CPC UCW
LLYFRGELL
LIBRARY
ABERYSTWYTH

AF616380

CHROMOSOMES TODAY

VOLUME 6

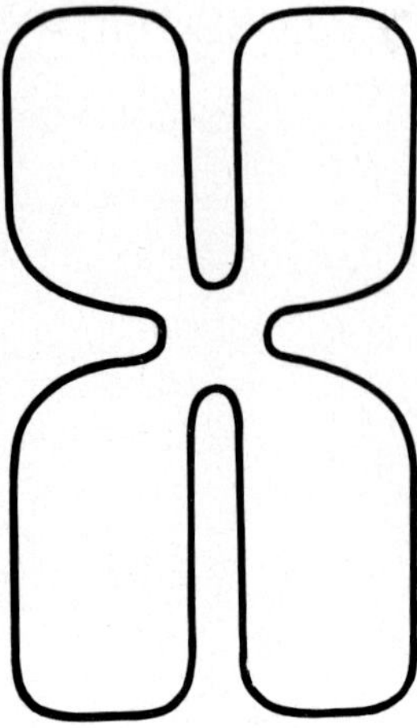

HELSINKI CHROMOSOME CONFERENCE

ORGANIZED BY
The University of Helsinki
The Folkhälsan Institute of Genetics

UNDER THE AUSPICES OF
Ministry of Education, Finland
The Sigrid Jusélius Foundation
Finnish Cultural Foundation
European Cell Biology Organization

This was the sixth of the International Chromosome Conferences inaugurated by C.D. Darlington and K.R. Lewis, held in Oxford in 1964, 1967 and 1970, in Jerusalem in 1972, and in Leiden in 1974. The proceedings of the conferences have appeared under the title Chromosomes Today, volumes 1 - 5.

CHROMOSOMES TODAY

VOLUME 6

Proceedings of the Sixth International Chromosome Conference held in Helsinki, Finland, August 29-31, 1977.

Editors
A. DE LA CHAPELLE
and
M. SORSA
Department of Medical Genetics and Department of Genetics, University of Helsinki

1977

ELSEVIER/NORTH-HOLLAND BIOMEDICAL PRESS
AMSTERDAM · NEW YORK · OXFORD

ISBN Elsevier/North-Holland: 0-444-80026-3

Published by:
Elsevier/North-Holland Biomedical Press
335 Jan van Galenstraat, PO Box 211
Amsterdam, The Netherlands

Sole distributors for the USA and Canada:
Elsevier North-Holland Inc.
52 Vanderbilt Avenue
New York, N.Y. 10017

Printed in The Netherlands

PREFACE

The tradition, started by Professor C.D. Darlington at Oxford in 1964, of bringing together scientists working in the various fields of chromosome research, is still alive. At the end of August 1977, the 6th International Chromosome Conference was held in Finland at the Korpilampi Forest Lake Hotel, a modern conference centre in the neighbourhood of the City of Helsinki.

The proceedings of this Conference, appearing as the 6th volume of CHROMOSOMES TODAY, differs from its predecessors in at least one striking way. It does not contain all the papers presented at the Helsinki Chromosome Conference, nor does it contain abstracts of all the contributions. While this change from earlier volumes may certainly be criticized, we should like to defend it for the reasons given below.

The rise in the number of active participants in the International Chromosome Conferences has been steep, from about 50 in 1964 to around 300 in 1977. The number of abstracts submitted to the Helsinki Chromosome Conference was so great that only a fourth could be accepted for oral presentation, even though two parallel sessions were arranged throughout most of the three conference days. Many scientists seem to agree that increasing the number of days or parallel sessions is not the solution to this problem. As a consequence, we encouraged scientists to present their data in poster form. This suggestion was favourably received, and the number of posters was actually twice that of the spoken papers. In fact, posters proved to be very popular as a means of scientific communication.

The remaining problem is the publication of the papers not included in the Conference Proceedings. In the present case the abstracts of all accepted papers and posters were printed in the Book of Abstracts (ISBN 951-45-1130-1) which can be purchased from the Organizers of the Conference. The Book of Abstracts thus provides a forum for the dissemination of the principal results presented at the Conference. Furthermore, it provides proof to employers and granting agencies that participants have actually contributed to the Conference.

Some of the Conference participants may wonder why all the eminent papers presented at the session entitled Past and Future of Chromosome Research are not included in the volume. The reason is that this session was videotaped and the edited tape can be obtained from the Organizers of the Conference.

Now let us look at another aspect of the present volume of CHROMOSOMES TODAY. We frankly considered the possibility of not publishing it at all - a major reason for this being the harsh terms put forward by publishing houses, especially the eventual high price of the volume. A second reason was the looming prospect that publication might, as in so many similar instances, take place some two years after the conference. Finally, we raised the question whether we should add to the general plague of "book pollution".

Our decision to go ahead after all was made easier by the following developments. Firstly, Elsevier/North-Holland, using the camera-ready technique, had the capacity for rapid publication at a reasonable price without any significant loss of quality. Secondly, so many outstanding contributions were submitted to the conference that it began to look like a potential blunder not to arrange for their printing. Thirdly, the themes appeared to cover an unusually wide selection of the various fields of chromosome research, in particular with reference to methodology. Thus the volume might become, when available a few months after the conference, something of a up-to-date textbook on current topics and achievements of chromosome research. To reach this goal and to keep to the necessary time-limit was our constant guideline when selecting and arranging the papers in this volume. We should like to add that speed was probably responsible for many of our mistakes in this endeavour.

It is not without satisfaction that we present the sixth volume of CHROMOSOMES TODAY only a few months after the Helsinki Chromosome Conference. We note with gratitude that the authors, without exception, understood the absolute necessity of keeping to the timetable. Cooperation with the Publishers, especially with Dr Jan Geelen, was always pleasant. We are impressed by the high degree of managerial and technical skill displayed at Elsevier/North-Holland. Our eminent secretary, Ms Terttu Kaustia, contributed significantly to the production of this volume. Finally, we thank our colleagues in the Organizing Committee of the Conference, Drs Veikko Sorsa, Svante Stenman and Peter Tigerstedt, who, even though not directly connected with the editing of this book, have helped us greatly in all respects.

Helsinki, October 1977

Albert de la Chapelle

Marja Sorsa

CONTENTS

CONFERENCE MEMBERS

AKSTEIN, Edna, Genetic Institute, The Chaim Sheba Med. Ctr., Tel-Hashomer, Israel
AL-RUBEAI, M., Dept. of Plant Biology, Queen Mary College, Mile End Road, London E1 4NS, U.K.
ALBERS, Focke, Bot. Inst., U. Bot. Garten, 23 Kiel, Dusternbrooker Weg 17, B.R.D.
ALONSO, Carlos, Dept. of Biochemistry, Universidad Autonoma CX, Madrid, Spain
ALVESALO, Lassi, Inst. of Dentistry, Lemminkäisenkatu 2, 20520 Turku 52, Finland
ARIENS, A.T., Ross van Lenneplaan 19, Sittard, The Netherlands
ARNASON, Ulfur, Inst. of Genetics, Sölvegatan 29, S-22362 Lund, Sweden
ARRIGHI, Frances, M.D. Anderson Hospital, Houston, Texas 77025, U.S.A.
ASHBURNER, Michael, Dept. of Genetics, University of Cambridge, Milton Road, Cambridge CB4 1XH, U.K.
AULA, Pertti, Children's Hospital, Stenbäckinkatu 11, 00290 Helsinki 29, Finland

BAGDASARIAN, Michael, Inst. of Biochemistry & Biophysics, Polish Academy of Sciences, Ul. Rakowiecka 36, 02-532 Warszawa, Poland
BAJER, A., Dept. of Biology, College of Liberal Arts, University of Oregon, Eugene, Oregon 97403, U.S.A.
BARAT, Monique, Lab. de Biologie Gen., Bat. 400, Université Paris Sud, 91405 Orsay, France
BARSACCHI-PILONE, G., Inst. of Histology & Embryology, Via A. Volta 4, 56100 Pisa, Italy
BAYLISS, M.W., ICI Corporate Laboratory, P.O. Box 11, The Heath, Runcorn, Cheshire WA7 4QE, U.K.
BEENSEN, Volkmar, Inst. of Anthropology & Human Genetics, Friedrich-Schiller University, Jena 69, Kollegiengasse 10, D.D.R.
BENIRSCHKE, Kurt, Dept. Reprod. Medicine, University of California, La Jolla, San Diego, Calif. 92037, U.S.A.
BENNICK, Anne, Norsk Hydro's Inst. Cancer Research, The Norwegian Radium Hospital, Montebello, Oslo 3, Norway
BER, Rosalie, Dept. of Tumor Biology, Karolinska Institutet, S-104 01 Stockholm 60, Sweden
BERGER, Roland, Lab. de Cytogénétique, Centre Hayem, Hopital Saint-Louis, 75010 Paris, France
BERNHEIM, Alain, Lab. de Cytogénétique, Centre Hayem, Hopital Saint-Louis, 75010 Paris, France
BIANCHI, Nestor, Inst. Multidisciplinario de Biologia Celular (IMBICE), C.C. 403, La Plata, 1900 Argentina
BIEDLER, June, Sloan-Kettering Institute, 145 Boston Post Rd, Rye, N.Y. 10580, U.S.A.
BIJLSMA, Jan, Lab. of Anthropobiology, Sarphatistraat 217, Amsterdam, The Netherlands
BLOCK, Karin, Inst. of Genetics, Sölvegatan 29, S-23262 Lund, Sweden
BOGDANOV, Yuri, Inst. of Molecular Biology, Academy of Sciences, Vavilov St. 32, Moscow 117312, U.S.S.R.
BOLUND, Lars, Aarhus University, DK-8000 Aarhus C, Denmark
BORGSTRÖM, Georg, Folkhälsan Inst. of Genetics, P.O. Box 819, 00101 Helsinki 10, Finland
BORZAN, Zelimir, Faculty of Forestry, Dept. of Forest Genetics, P.O. Box 178, 41001 Zagreb, Yugoslavia
BOSMA, Anna A., Vakgroep Funktionele Morfologie, Faculteit der Diergeneeskunde, Bekkerstraat 141, Utrecht, The Netherlands
BOZCUK, Nihat, Hacettepe University, Inst. of Biology, Beytepe Campus, Ankara, Turkey
BRIGHTON, Christine, Cytogenetics Section, Jodrell Laboratory, Royal Botanic Gardens, Kew, Richmond, Surrey TW9 3A3, U.K.
BRØGGER, Anton, Inst. for Cancer Research, The Norwegian Radium Hospital, Montebello, Oslo 3, Norway
BUCKTON, K.E., M.R.C. Clinical Population Cytogenetics Unit, Western General Hospital, Crewe Road, Edinburgh, Scotland
BüHLER, Erica, University Childr. Hosp., Dept. of Genetics, CH-4005 Basel, Switzerland

BURG, Kornel, Biological Research Center, Hungarian Academy of Sciences, Institute of Genetics, H-6701 Szeged, P.O. Box 521, Hungary
BURGERHOUT, Wim, Erasmus Univ. Medical Faculty, Dept. of Cell Biology & Genetics, P.O. Box 1738, Rotterdam, The Netherlands
BÜRKI, Kurt, Swiss Inst. for Experimental Cancer Research, CH-1066 Epalinges, Switzerland
BUYS, Charles, Dept. of Human Genetics, Groningen State University, Antonius Deusinglaan 4, Groningen, The Netherlands

CAPANNA, E., Inst. Vertebrate Zoology, c/o Inst. Comparative Anatomy, Via Borelli 50, I-00161, Italy
CASPERSSON, Torbjörn, Institut för Cellforskning, Karolinska Institutet, 104 01 Stockholm 60, Sweden
CHAKI, Rina, Genetic Institute, The Chaim Sheba Med. Ctr., Tel-Hashomer, Israel
CHIARELLI, B., Inst. of Anthropology, Via Acc. Albertina 17, 10123 Torino, Italy
COHEN, Maurice, California Inst. of Technology, Pasadena, Calif. 91109, U.S.A.
COMINGS, David, Dept. of Medical Genetics, City of Hope National Medical Cneter, Duarte, Calif. 91010, U.S.A.
COUTURIER, Jerome, Institut de Progenèse, 15 rue de l'Ecole de Médecine, 75006 Paris, France
CREMER, C., Inst. of Human Genetics, Freiburg University, 78 Freiburg im Breisgau, Albertsstrasse 11, B.R.D.
CRICK, Francis, The Salk Institute, P.O. Box 1809 San Diego, Calif. 92112, U.S.A.

DAKER, M.G., 34 Gibbs Green, Edgware, Middlesex, U.K.
DANCIS, Barry, Dept. of Biology, Temple University, Philadelphia, Pa. 19122, U.S.A.
DARLINGTON, C.D., Botany School, South Parks Road, Oxford OX1 3RA, U.K.
DE GROUCHY, Jean, Hôpital des Enfants Malades, 149 rue de Sèvres, Paris 15, France
DE JONG, J.H., Genetical Dept., University of Amsterdam, Kruislaan 318, Amsterdam, The Netherlands
DE LA CHAPELLE, Albert, Dept. of Medical Genetics, University of Helsinki, Haartmaninkatu 3, 00290 Helsinki 29, Finland
DELHANTY, J.D.A., Dept. Human Genetics, The Galton Laboratory, University College, London WC1, U.K.
DERKSEN, Jan, Dept. of Histology, Karolinska Institutet, S-104 01 Stockholm 60,Sweden
DOVER, Gabriel, Dept. of Genetics, University of Cambridge, Milton Road, Cambridge CB4 1XH, U.K.
DUMARS, K.W., Dept. of Child Health, Welsh National School of Medicine, Heath Park, Cardiff CF4 4XN, South Glamorgan, U.K.
DUTRILLAUX, Bernard, Institut de Progénèse, C.N.R.S., 15 rue de l'Ecole de Médecine, Paris 75006, France

EDGREN, Johan, IV Dept. of Medicine, Unioninkatu 38, 00170 Helsinki 17, Finland
EDSTRÖM, Jan-Erik, Dept. of Histology, Karolinska Institutet, Stockholm 60, Sweden
ENGELHARDT, Peter, Dept. of Genetics, University of Helsinki, Salomonkatu 17 A, 00100 Helsinki 10, Finland
EVANS, H.J., MRC Cytogenetics Unit, Western General Hospital, Crewe Road, Edinburgh EH4 2XO, Scotland

FISCHER, Patricia, Inst. for Cancer Research, University of Vienna, Borschkegasse 8 A, A-1090 Wien, Austria
FONATSCH, Christa, Inst. of Genetics, Medizinische Hochschule, P.O. Box 61 01 80, 3000 Hannover 61, B.R.D.
FORD, C.E., Sir William Dunn School of Pathology, University of Oxford, South Parks Road, Oxford OX1 3RE, U.K.
FREDGA, Karl, Inst. of Genetics, University of Lund, Sölvegatan 29, S-223 62 Lund, Sweden
FREEMAN, Viola, Dept. of Pediatrics, McMaster University Med. Center, 1200 Main St. West, Hamilton L8S 4J9, Ontario, Canada
FREY, Marja-Liisa, Dept. of Medicine, Div. of Zoology 3, University of Turku, Lemminkäisenkatu 1, SF-20520 Turku 52, Finland

FRIBERG, Kristina, Inst. for Medical Cell Research & Genetics, Karolinska Institutet, S-104 01 Stockholm 60, Sweden
FRIEBE, Bernd, D-1000 Berlin 47, Severingstrasse 5, B.R.D.
FRIEDLANDER, M., Dept. of Biology, University of the Negev, Beer Sheva, Israel

GADELLA, Theodorus, Ds. Pasmastraat 7, Bunnik, The Netherlands
GAHMBERG, Nina, II Dept. of Pathology, University of Helsinki, Haartmaninkatu 3, 00290 Helsinki 29, Finland
GARVER, James, Inst. Anthropogenetics, Wassenaarseweg 72, Leiden, The Netherlands
GEBAUER, J., Inst. Human Genetics, Göttingen University, Nikolausbergerweg 5 A, D-3400 Göttingen, B.R.D.
GEBHART, Erich, Inst. Human Genetics, Erlagen University, Bismarkstrasse 10, D-8520 Erlagen, B.R.D.
GEELEN, Jan, Elsevier/North Holland Biomedical Press, P.O. Box 1527, Amsterdam, The Netherlands
GERAEDTS, J.P.M., Inst. Anthropogenetics, Sylvius Laboratoria, Wassenaarseweg 72, Leiden, The Netherlands
GERSTEL, D.U., Dept. of Crop Science, North Carolina State University, Raleigh, N.C. 27607, U.S.A.
GHOSH, Sibdas, Institut f. Zellforschung, Deutsches Krebsforschungszentrum, P.O. Box 101949, D-6900 Heidelberg, B.R.D.
GONZALES, Jacques, Lab. d'Histologie, Embryologie Cytogénétique, Hôpital Pitié-Salpetrière, 105 Boulevard de l'Hôpital, 75013 Paris, France
GOUW, W.L., R.U. Antropogenetisch Inst., Ant. Deusinglaan 4, Groningen, The Netherlands
GRANBERG-ÖHMAN, Ingrid, Huddinge Hospital, S-14186 Huddinge, Sweden
GRIGG, G.W., CSIRO, Molecular & Cell Biology Unit, P.O. Box 90, Epping, N.S.W. 2121, Australia
GRIMM, Tiemo, Inst. Human Genetics, Göttingen University, Nikolausberger Weg 5 A, D-3400 Göttingen, B.R.D.
GRIPENBERG, Ulla, Dept. of Genetics, University of Helsinki, P.Rautatiekatu 13, 00100 Helsinki 10, Finland
GROPP, A., Inst. of Pathology, Medizinische Hochschule, Ratzeburger Allee 160, Lübeck 24, B.R.D.

HAAPALA, Olli, Dept. of Biomedicine, University of Tampere, 33520 Tampere 52, Finland
HADLACZKY, Gyula, Biological Research Ctr., Hungarian Academy of Sciences, Inst. of Genetics, H-6701 Szeged, P.O. Box 521, Hungary
HAGELTORN, Matts, Dept. of Animal Genetics, Royal Veterinary College, S-75007 Uppsala 7, Sweden
HAGEMANN, R., Dept. of Genetics, Martin-Luther University, DDR-402 Halle/S., Domplatz 1, D.D.R.
HAGEMEIJER HAUSMAN, Anne, Dept. of Cell Biology, Erasmus University, P.O. Box 1738, Rotterdam, The Netherlands
HAGLUND, Ulla, Inst. for Medical Cell Research & Genetics, Karolinska Institutet, S-104 01 Stockholm 60, Sweden
HAGMAN, Max, Inst. of Forest Research, Kornetintie 8, 00380 Helsinki 38, Finland
HALKKA, Liisa, Dept. of Genetics, University of Helsinki, P.Rautatiekatu 13, 00100 Helsinki 10, Finland
HALKKA, Olli, Dept. of Genetics, University of Helsinki, P. Rautatiekatu 13, 00100 Helsinki 10, Finland
HAMERTON, John, Division of Genetics, Health Sciences Centre, Winnipeg, Manitoba R 3E 023, Canada
HANSMANN, J., Inst. f. Humangenetik, Universität Göttingen, 34 Göttingen, Nikolausberger Weg 52, B.R.D.
HANSTEEN, I.L., St. Joseph's Hospital, N 3900 Porsgrunn, Norway
HARTUNG, Michèle, 4 Parc Jean Mermoz, Marseille 13008, France
HAUSCHTECK-JUNGEN, E., Strahlenbiologisches Inst., August Forel-str. 7, 8029 Zürich, Switzerland
HAYASHI, Ken, Chromosome Lab., Univ.-Frauenklinik, D-74 Tübingen, Schleichstrasse 4, B.R.D.

HECHT, Fred, Genetics Unit, Univ. of Oregon, Health Sciences, Child Development Center, Portland, Oregon 97207, U.S.A.
HENS, Luc, Laboratory of Anthropogenetics, Free Univ. of Brussels, 1050 Brussels, Belgium
HESS, Oswald, Inst. f. Allgemeine Biologie, Univ. Düsseldorf, D-4000 Düsseldorf, Ulembergstrasse 127-129, B.R.D.
HESS, Claudia, Strahlenbiologisches Inst., August Forel Str. 7, CH-8029 Zürich, Switzerland
HILWIG, Ingeborg, Hoechst AG, Genwebezuchtung H 811, 623 Frankfurt Am Main 80, B.R.D.
HIMBERG, Mikael, Porthaninkatu 3, SF-20500 Turku, Finland
HOLM, Preben, Dept. of Physiology, Carlsberg Laboratory, 2500 Valby, Copenhagen, Denmark
HOLMGREN, Paul, Dept. of Genetics, Umeå University, S-90187 Umeå, Sweden
HONGELL, Karin, Dept. of Genetics, University of Helsinki, P.Rautatiekatu 13, 00100 Helsinki 10, Finland
HOSSFELD, D.K., Medical University Clinic, Tumor Research, 4300 Essen 1, Hufelandstrasse 55, B.R.D.
HUSTINX, Th., Inst. of Anthropogenetics, Nijmegen Univ., Nijmegen, The Netherlands

JALBERT, P., Cytogenetics Lab., Centre Hospitalier Régional et Universitaire, B.P. 217X, 38043 Grenoble Cedex, France
JAMRICH, Milan, Inst. of Molecular Genetics, 69 Heidelberg, im Neunheimerfeld 230, B.R.D.
JENSEN, R.A.C., Percy FitzPatric Inst., University of Cape Town, Rondebosch, Cape, 7700 Rep. of South Africa
JOHANNISSON, Reiner, Inst. f. Allgemeine Biologie, Univ. Düsseldorf, Geb. 26.02/02, D-4000 Düsseldorf, B.R.D.
JONES, Keith, Jodrell Laboratory, Royal Botanic Gardens, Kew, Richmond, Surrey TW9 3DS, U.K.
JOY, Peter, Hankkija Plant Breeding Inst., SF-04300 Hyrylä, Finland

KAISER-McCAW, Barbara, Univ. of Oregon Health Sciences Center, Crippled Children's Division, Child Development Center, Genetics Unit, Portland, Oregon 97207, U.S.A.
KALLIO, Hanna, The Women's Clinic, Histology Lab., Haartmaninkatu 2, 00290 Helsinki 29, Finland
KALUZEWSKI, Bogdan, Lab. of Genetics, Inst. of Endocrinology, Medical Academy of Lodz, 3 Sterling St., 91-425 Lodz, Poland
KAVENOFF, Ruth, Dept. of Chemistry B-017, Univ. of California, San Diego, La Jolla, Calif. 92093, U.S.A.
KERÄNEN, Sirkka, Dept. of Virology, Univ. of Helsinki, Haartmaninkatu 3, 00290 Helsinki 29, Finland
KESSLER, Bezalel, The Volcani Center, Aro, Bet-Dagan, Israel
KIHLMAN, B.A., Dept. of Genetics & Plant Breeding, Royal Agricultural College of Sweden, S-750 07 Uppsala 7, Sweden
KIRSCH-VOLDERS, M., Lab. Anthropogenetics, VUB, 2 Pleinlaan, B-1050 Brussels, Belgium
KIVI, Erkki, Hankkija Plant Breeding Institute, SF-04300 Hyrylä, Finland
KLASTERSKA, Irena, Stockholm Univ., Wallenberg Lab., Lilla Frescati, S-104 05 Stockholm 50, Sweden
KLOETZEL, Peter-Michael, Inst. f. Allgem. Biologie, Universitätstr. 1, University of Düsseldorf, 4000 Düsseldorf, B.R.D.
KNUUTILA, Sakari, Rinnekoti Institution for the Mentally Retarded, Research Dept., 02980 Espoo 98, Finland
KOCH, Gerhard, Inst. f. Humangenetik der Univ., 852 Erlangen, Bismarckstr. 10, B.R.D.
KORF, Bruce, The Rockefeller Univ., 1230 York Ave., New York, N.Y. 20021, U.S.A.
KORSNES, Lars, Norsk Hydro's Inst. for Cancer Research, The Norwegian Radium Hospital, Montebello, Oslo 3, Norway
KOVANEN, Riitta, The Women's Clinic, Histology Lab., Haartmaninkatu 2, 00290 Helsinki 29, Finland
KOROCHKINA, Lyuba S., Inst. of Cytology & Genetics, Academy of Sciences, Siberian Dept., Pravda St. 9, Novosibirsk 630090, U.S.S.R.
KUSHNIR, Uri, Dept. of Plant Genetics, Weizmann Inst. of Science, Rehovot, Israel

LACADENA, Juan-Ramon, Dept. of Genetics, Faculty of Biology, Univ. Complutense, Madrid-3, Spain
LAFOURCADE, Jacques, Hôpital de la Salpetrière, 47 Boulevard de l'Hôpital, 75634 Paris, France
LAMBERT, A.M., Inst. de Botanique, Lab. de Phytogénétique, 28 rue Goethe, 67083 Strasbourg Cedex, France
LATT, Samuel, Harvard Medical School, Children's Hospital Med. Center, 300 Longwood Ave., Boston, Mass. 02115, U.S.A.
LAURENT, Colette, Lab. de Cytogénétique, Inst. Pasteur de Lyon, 77 rue Pasteur, 69365 Lyon Cedex 2, France
LAWLER, Sylvia, Royal Marsden Hospital, Fulham Road, London S.W. 3, U.K.
LECHER, P., Lab. Zoology, University of Clermont-Ferrand II, 63170 Aubière, France
LEISTI, Jaakko, Rinnekoti Institution for the Mentally Retarded, 02980 Espoo 98, Finland
LEJEUNE, Jerome, Lab. de Génétique Fondamentale, Inst. de Progenèse, 15 rue de l'Ecole de Médécine, Paris 75006, France
LEVAN, Albert, Inst. of Genetics, Univ. of Lund, Sölvegatan 29, S-223 62 Lund, Sweden
LEVAN, Göran, Inst. of Genetics, Univ. of Lund, Sölvegatan 29, S-223 62 Lund, Sweden
LIMA-DE-FARIA, A., Inst. of Molecular Cytogenetics, University of Lund, Tornavägen 13, S-223 63 Lund, Sweden
LIMON, Janusz, Genetic Lab., Inst. of Medical Biology, 80210 Gdansk, Poland
LINDER, Robert, 2 rue de Westhalten, 68250 Rouffach, France
LINDSTEN, Jan, Lab. of Clinical Genetics, Karolinska Hospital, Stockholm, Sweden
LINNERT, G, Inst.f.Angewandte Genetik, Albrecht-Ther-Wik 6, D-1000 Berlin 33, B.R.D.
LUBSEN, Nicolette, Dept. of Genetics, Univ. of Nijmegen, Toernooiveld, Nijmegen, The Netherlands

MANDAHL, Nils, Inst. of Genetics, Univ. of Lund, Sölvegatan 29, S-223 62 Lund, Sweden
MAX, Christina, Inst. of Genetics, Univ. of Lund, Sölvegatan 29, S-223 62 Lund, Sweden
McQUADE, Henry, Dept. of Radiology, Univ. of Missouri Med. Center, Columbia, Missouri 65201, U.S.A.
MEHDIPOUR, Parvin, Bismarckstrasse 10, D-8520 Erlangen, B.R.D.
MIKKELSEN, Margareta, John F. Kennedy Inst., Gl. Landevej 7, 2600 Glostrup, Denmark
MITELMAN, Felix, Dept. of Clinical Genetics, Lund Univ. Hospital, S-22185 Lund, Sweden
MOREAU, Nicole, 53 Avenue Rockefeller, 69003 Lyon, France
MOSES, Montrose, Dept. of Anatomy, Duke Univ. Medical Center, Durham, N.C.27710, U.S.A.
MUKHERJEE, R.N., International Atomic Energy Agency, Kärtner Ring 11, P.O. Box 590, A-1011 Wien, Austria
MüLLER, Hansjakob, Basler Kinderspital, Römergasse 8, 4005 Basel, Switzerland
MüNTZING, Arne, Inst. of Genetics, Univ. of Lund, Sölvegatan 29, S-223 62 Lund, Sweden

NAGL, Walter, Dept. of Biology, Division of Cytology, The University, P.O. Box 3049, D-6750 Kaiserslautern, B.R.D.
NAKAI, Sayaka, National Inst. of Radiological Sciences, 4-9-7 Anagawa, Chiba 280, Japan
NARAYAN, R.K.T., Dept. of Agricultural Botany, Univ. College of Wales, Aberystwyth, Wales, SY23 3DD, U.K.
NARDI-DELLATOGNA, Irma, Inst. of Histology & Embryology, Via A. Volta 4, 56100 Pisa, Italy
NATARAJAN, A.T., Dept. of Radiation Genetics, Wassenaarseweg 72, Leiden, The Netherlands
NEVO, Yaffa, Dept. of Plant Genetics, The Weizmann Inst. of Science, Rehovot, Israel
NEVSTAD, N.P., Norsk Hydro's Inst. Cancer Research, The Norwegian Radium Hospital, Montebello, Oslo 3, Norway
NIELSEN, Karin, Inst. of Genetics, Univ. of Lund, Sölvegatan 29, S-223 62 Lund, Sweden
NIENSTEDT, Irma, Dept. of Genetics, Univ. of Turku, SF-20500 Turku 50, Finland
NORDENSON, Ingrid, Kungsgårdsvägen 13, S-90250 Umeå, Sweden

OKSALA, Tarvo, Dept. of Genetics, Univ. of Turku, SF-20500 Turku 50, Finland
OLIVIERI, Gregorio, Inst. of Genetics, Città Universitaria, Roma, Italy
ÖSTERGREN, G., Inst. of Genetics, Agricultural College of Sweden, Uppsala 7, Sweden
ÖZARSLAN, S.H., Istanbul Univ., Fen Fakultesi, Vezneciler-Istanbul, Turkey

PAGES, Montserrat, Dept. of Biochemistry, Universidad Autonoma CX, Madrid, Spain
PANITZ, Reinhard, Inst.f.Kulturpflanzenforschung, 4325 Gatersleben, D.D.R.
PAPES, Drazena, Dept. of Botany, Univ. of Zagreb, Roosveltov Trg 6/III, P.p. 933, YU-41001 Zagreb, Yugoslavia
PARRINGTON, J.M., M.R.C. Human Biochemical Genetics, The Galton Lab., University College of London, London NW1 2HE, U.K.
PARVINEN, Martti, Dept. of Anatomy, Univ. of Turku, 20520 Turku 52, Finland
PASQUALI, Francesco, Gruppo Euratom, Via Forlanini 14, 27100 Pavia, Italy
PAVLOVA, Margarita, Inst. of Hydrobiology, Ukranian SSR Acad. of Science, 44 Vladimirskaya, P.O. Box 252003, Kiev, U.S.S.R.
PEARSON, Peter, Instituut voor Anthropogenetica, Wassenaarseweg 72, Leiden, The Netherlands
PFEIFER, Susan, Dept. of Virology, Univ. of Helsinki, Haartmaninkatu 3, 00290 Helsinki 29, Finland
PHILIP, Preben, Division of Haematology, Dept. of Medicine A, Rigshospitalet, Blegdamsvej 9, DK-2100 Copenhagen, Denmark
POHLMAN, Joachim, Inst. f. Allgem. Botanik, Jungiusstrasse 6, 2000 Hamburg 36, B.R.D.
PORTIN, Petter, Dept. of Genetics, Univ. of Turku, SF-20500 Turku 50, Finland
PRIILINN, Oskar, Inst. of Experimental Biology, Acad. of Science of the Estonian SSR, 203051 Estonian SSR, Harku, U.S.S.R.
PUNNETT, Hope, St. Christopher's Hospital, 2600 N. Lawrence St., Philadelphia, Penn. 19133, U.S.A.

QUACK, Bernadette, Laboratoire de Cytogénétique, Centre Hospitalier, 73011 Chambery, France

RAINER, Srecko, 68 Goce Delceva, 61000 Ljubljana, Yugoslavia
RAO, S.R.V., Dept. of Zoology, Univ. of Delhi, Delhi 110009, India
RASMUSSEN, Søren, Carlsberg Lab., Dept. of Physiology, Gl. Carlsbergvej 10, DK-2500 Copenhagen, Denmark
REES, Hubert, Dept. of Agricultural Botany, U.C.W. School of Agricultural Science, Aberystwyth Dyfed, 5423 3DD, Wales, U.K.
REEVES, Brian, Royal Marsden Hospital, Fulham Road, London SW3 6JJ, U.K.
REHDER, Helga, Abt. f. Pathologie der Medizinischen Hochschule Lübeck, 24 Lübeck, Ratzeburger Allee 116, B.R.D.
RIBBERT, D., Inst. of Zoology, Univ. of Münster, 44 Münster (Westf.), Badestr.9, B.R.D.
RILEY, Ralph, Plant Breeding Inst., Trumpington, Cambridge CB2 2LQ, U.K.
RINGERTZ, Nils, Inst. for Medical Research, Medical Nobel Inst., Karolinska Institutet, 104 01 Stockholm 60, Sweden
RIS, Hans, Dept. of Zoology, Univ. of Wisconsin, Madison, Wisc. 53706, U.S.A.
ROMMEL, M., Internationale Agrarwirtschaft, Steinstr. 19, 343 Witzenhausen, B.R.D.
ROUSI, Arne, Dept. of Botany, Univ. of Turku, SF-20500 Turku 50, Finland
ROWLEY, Janet, The Franklin McLean Memorial Research Inst., 950 East 59th Street, Chicago, Illinois 60637, U.S.A.
RYDLANDER, Lars, Dept. of Histology, Karolinska Institutet, S-10401 Stockholm, Sweden
RÖPER, Wolfram, Inst. f. Allgemeine Botanik, Jungiusstr. 6, D-2000 Hamburg 36, B.R.D.

SACHSSE, Walter, Inst. of Genetics, Univ. of Mainz, D-6500 Mainz, Saarstr. 21, B.R.D.
SAKSELA, Eero, The Women's Clinic, Haartmaninkatu 2, SF-00290 Helsinki 29, Finland
SANDBERG, Avery, Roswell Park Memorial Inst., 666 Elm St., Buffalo, N.Y. 14203, U.S.A.
SAURA, Anja, Dept. of Genetics, Univ. of Helsinki, Salomonkatu 17 A, SF-00100 Helsinki 10, Finland
SAVAGE, J.R.K., MRC Radiobiology Unit, Harwell, Didcot, Berks., U.K.
SAVONTAUS, M.-L., Dept. of Genetics, Univ. of Turku, SF-20500 Turku 50, Finland
SCHERES, J.M.J.C., Dept. of Human Genetics, Division of Cytogenetics, Faculty of Medicine, Univ. of Nijmegen, Nijmegen, The Netherlands
SCHLAMMADINGER, Joseph, Max-Planck-Inst. f. Biophysikalische Chemie, Abt. 06, Postfach 968, D-3400 Göttingen-Nikolausberg, B.R.D.
SCHLEGEL, Rolf, Martin-Luther Univ., Halle-Wittenberg, DDR-4104 Hohenthurm, Berliner Str. 2, D.D.R.
SCHMID, Werner, Dept. of Paediatrics, Children's Hospital, Steinwiesstrasse 75, Zürich, Switzerland

SCHRÖDER, Jim, Folkhälsan Inst. of Genetics, P.B. 819, 00101 Helsinki 10, Finland
SCHVARTZMAN, Jorge, Dept. of Cytology, Inst. of Cellular Biology, Velazquez 144, Madrid-6, Spain
SCHWANITZ, Gesa, Inst. of Human Genetics, Bismarckstr. 10, D-8520 Erlangen, B.R.D.
SCHWARZACHER, Hans, Inst. of Histology & Embryology, The University, A-1090 Wien, Schwarzspanierstrasse 17, Austria
SCHWEIZER, D., Dept. of Cytogenetics, University, Rennweg 14, A-1030 Wien, Austria
SCHWINGER, E., Inst. f. Gerichtliche Medizin, 55 Bonn, Stiftsplatz 12, B.R.D.
SCOTT, D., Cell Culture Dept., Patterson Labs., Christie Hospital, Manchester M20 9BX, U.K.
SEABRIGHT, Marina, Wessex Regional Cytogenetics Unit, General Hospital, Salisbury, Wilts., U.K.
SELANDER, Ritva-Kajsa, Folkhälsan Inst. of Genetics, P.B. 819, 00101 Helsinki 10, Finland
SENTEIN, Paul, Lab. of Histology, 2 rue Ecole de Médécine, 34000 Montpellier, France
SERRA, A., Università Cattolica, Facolta de Medicina, Inst. di Genetica Umana, 00168 Roma, Italy
SHARMA, A.K., Dept. of Botany, Univ. College of Science, University of Calcutta, 35 Ballygunge Circular Road, Calcutta-700019, India
SHARMA, Archana, Dept. of Botany, Univ. College of Science, University of Calcutta, 35 Ballygunge Circular Road, Calcutta-700019, India
SHARP, James, King's College Hospital, Denmark Hill, London SE5 8RX, U.K.
SHAW, David, Cytogenetics Dept., Plant Breeding Inst., Trumpington, Cambridge CB2 2LQ, U.K.
SLATER, Rosalyn, Lab. of Anthropobiology, Menselijke Erfelijkheidsleer, Sarphatistr. 217, Amsterdam-C, The Netherlands
SNAIDER, Tamara, Inst. of Experimental Biology, Academy of Sciences of the Estonian SSR, 203051 Estonian SSR, Harku, U.S.S.R.
SÖDERSTRÖM, Karl-Ove, Inst. of Biomedicine, Dept. of Anatomy, University of Turku, SF-20520 Turku 52, Finland
SOLDATOVIC, Bogosav, Inst. for Biological Research, 29 Novembra 142, Belgrad, Yugoslavia
SORSA, Marja, Dept. of Genetics, University of Helsinki, Salomonkatu 17 A, SF-00100 Helsinki 10, Finland
SORSA, Veikko, Dept. of Genetics, University of Helsinki, Salomonkatu 17 A, SF-00100 Helsinki 10, Finland
SPECKMANN, G.J., Foundation for Agric. Plant Breeding, P.O. Box 117, Wageningen, The Netherlands
STAHL, A., Lab. d'Histologie, Faculté de Médécine, Boulevard Jean-Moulin, 133385 Marseille Cedex 4, France
STARK, Leslie, 15 Barolin St., Bundaberg 4670, Australia
STEFFENSEN, Dale, Dept. of Genetics & Development, 515 Morrill Hall, Urbana, Illinois 61810, U.S.A.
STENGEL-RUTKOWSKI, Sabine, Genetische Beratungstelle/Kinderpoliklinik, Universität München, Pettenkoferstr. 8a, D-8000 München 2, B.R.D.
STENMAN, Svante, III Dept. of Pathology, University of Helsinki, Haartmaninkatu 3, SF-00290 Helsinki 29, Finland
STENSTRAND, Kristina, Inst. of Radiation Protection, P.O. Box 268, 00102 Helsinki 10, Finland
STEPHAN, Gunther, Inst. f. Strahlenhygiene, 8042 Neuherberg, Post Schleissheim, Ingolstädter Landstrasse 1, B.R.D.
STOLL, Claude, Inst. de Puericulture, Hospices Civils, 67000 Strasbourg, France
STOLTZMANN, V., Inst. of Biochemistry & Biophysics, Polish Academy of Sciences, Ul. Rakowiecka 36, 02-532 Warszawa, Poland
SUMNER, A.T., MRC Cytogenetics Unit, Western General Hospital, Edinburgh EH4 2XU, Scotland
SUOMALAINEN, Esko, Dept. of Genetics, University of Helsinki, P.Rautatiekatu 13, SF-00100 Helsinki 10, Finland
SUOMALAINEN, Hannu, Dept. of Medicine, Div. of Zoology 3, Lemminkäisenkatu 1, 20520 Turku 52, Finland

SWOLIN, Birgitta, Central Lab. of Clinical Chemistry, Sahlgrens Hospital, S-41345 Göteborg, Sweden

TAKENOUCHI, Y., Biological Laboratory, Sapporo College, South 22, West 12, Sapporo 064, Japan
TANGUAY, R., Génétique Humaine, Le Centre Hospitalier, 2705 Boulevard Laurier, Université Laval, Québec G1V 4G2, Canada
TATES, A.D., Dept. of Radiation Genetics & Chemical Mutagenesis, Wassenaarseweg 72, Leiden, The Netherlands
THERMAN, Eeva, The Univ. of Wisconsin, Dept. of Medical Genetics, Genetics Bldg., Madison, Wisconsin 53706, U.S.A.
TIGERSTEDT, P.M.A., Dept. of Plant Breeding, University of Helsinki, Viikki, SF-00710 Helsinki 71, Finland
TOLKSDORF, Marlis, Univ. Children's Clinic, 23 Kiel, Fröbelstr. 15/17, B.R.D.
TRAUT, Walther, Abt. Biologie, Ruhr-Univ., D-4630 Bochum, B.R.D.
TREPTE, Hans-Heinrich, I Inst. of Zoology, Berlinerstr. 28, D-3400 Göttingen, B.R.D.
TURC, Claude, Lab. of Cytogenetics, Faculté de Médécine, 7 Blvd Jeanne d'Arc, 21033 Dijon Cedex, France

UCHIDA, Irene, Dept. of Pediatrics, McMaster Univ. Medical Center, 1200 Main St.West, Hamilton L8S 4J9, Ontario, Canada
UTAKQJI, T., Cancer Institute, Kami-Ikeburo 1-37-1, Toshima-ku, Tokyo 170, Japan

VAN BUUL, P.P.W., Sylvius Labs., Dept. of Radiation Genetics & Chemistry, Wassenaar-seweg 72, Leiden, The Netherlands
VAN DEN BERGHE, Herman, Div. Human Genetics, Minderbroedersstr. 12, B-3000 Leuven, Belgium
VAN DER HAGEN, Carl-Birger, Inst. of Medical Genetics, Univ. of Oslo, Blindern, Oslo 3, Norway
VAN DER PLOEG, Mels, Dept. Histochem. Cytochem., Sylvius Lab., 72 Wassenaarseweg 72, Leiden, The Netherlands
VAN EGMOND-COWAN, I., Inst. of Anthropogenetics, Wassenaarseweg 72, Leiden, The Netherlands
VAN HEEMERT, C., Euratom Inst. for Atomic Sciences, P.O. Box 48, Wageningen, The Netherlands
VAN HEMEL, Jan, Transitorium 3, Padnalaan S, Utrecht, The Netherlands
VAN WENT, Greta, National Inst. of Public Health, Dept. of Pharmacology, P.O. Box 1, Bilthoven, The Netherlands
VARLEY, J., Zoology Dept., Univ. of Leicester, Leicester LE1 7RH, U.K.
VARSHAVSKY, A.J., Inst. of Molecular Biology, Academy of Sciences of the USSR, Vavilov St. 32, Moscow B-312, U.S.S.R.
VETTERLEIN, Monika, Inst. for Cancer Research, Univ. of Vienna, Borschkegasse 8 A, A-1090 Wien, Austria
VIINIKKA, Yrjö, Dept. of Genetics, Univ. of Turku, 20500 Turku 50, Finland
VON KOSKULL, Harriet, III Dept. of Pathology, University of Helsinki, Haartmanink. 3, SF-0290 Helsinki 29, Finland
VOULLAIRE, Lucille, Cytogenetics Lab., Queen Victoria Hospital, Lonsdale St., Melbourne, Australia
VROLIJK, Hans, Inst. of Anthropogenetics, Wassenaarseweg 72, Leiden, The Netherlands

WAHRMAN, Jacob, Dept. of Genetics, The Hebrew Univ. of Jerusalem, Israel
WAKSVIK, Helga, Norsk Hydro's Inst. of Cancer Research, The Norwegian Radium Hospital, Montebello, Oslo 3, Norway
WALLACE, Clive, Dept. of Anatomy, Medical School, Univ. of the Witwaterstrand, Johannesburg 2001, Rep. of South Africa
WAYNE, A.W., Dept. of Haematology, King's College Hospital, Denmark Hill, London SE5 8RX, U.K.
WEBB, G.C., Dept. of Population Biology, Research School of Biological Sciences, P.O. Box 475, Canberra City, A.C.T. 2601, Australia
WEIMARCK, Anna, Inst. of Genetics, Sölvegatan 29, S-223 62 Lund, Sweden
WEIMARCK, Gunnar, Dept. of Plant Taxonomy, Ö. Vallgatan 20, S-223 61 Lund, Sweden

WENNSTRÖM, Johan, Kurjentie 5 F, SF-06100 Porvoo 10, Finland
WESTERMAN, M., Dept. of Genetics, Latrobe Univ., Bundoora, Victoria 3083, Australia
WESTIN, Jan, Section of Haematological Oncology, Dept. of Medicine II, Sahlgren's Hospital, S-413 45 Göteborg, Sweden
WOLF, Erich, Lab. of Cytogenetics, Ortlerweg 18, D-1000 Berlin 45, B.R.D.
WULF, H.C., Afd. D., The Finsen Inst., Strandboulevarden 49, Copenhagen Ø, Denmark

YOSHIDA, Y., Biological Lab., Tachikawa College of Tokyo, 3-6-33 Azuma-chô, Akishima-shi, Tokyo, 196 Japan

ZAINIEV, Gafur, Inst. of Cytology & Genetics, Siberian Dept., Acad. of Sciences of the USSR, Novosibirsk, U.S.S.R.
ZANG, Klaus, Inst. of Human Genetics, University Clinics, 6650 Homburg (Saar), B.R.D.
ZECH, Lore, Inst. for Medical Cell Research, Karolinska Institutet, S-104 01 Stockholm 60, Sweden
ZETTERBERG, Anders, Inst. for Medical Cell Research, Karolinska Institutet, S-104 01 Stockholm 60, Sweden

Chromosomes Today Volume 6, A. de la Chapelle and M. Sorsa eds.
© 1977 Elsevier/North-Holland Biomedical Press. Amsterdam, The Netherlands

THE CHROMOSOME REVOLUTION

C.D. Darlington
Botany School
Oxford

I

I am grateful to our hosts in Helsinki for giving me this opportunity of speaking today. For I want to persuade you that, if we look back at our journey with the chromosomes over the last hundred years and the story it tells us, we shall find that its meaning has changed and is changing with every advance we have made.

The chromosomes unfolded themselves to the eyes of our predecessors in the thirty years between the death of Darwin and the discovery of Drosophila. During that time a small body of pioneers won from the chromosomes a new view of life. It was that organisms do not, as had been supposed, manage their affairs for themselves. On the contrary, the chromosomes somehow manage these things for them.

The notion that the organism might be managed by something very small within its cells, of course, began earlier. It grew up with the microscope. A year to remember is 1875. Then Strasburger argued that the nucleus organised the division of cells, and Hertwig argued that the nuclei were what mattered in the fusion of cells. Three years later Claude Bernard was therefore able to say that the nucleus was the centre of construction and generation in the cell[1, 2, 3, 4].

The vegetative and the sexual problems raised by the activity of the nucleus came into focus separately but almost at once. In a great review in 1888, Waldeyer proposed that Boveri's "chromatic elements" in the nucleus and Flemming's split threads in mitosis were the same things and should be given the same name: the chromosomes[5]. This proposal of Waldeyer looked like a mere definition. But it was much more. It united the theories of the pioneers who had preceded him. It implied that nuclei, mitoses and chromosomes are all connected; and they are always connected also with the division (or multiplication) of cells. The chromosomes must therefore maintain their individuality or molecular continuity from one mitosis to the next. They must also be in a position to control heredity, variation and development in all higher organisms. And they must have been so throughout evolution.

Forty years later, as I remember, these far-reaching implications, seemed to have been overwhelmingly vindicated. But they were still being bitterly contested by the great majority who had never seen the chromosomes. The continuing dispute was useful for it showed us both the scope and the limits of the chromosomes. For example, it allowed Belar to assert that, although there might be bacteria and algae outside the rule of the chromosomes, all Protista, as well as all higher plants and animals, lay within it[6].

The name of the chromosomes thus contained within it both a programme of work and a manifesto for the workers, for us. This was even more obvious when Weismann broached the sexual side of the chromosome business.

To Weismann, arguing a transmission of heredity free from the body and the organism[7], the chromosomes were a gift from heaven. He did not wait to name them or even to see them before endowing them with all (and more than all) the properties we allow them. Some of his ideas and his names of germplasm, idants and little ids, have confused us ever since. But a greater difficulty for us is that he had two ideas which (in my opinion) were so vast, so intuitive, and so long before their time, that he was unable to pull them apart. Together they constitute the third step in our argument.

Weismann's first idea concerned a visible process. A reduction of chromosomes, he claimed, was universally demanded in all sexual reproduction in both male and female lines. Much later (in 1905) this prediction was to be sealed by the naming of meiosis. Weismann's second idea concerned properties of this meiosis that were invisible and remote, namely, its evolutionary origin and its causal connections. Here he was using the chromosomes to contradict a fallacy still widely held today. This was the idea that sexual reproduction owed its origin to the egg's instant need for the stimulus of the sperm. In its place he proposed that a long-term evolutionary advantage was the prime mover. This advantage lay in amphimixis, a mixing and unmixing of the materials of heredity which exposed them to natural selection. Thus the chromosomes from being the basis of heredity, became also the basis of evolution. In one stroke Weismann claimed to have taken away the ground from under Lamarckian inheritance and Darwinian pangenesis[8].

Weismann's theory of meiosis extended and implemented Darwin's theory of natural selection and like it reached as far as biology can reach. I remind you of it because, in the present chaos and frequent triviality of biological thought, today, ninety years later, I do not know how many biologists (i) know of it, (ii) accept it, or (iii) grasp its meaning. In the thirty crucial years which followed, however, it was this theory which put the chromosomes, meiosis and evolution together in one piece. How did this come about?

The first step was taken when Rückert (in 1892) saw the chromosomes at meiosis in the shark's egg. There they were, all paired in diplotene. Each pair showed one to seven points of contact or crossing over (Uberkreuzung). He could not see the chromatids but these points were evidently what we call chiasmata. Now, Rückert wrote, "since the chromosomes on our present understanding must be the bearers of heredity", and since, on Weismann's view, there must be a mixing of different qualities in heredity, these points of contact must be the places where such mixing (Substanzaustausch) takes place[9]. This inference was, to Rückert, a necessary corollary (Ergänzung) of Boveri's hypothesis of the individuality and continuity of the chromosomes. For since *Ascaris*, for example, has only one pair of chromosomes, any other assumption "leads to palpable absurdities". This was the fourth great speculative step in developing the chromosome theory. From it arose the sequence of discoveries on which our genetics is based: (i) the elucidation of the stages of meiosis by Winiwarter in 1901, (ii) the clinching of the connection with mendelism by Sutton in 1902, (iii) the recognition of the points of crossing over as chiasmata by Janssens in 1909 [10, 11, 12].

At this point what Weismann meant by mixing could be equated with what we mean by "genetic recombination". To complete the argument it remained only to connect the unmixing of the chromosomes at meiosis with the devices which promote their mixing by outbreeding. These were unravelled when East and his successors took up the threads of Darwin's last enquiry and discovered the evolution of the breeding system in plants.

II

With mendelism in 1900 there came an exchange of ideas between the new men whose chromosome theory had been established largely by "experiments of nature", as Wilson put it[4] and the even newer men, the mendelian breeders who relied on experiments with organisms; visible experiments with visible things. It was an exchange limited by distrust from the visible organism side. In 1909 Johannsen proposed his terms *gene* and *genotype*, excluding any possible basis in the chromosomes. In the same year Janssens introduced his crossing over and his chiasmata with an invitation to the breeders to take note. In 1911 Morgan took note. He adopted the chromosome theory[13]. But the chromosomes themselves he set on one side. It was left to Muller, taking his own line in 1914, to nail Johanssen's genes into place in or on the chromosomes - where they have remained[14].

To us today this looks like a reconciliation and a fusion of ideas. But to Morgan it was not so. In "The Theory of the Gene" in 1926[15] "characters" had at last become "genes" but genes, like atoms, were invisible. The theory was one of linkage groups. It rested on "experimentally determined genetic evidence", that is on evidence that could disregard the chromosomes. Later Morgan admits, or claims, that "The cytologist has given us an account of the chromosomes which fulfills the requirements of genetics". Hence the chromosomes were useful "to supply a mechanism for the theory of heredity", a theory based solely on breeding experiments. The chromosomes themselves were not experimental and therefore not genetic. They could not create or solve problems of their own since they always did what "genetics", that is to say, organisms, required them to do. To suppose, with Janssens, that meiosis had a value in evolution was therefore, as Morgan told me, a teleological speculation.

From this, as we may say, _organismal_ point of view, the mechanical question as to how the chromosomes did their job was a matter of indifference. And for twenty years it remained so. Similarly, the evolutionary question as to why crossing over was suppressed in one sex could not be asked and need not be answered. And this property, crucial as I would say for experiment and for nature, was relegated by Morgan to a footnote (15: p.12).

Morgan, it seems, felt bound to cut the chromosomes off from the chromosome theory. Why? Probably because in this way he would least offend his own belief, which was also the established or conventional belief, in the primacy of organisms. In consequence he was forced into an argument which has become in our eyes historically false. It was genetically false because the chromosome theory of heredity was understood (witness Wilson in 1900) before mendelism was rediscovered. It was physiologically false because the chromosome theory of development had been understood twenty years earlier. It was evolutionarily false because chromosomes came, if not before organisms in time, at least along with them.

From my point of view, at the time, the theory of heredity was based on the study of chromosomes at all stages of the lives of the whole world of plants and animals; it had been confirmed, developed and implemented by the _Drosophila_ experiments. Now we had to return to the chromosomes and see what they did in the whole range of living nature. We had to examine systematically the total results of their activities. In this way we might discover that the connections between evolution and heredity were not what experimental breeders might suppose.

In my attempt to do this, I assumed that all chiasmata resulted from crossing over and all chromosome pairing resulted from chiasma formation. These assumptions pushed the economy of hypothesis too far. But they proved to be a necessary step in arriving at a number of generalisations which could hardly have been reached at all from breeding experiments [13, 16, 17].

(i) That meiosis has two contrasted forms, with and without chiasmata, giving the option of two tracks in heredity.

(ii) That all sexually reproducing species have chiasmata at least in one track.

(iii) That chiasmata are always the consequence of crossing over between chromosomes at the haploid level.

(iv) That all chromosomes in all sexually reproducing organisms are therefore subject to recombination by crossing over.

(v) That chromosomes are therefore inherently divisible into units connected with what Muller described as genes.

(vi) That chromosomes are therefore inherently capable of exchanges at the haploid level.

(vii) That these exchanges at the haploid level underlie both the continuities and the discontinuities of species at haploid as well as at diploid levels.

We can at once see points at which these generalisations call out to be qualified, elaborated or explained. We cannot doubt, moreover, that the evolution of heredity has gone or may go beyond any limits we now know. Moreover its molecular basis goes back in time beyond the chromosomes and mitosis as we now understand them[18]. But this experience tells us that each new turn in our understanding is likely to enlarge as well as correct what has gone before.

III

To understand the place of the chromosomes in nature we have to assemble highly contrasted sources of evidence. Plants and animals, breeding and microscopy, experiment and nature, all differ in what they tell us, notably as between the integral and the analytical. But besides all these differences there is the deep contrast between the organisms which we know as the whole of ourselves and the chromosomes which even for us are only parts.

We can accustom ourselves in theory to the chromosomes controlling our heredity and development. We may also see that by the invention of meiosis the chromosomes brought organisms and species into a relationship from which all later evolution sprang. But the naturalist, if he has a theory, will still hanker after organisms and species as the real self-governing things. He can see his groups of animals or plants or human beings breeding. But he cannot see chromosomes breeding. Do they in fact breed? That is not the way he looks at them[19].

The history and geography of our subject show us how this gulf opened between the visible whole and the invisible parts. Cells and chromosomes, genes and nucleic acids, the analytical hierarchy of genetics, burst on the world out of German microscopy and chemistry. They made their western impact on naturalists who had thought first and last of organisms. Look at the reactions of the two Anglo-Saxons who bore the painful brunt of the new ideas. Bateson accepted mendelism because it established discontinuity between organisms and in doing so set aside both Lamarckism and biometry. But he rejected the chromosomes because they did not do what his organisms needed them to do[13, 20]. Morgan, on the other hand, rejected mendelism because it did not allow what his Lamarckian view of organisms and evolution required. Only _Drosophila_ and his pupils, who were Wilson's pupils, forced him to accept mendelism. He then accepted the chromosomes into the bargain on condition of their subordinate status, an almost Lamarckian status, of doing what organisms required them to do. To the end he was clinging (as Julian Huxley later did) to human culture as the surviving example of an inheritance of acquired characters, characters acquired by organisms[21, 22, 23, 24].

If Bateson and Morgan found the chromosomes hard to avoid, their successors, the statistical neo-Darwinians, found it much easier. For Fisher in 1930, there were "dark-staining bodies or chromosomes which are to be seen in the nuclei of cells at certain stages of cell division"[25]. Haldane, working beside me, could not be so disparaging[26]. But for both the chromosomes were pieces from the Morgan model. Bemused by their apparent deductive success in using this model to vindicate Darwin's theory, they took it to be final. The chromosomes performed the duties required by experimental breeding which was, as Haldane like Morgan put it to me, the genuine "genetics". They dared not notice the evidence that the Darwinian evolution of organisms is tripped and trapped in nature, not just by Sewall Wright's drift, but by a hundred unexpected accidents of chromosome perversity[27].

We have indeed for too long allowed these accidents to be brushed under the carpet. We have been ashamed to admit that the chromosomes do a great deal of what they do as organisms in their own right. Accessory and sex chromosomes, diminishing and expelled chromosomes, heterochromatic, inert, and unstable chromosomes, interchange and inversion complexes, centromeres, organisers and controlling elements, all evolve by their own rules often sacrificing the interests of their hosts.

This rich variety of behaviour we can see at work in a world in which genes and chromosomes, meiosis and fertilisation, organisms and communities, interact and are adapted to interact. The chromosomes therefore, carrying their pedigrees on their backs, maintain, not only the equipment for the present, but also the evidences of the past and the threats or promises for the future. All of these concern both organisms and chromosomes. For chromosome variations are available both to hold a species together and to split it apart. And in both these crucial operations the chromosomes are often bound to take the initiative. They do so in the sense in which de Vries supposed his "mutations" to take the initiative in evolution.

How important this initiative may be is best shown by the evolutionary trends in which chromosome changes proceed without any organismal connection. For example, in sex chromosomes and interchange complexes, with inactivation, and in mere size and polyploidy[4, 11]. These trends, like those which interested Lamarck at the level of organisms, can now be seen to arise from positive feedback in natural selection[19].

Thus looking at the chromosomes as organisms we have to see their relations with their hosts in three stages: (i) chromosomes instruct their organisms as individuals — in heredity and in experiment; (ii) organisms, through natural selection, guide their chromosomes in evolution and in nature; (iii) chromosomes limit their directions of evolution by their own integral character. But, however we look at them, this integral character of the chromosomes is not deducible from their analytical components, the genes, any more than it is deducible from the character of organisms themselves.

In asking you to think of chromosomes as organisms I am not of course suggesting that we should abandon the reductionist principles on which we have worked so long. I am suggesting that we have to apply these principles to a complexity of structures and processes which are comparable in their history with those of the macroscopic organisms in which they live. With the chromosome revolution we have to enter a new world much of which still lies beyond our horizon.

REFERENCES

1. Strasburger, E. (1875). Ueber Zellbildung und Zelltheilung Dabis, Jena.
2. Bernard, C. (1878). Leçons sur les Phénomènes de la Vie. Paris.
3. Roux, W. (1883). Über die Bedeutung der Kerntheilungsfiguren. Engelmann, Leipzig.

4. Wilson, E.B. (1900, 1925). The Cell (3rd Ed. Macmillan, N.Y. note: fig. 36. Reprint 1966 Introd. H.J. Muller).
5. Waldeyer, W. (1888). Über Karyokinese und ihre Beziehungen zu den Befruchtungsvorgängen. Arch. mik. Anat: 32, 1-122. (Trans. Q. J. mic. Sci., N.S. 30, 159-281.
6. Belar, K. (1926). Der Formwechsel der Protistenkerne. Ergeb. Fortschr. Zool. 6, 1-323.
7. Weismann, A. (1883). Über die Vererbung (erworbener Eigenschaften) Fischer, Jena.
8. Weismann, A. (1887). On the significance of the polar globules. Nature 36, 607-609.
9. Rückert, J. (1892). Zur Entwickelungsgeschichte des Ovarialeies der Selachiern. Anat. Anz. 7, 107-158.
10. Sutton, W.S. (1902). On the morphology of the chromosome group in Brachystola magna. Biol. Bull. 4, 24-39.
11. Darlington, C.D. (1965). Cytology. Churchill, London. (note: fig. 110).
12. Janssens, F.A. (1909). Spermatogénèse dans les Batraciens V. La Théorie de la Chiasmatypie, Nouvelle Interprétation des Cinèses de Maturation. Cellule 25, 387-411.
12a. Winiwarter, Hans von (1901). Recherches sur l'ovogenèse et l'organogenèse de l'ovaire des Mammifères (Lapin et Homme). Arch. de Biol. 17, 33-200.
13. Darlington, C.D. (1964). Genetics and Man. Allen & Unwin, London.
14. Muller, H.J. (1914). A gene for the fourth chromosome of Drosophila. J. Exp. Zool. 17 (3).
15. Morgan, T.H. (1926). The Theory of the Gene. Yale U.P. New Haven.
16. Darlington, C.D. (1973). The place of the chromosomes in the genetic system. Chromosomes Today 4, 1-13 (Jerusalem).
17. Darlington, C.D. (1976). The great events in chromosome evolution. Current Chromosome Research 1-5, Kew: Elsevier.
18. Catcheside, D.G. (1977). The Genetics of Recombination. Arnold, London.
19. Darlington, C.D. (1977). The Little Universe of Man. Allen & Unwin, London.
20. Coleman, W. (1970). Bateson and Chromosomes. Centaurus 15, 228-314.
21. Morgan, T.H. (1932). The Scientific Basis of Evolution. Norton, N.Y.
22. Muller, H.J. (1943). E.B. Wilson. Am. Nat. 77, 5-37.

23. Muller, H.J. (1946). T.H. Morgan. Science 103, 530-551.
24. Shine, Ian. & S. Wrobel. Thomas Hunt Morgan. Kentucky U.P., Lexington.
25. Fisher, R.A. (1930). The Genetical Theory of Natural Selection. Oxford U.P.
26. Haldane, J.B.S. (1932). The Causes of Evolution. Longman, London.
27. Darlington, C.D. (1971). Axiom and process in genetics. Nature 234, 521-5.

CHROMOSOME STRUCTURE

Chromosomes Today Volume 6, A. de la Chapelle and M. Sorsa eds.

CHROMOSOME STRUCTURE - AN INTRODUCTION

Biophysical techniques in chromosome analysis

TORBJÖRN CASPERSSON

Institute for Tumor Pathology, Karolinska Institutet,
S 104 01 Stockholm 60 (Sweden)

The supplementation of conventional microscopy by different other biophysical techniques, in the first line measuring procedures, all aiming at the use of the microscope as a quantitative tool, proved to be of value for cell and chromosome studies already long ago. The very first can be said to have been the quantitative polarization method which was developed quite early, but it found only few applications in cell biology. In the mid-thirties the first ultramicrospectrophotometric procedures were developed directly for cell biological purposes. In the forties and early fifties they were joined by X-ray based measuring cytochemical techniques and by fluorometry and also microinterferometry. The amount and distribution of many different cell constituents can be determined with these techniques. For work on nuclear components spectrophotometry and fluorometry proved to be the most useful ones.

The widespread interest in the organization of chromosomes and chromatin gave the most important stimulus in the early days to the development of new biophysical procedures. The thirties were, of course, a period when through Darlington's inspired and inspiring work and through the magnificent earlier work of the Morgan school, supplemented by Hans Bauer's and T. Painter's findings of polytene chromosomes in the tissues of Diptera, much interest was focused on the one side on the metaphase chromosome and its structure, and on the other, on the salivary gland type chromosomes, where it seemed that the functioning individual gene could be studied in loco - a fascinating aspect.

It is perhaps not so easy to fathom nowadays that the early thirties was a time when cytochemistry was still in such a state that claims could be heard that chromosomes consisted only of proteins and lipids - there are Feulgen-positive lipids - nothing was known about the function of nucleic acids, the best guess seemed to be that they were just waste products of the cell´s metabolism and, finally, Nature´s mysterious ways of synthesizing proteins just appeared as black magic.

In the middle thirties, however, measuring instrumentation was developed and the first UV-absorption spectra could be taken from individual metaphase chromosomes and even from individual bands and interbands of Drosophila salivary gland chromosomes. Mitotic and also meiotic chromosomes were studied in many different objects. The geneticist Jack Schultz from the Morgan group worked with us in Stockholm on Drosophila polytene chromosomes at that time looking for changes in the DNA-amounts in individual bands correlated with gene effects and he could show such changes in certain variegated Drosophila stocks.

Only a few years later, 1940/41, the techniques were so advanced that very detailed spectra could be measured even in small nuclear elements so that information could be obtained not only about the nucleic acids present but also proteins and even certain amino acids. During the first years of the second world war so much information had been collected that it was evident that nucleic acids play a crucial role in the cellular protein synthesis.

This made our group in Stockholm during most of the years of the war to concentrate the cytochemical work on the relation between RNA and cytoplasmic protein synthesis, but shortly after the end of the war cytochemical chromosome work was again running full speed. We had the privilege then to have as long time staying guests such grand men of cytology as L. Geitler and Hans Bauer. Geitler, among other things, tried to find a cytological material where UV-dichroism measurements could be made at the time of expected gene reproduction - the thought was to look for possible orientation of the DNA at such a time. One of Bauer´s main contributions from this time was the development of cultivation procedures for Chironomus. The reason was Schultz´ work, mentioned above which made it appear possible cytochemically to observe changes in individual polytene chromosome bands correlated to gene function. The very small size of the Drosophila chromosomes, however, made the measurements very difficult and the much larger chromosomes of Chironomus should offer a much more favorable situation for studies of individual bands. Bauer´s younger collaborator Wolfgang Beerman also spent a year with us in the early fifties continuing Bauer´s work, what led up to Beermans later fine series of studies of the structure of Chironomus chromosomes.

During the following decade many laboratories worked cytophotometrically with DNA-determinations and this contributed to the development of our present view of the role and mode of function of the DNA.

Little was done on the cytochemistry of the chromosome structures during this rather long period in the fifties and the sixties. This was, however, a time when cancer research expanded very fast all over the Western world and the work on carcinogenesis led little by little to a renewed interest in the behaviour and structure of individual chromosomes. Of natural reasons the emphasis was now on the mammalian chromosomes. Other inducements to chromosome research came during the same period from the atomic bomb tests and the expanding work on nuclear energy, mirrored for instance in the long standing work of the United Nations´multinational committe for the study of biological radiation effects.

If I may again take an example from our own research group, our reaction to these trends about 10 years ago was to try quite hard to bring our different cytochemical light-optical techniques to work well also with individual mammalian chromosomes in spite of their small size. This was a natural link in our then running carcinogenesis work - an effort to identify very early nuclear changes during transformation, changes which could not possibly be detected by simple DNA-determinations in whole nuclei. We had rather good success with the improvement of the technical procedures but hit then straight on the chromosome identification problem in the mammalian cells. Obviously any cytochemical work on individual chromosome parts would have little meaning if one could not recognize and identify the chromosome and/or chromosome part studied. We tried hard to lick this problem by aid of different variations of high resolution UV-microspectrophotometry measuring at first DNA-amounts and then DNA-distribution in individual mammalian chromosomes. The results were quite promising - we even got clear indications of the existence of a banding system in chromosomes - but the technical work involved was necessarily so very complex and demanding that it was difficult to envisage practical application of such methods.

Then, starting with the thought that of simple statistical reasons there must be considerable differences in the base composition of individual chromosome parts as the complexity of the genetic information is carried by differences in the base sequences in the DNA we looked for specific base-reacting compounds, which contained fluorescent groups as indicators. There were special reasons for our choice of fluorometry in this work - the main one being the ease with which the distribution of fluorescent substances can be measured and recorded also in structures of a size close to the limit of the microscope's resolution. Many substances were tried and this led to the development of the quinacrine and quinacrine mustard banding techniques for chromosome identification.

The only comment to be made here about this method is that I am sure it would not have been developed at that time if there had not been directly available in the laboratory quite elaborate microfluorometers which gave clear objective information about the existence of banding patterns in the chromosomes where by eye we could only see fuzzy irregularities in the preparations of those days. In order to collect such a large observational material on human chromosomes that it could be analyzed statistically by computerized procedures a special fast semiautomated fluorescence profile recorder was built by aid of which measurements from nearly 30 000 chromosomes were recorded and could be analyzed. This gave the solid basis for the fluorescence chromosome identification system in man.

The fields of application in medicine and biology of the nowadays available several different banding techniques, mutually supplementing each other in a very efficient way, are at present growing very fast. However, it will never be possible fully to exploit the great potentialities of these already available techniques for work on such most important and at present much discussed fields as the analysis of the effects of environmental clastogens and of carcinogens of different origins with the at present generally used routine procedures - they are too slow and also too wearisome for the operator. Here again biophysical auxiliary tecniques can come in and speed up and facilitate the work.

The efforts in several well equipped laboratories to develop fully automated computerized chromosome analytical techniques have been extensively discussed in the last years´literature. Regrettably enough they do not look very promising at this moment, especially not for the detailed search for and analysis of chromosome aberrations which is a necessary and an important part of the work on the fields mentioned above.

In our laboratory we have during the last few years been trying other simpler ways, namely to keep the aberration analysis by visual inspection, but making it faster and more convenient by different technical tricks. Crucial in these instruments is the replacement of the time consuming photographic work - partly or entirely - by TV-based procedures in which electronic manipulation of the picture on the monitor greatly facilitates and speeds up the work. Such units have been built for as well stained specimens as for fluorochromed ones. In the former one can dispense entirely with photography what means a great saving in time. In the latter we still keep the primary photography of the specimen and work with TV-techniques in the negative. In our experience from large scale aberration analysis work one can make very great gains in time and efficiency with such auxiliary devices. Similar work is going on in other laboratories. Our efforts are cited here only because they give another example of the usefulness of various biophysical tools for chromosome research in general It is to be expected that in the future these procedures can be considerably improved upon.

Biophysical procedures, supplementing the microscope in various ways, have thus helped the chromosome analytic work along during a period of more than 40 years and still new developments on these fields are to be expected in the future.

Chromosomes Today Volume 6, A. de la Chapelle and M. Sorsa eds.
© 1977 Elsevier/North-Holland Biomedical Press, Amsterdam, The Netherlands

MAMMALIAN CHROMOSOME STRUCTURE

DAVID E. COMINGS, M.D.
City of Hope National Medical Center, Department of Medical Genetics, 1500 East Duarte Road, Duarte, California, 91010 U.S.A.

The subject of mammalian chromosome structure continues to be an exciting and controversial domain of scientific endeavour. Some of the problems that were most controversial a number of years ago, such as uninemy versus polynemy and the structure of the 250 Å fiber are now solved. That the chromatid of mitotic chromosomes is a uninemic structure is no longer a subject of significant contention. Likewise, the voluminous explosion of literature on the nu body has solved all but a few details about the structure of the chromatin in its extended "nucleosomes on a string" configuration or in its compact configuration forming the 250 Å fiber. Despite this progress significant areas of controversy regarding the tertiary structure of the chromatin in metaphase chromosomes still remain. The following are some aspects of this problem I would like to emphasize.

1. The structure of *Drosophila* polytene chromosomes and mammalian metaphase chromosomes may be more similar than they appear. A number of years ago Crick suggested the active genetic material of *Drosophila* polytene chromosomes was in the interband regions rather than the bands. Recent studies by Jamrich et al.[2] which show RNA polymerase to be located in the interband regions and puffs, is in favor of this concept. The observation that ^{3}H-uridine autoradiography of polytene chromosomes shows a low level of grains over the interbands[3,4] also agrees with this proposal. The analogy with mammalian bands and interbands is enticing. The fact that interband (R-band) DNA is early replicating and GC-rich and G-band DNA is late replicating and AT-rich has long been strong evidence in favor of R-bands containing active euchromatin and G-bands containing inactive heterochromatin (see Ref. 5 and 6 for review). The fact that the only three human chromosomes, 21, 13 and 18, that are tolerated in the trisomic state also contain the least amount of R-band DNA[7] supports this correlation. The recent demonstration that cDNA to poly A mRNA preferentially hybridizes to the R-bands[8] also supports this proposal. We have suggested that the increased amount of nonhistone protein associated with euchromatin interfers with the binding of Giemsa to R-band DNA and is partly responsible for the poor staining of R-bands seen in G-banding[9]. While there are obviously many more bands in *Drosophila* polytene chromosomes (about 5,000) than in human chromosomes (about 350), fine structure mapping of extended pro-metaphase chromosomes[10] or chromosomes extended by treatment with certain chemicals[11] indicate there are 2-5 sub-bands in each major G-band and the total number of metaphase

chromomeres is closer to 1,200. It would come as no surprise that if polytene chromosomes could be produced in humans there would be more bands than in *Drosophila*. Attempts to "condense" the polytene chromosomes can also be of interest. Holmquist and Steffensen[12] have continued to fill out the map originally published by Hannah[13] to show the location of intercalary heterochromatin in *Drosophila* polytene chromosomes. The pattern has intriguing resemblances to the gross banding pattern of mammalian chromosomes (Figure 1).

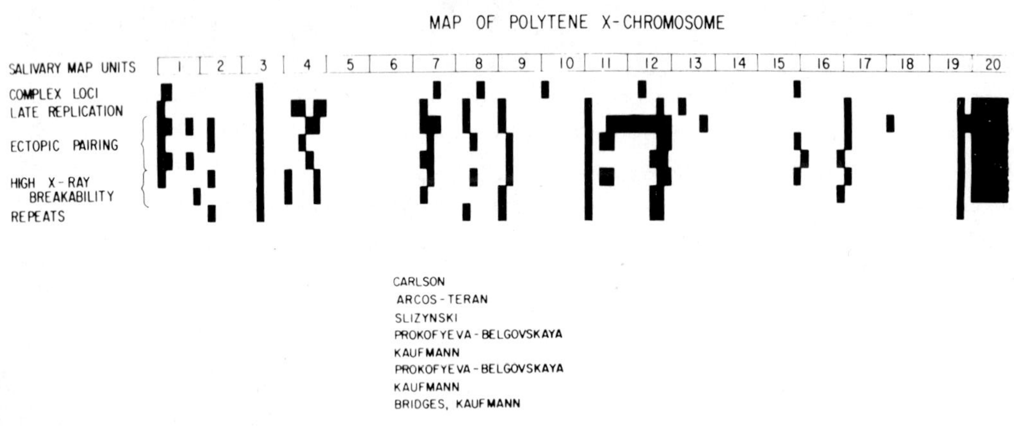

Fig. 1. Attributes of intercalary heterochromatin mapped on the polytene X-chromosome of *Drosophila melanogaster*. Band repeats, high x-ray breakability and high ectopic pairing are redrawn from Hannah[13]. More recent material is included on ectopic pairing[14], late-DNA replication[15] and complex loci[16,17]. Region 20 contains centric heterochromatin. Figure and Legend kindly supplied by Gerald Holmquist and Dale Steffenson.

2. The role of nuclear matrix in chromosomal formation. One of the few satisfactory ways to account for the bands in polytene chromosomes is to assume that each DNA strand contains a series of loops of varying sizes. The polytene structure would thus have the appearance shown in Figure 2[18-20].

This is consistent with the presence of domains of supercoiled DNA found to be present in eukaryotic nuclei[21-23]. Ide et al.[21] found that mammalian nuclei with supercoiled DNA also possessed a major nonhistone protein with a molecular weight of 68,000 daltons. This coincides with that of the major nuclear matrix protein[24-26]. Nuclear matrix proteins constitute 10-15% of the total nuclear nonhistone protein

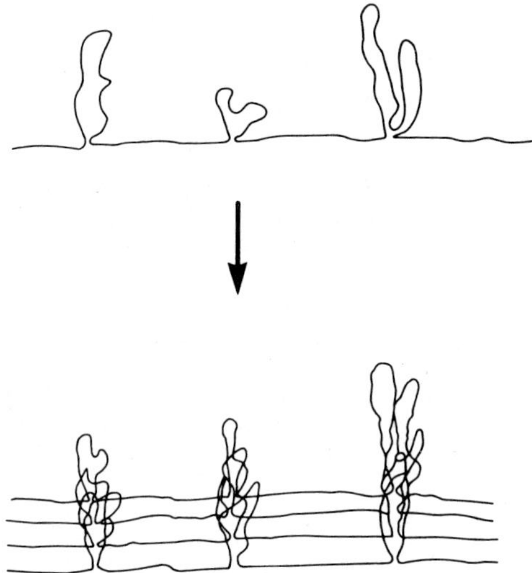

Fig. 2. Chromomeres of polytene chromosomes as loops in a linear DNA molecule.

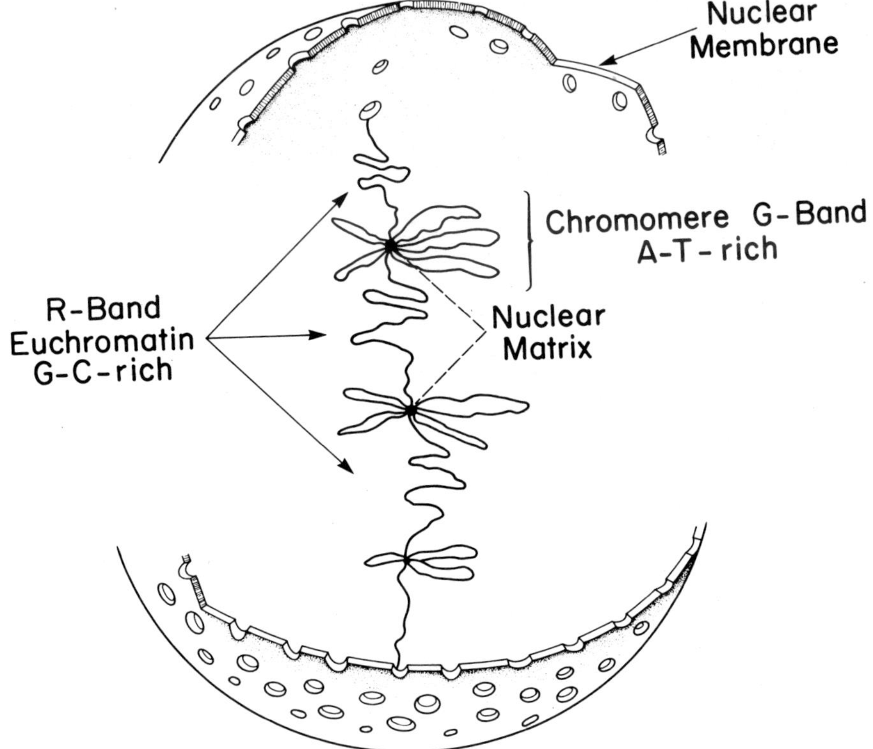

Fig. 3. Model of an interphase chromosome with telomeres attached to the nuclear membrane and G-band AT-rich DNA attached to the intranuclear matrix. This has been enlarged for diagramatic purposes. In reality, the two telomeres attach closer to each other on a smaller region of the nuclear membrane.

and a much higher percentage of the nonhistone proteins of well washed nuclei[27]. Our studies of the DNA binding properties of nuclear matrix indicate it preferentially binds to AT-rich DNA with a clear preference for the poly-T strand[28]. This is of particular interest since the G-band DNA is AT-rich and may contain stretches of very AT-rich DNA[29]. The nuclear matrix consists of the inner nuclear membrane, a portion of the nucleolus, and an extensive amount of intranuclear matrix. DNA is attached to all three of these elements[24-26]. Figure 3 is thus a reasonable model of interphase chromosomes with chromomeres of relatively AT-rich DNA bound to matrix to form G-bands and interband regions of genetically active chromatin to form the R-bands. Replicating DNA also appears to be attached to the nuclear matrix[29a]. The metaphase chromosomes represent the condensed state of this structure with several small chromomeres condensing to form visible G-bands (Figure 6).

3. Multiple longitudinal DNA fibers do not exist. A persistent feature of many models of chromosome structure is the presence of primarily longitudinal strands of DNA traversing through two or more chromomeres[30-33]. I believe there are several strong arguments against this. One is the fact that chromosome breaks follow a linear relationship to dose of X-ray (for review Ref. 34). This effectively precludes more than a single longitudinal DNA fiber. The correlation between the genetic map and the chromomere map is also inconsistent with multiple longitudinal fibers. Such fibers also introduce enormous complications in the production of simple break and reunion translocations. The concept of multiple longitudinal fibers has grown almost entirely from the fact that even a slight degree of tension on water spread chromosomes causes the fibers to line up parallel to the long axis of the chromatid and these are interpreted as a distinct set of longitudinal fibers.

4. The chromatid can be uncoiled. From EM studies of thousands of water spread chromosomes treated by a variety of different ways it is apparent that the chromatid can be preserved as is or extended to almost any degree. For example, as shown in figure 4a chromosomes were spread on the surface of 0.2 M HCl and 1.0 M NaCl for 120 seconds then picked up on grids for electron microscopy. The chromatids are spread to a total length of up to 50 microns and where they are not dispersed the chromatid has a width of 0.12 μm. In b) they were spread on 0.05% SDS and here show a banding pattern with chromomeres and a chromatid width of 0.5 to 1.0 μm between the chromomeres and 1.3 to 1.8 μm at the chromomeres. In Figure 4 c) and d) the chromosomes have been spread on 0.65 M NaCl, 0.01 M EDTA, 0.01 M Tris pH 7.54 plus 5 μg/ml RNase A, and 5 μ/ml of RNase T1, incubated for 10 minutes at 37° and picked up on grids. Here the chromatid is extended up to 75 μm and the width is as low as 0.1 μm. Bak, Zeuthen and Crick[35] have reported that under one set of conditions the chromatids of human chromosomes decondense to a structure with a width of 0.35-0.43 μm. They claim this is a hollow tube but we have found no evidence for this in thin section or whole mount electron microscopy and there is no indication of this by scanning electron microscopy of chromosomes[33,36]. While it is true that

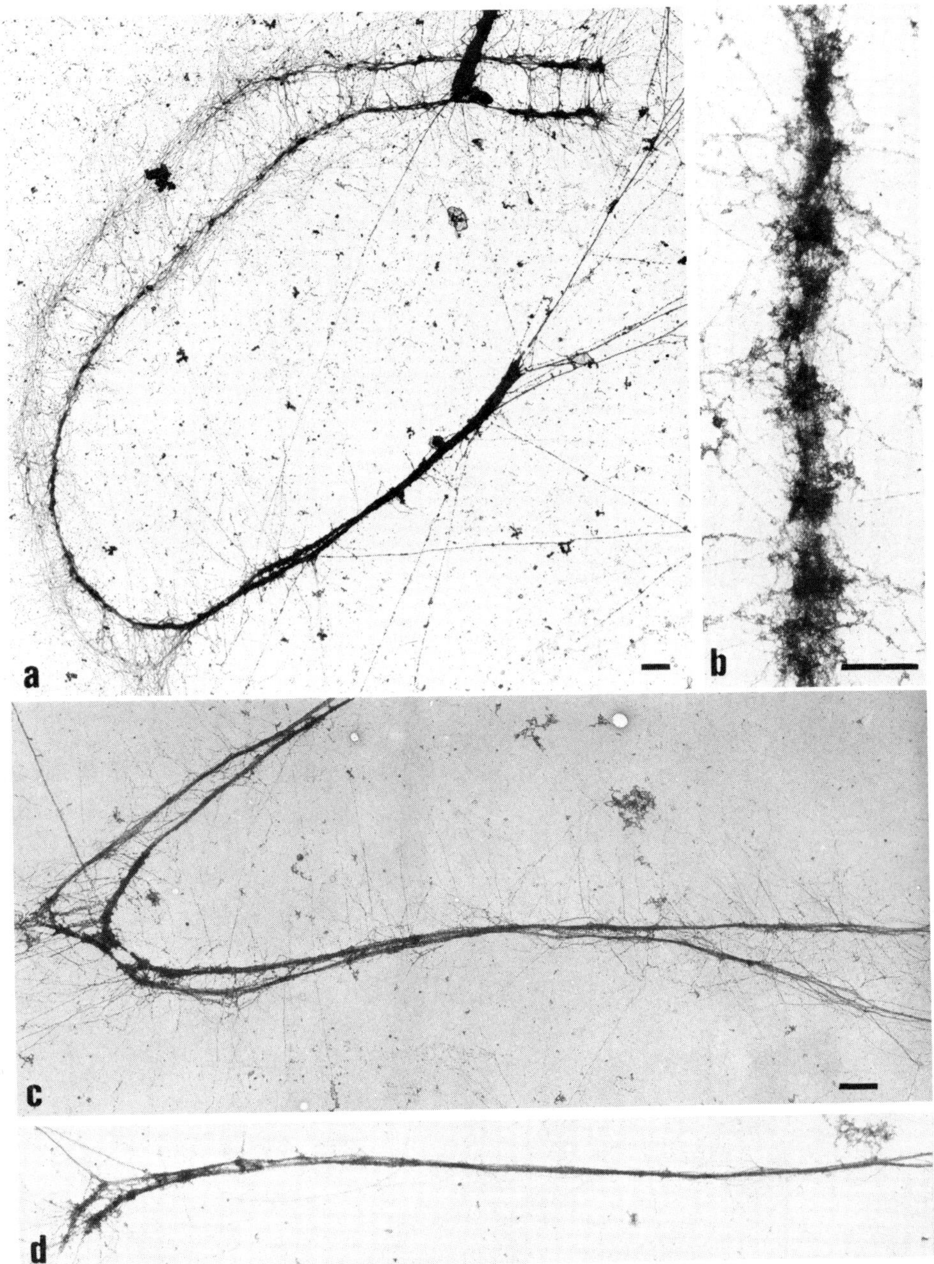

Fig. 4. a) Chinese hamster metaphase cells spread on 0.2 N HCl, 1.0 M NaCl for 120 seconds; b) Spread on 0.05% sodium dodecyl sulfate for 30 seconds; c) and d) Chinese hamster metaphase cells in McCoy's Media mixed with equal volume of distilled water than spread on 0.65 M NaCl, 0.01 M EDTA, 0.01 M Tris pH 7.4 + RNase A 5 μg/ml, RNase Ti 5 μg/ml, RNase Ti 5 μ/ml at 37° for 10 minutes. Chromosomes were then picked up on grinds, stained in 2% uranyl acetate, dehydrated to 100% ethanol, placed in amyl acetate and air dried. Photographs by Dr. T.A. Okada.

with any given set of conditions a rather constant degree of uncoiling is seen, almost any width structure can be produced with different conditions. As a further example, in 1962 Brooke et al.[37] reported uncoiling human chromosomes with 0.01 M KCl. Different degrees of uncoiling were produced including chromotids with a width of 0.25-0.35 μm.

One of the frustrating paradoxes of studying the higher order structure of metaphase chromosomes is that they must be decondensed to see the manner of the tertiary folding of the chromatin and yet this very decondensation tends to destroy the tertiary structure which is being sought and produces instead a bewildering array of artifacts.

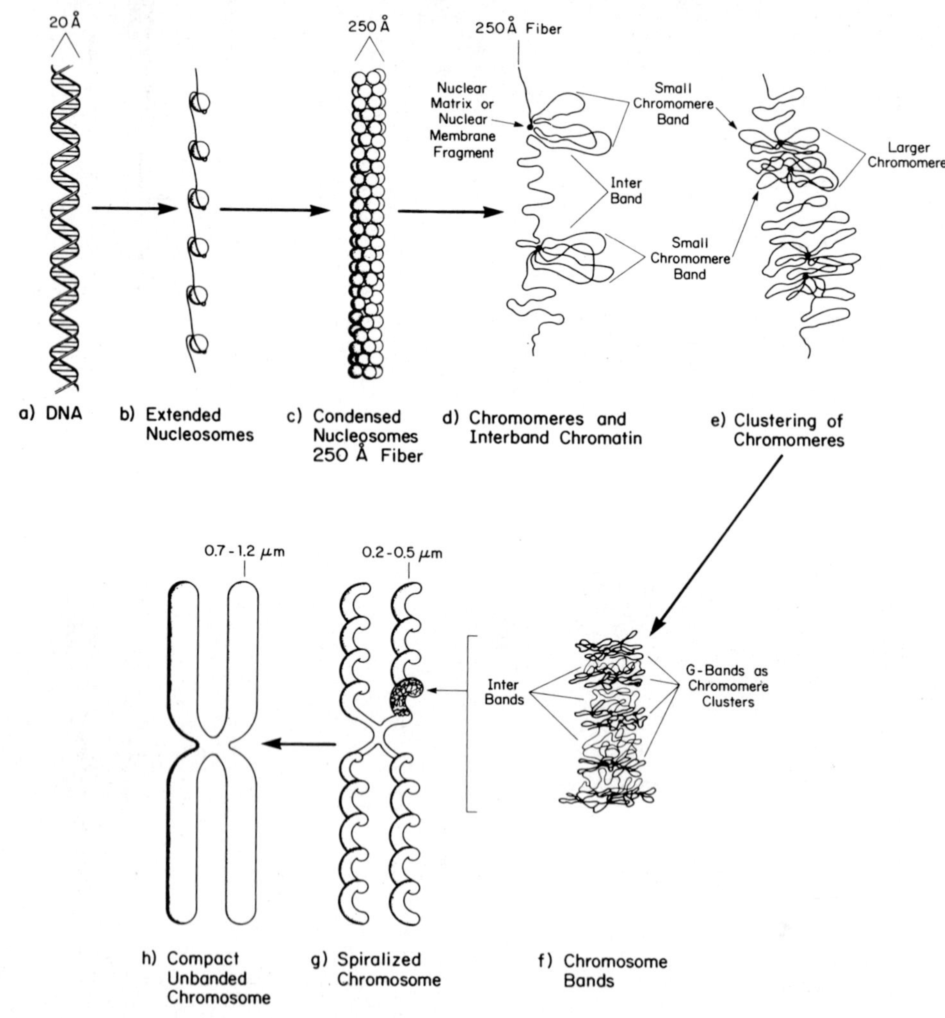

Fig. 5 - (See text)

5. Remnants of the nuclear matrix persist in metaphase chromosomes. If the interphase chromatin were arranged into chromomeres through the attachment of chromatin to the nuclear membrane and other parts of the nuclear matrix (as in Figure 3), it would be extrememly convenient for the conservation of chromosome structure if the matrix did not detach during mitosis. We have observed the presence of membrane fragments[38] and nuclear annulae[39] persisting in well spread metaphase chromosomes. Engelhardt and Pusa[40] and recently Maul[40] have also observed annulae in meiotic and metaphase chromosomes by thin section electron microscopy. By scanning electron microscopy membrane plates are sometimes seen at the teleomeres[42]. The degree to which the intra-nuclear matrix remains bound to the chromosome during mitosis requires further study.

Figure 5 summarizes numerous aspects of mammalian chromosome structure. a) begins with 20 Å DNA fiber which in b) is complexed with nucleosomes to form the extended "nucleosome on a string" configuration. In c) condensation of the nucleosomes results in a 100 to 300 Å thick fiber[43,44] forming the classical 250 Å chromatin fiber of metaphase and interphase chromatin. d) illustrates the proposal discussed in this paper that nuclear matrix binds to AT-rich segments of DNA present in G-band chromatin to form chromomere loops and inter-chromomere DNA. These condense, e) and f), to form G-bands and R-bands. h) illustrates a partially uncoiled chromosome producing spiralization. The dimensions of this spiral fiber are 0.2 to 0.5 μm based on measurements from the paper by Ohnuki[45]. This can be further despiralized and depending upon the conditions the resulting strand can be 0.35 - 0.43 μm[35], .25-.35 m[37] or down to 0.1 m (Figure 4). When this spiralized chromosome is fully condensed it forms the classical appearing chromosome (h).

REFERENCES

1. Crick, F. (1971) Nature 234, 25-27.
2. Jamrich, M.J., Greenleaf, A.L., and Bautz, E.K.F. (1977) Proc. Nat. Acad. Sci. 74, 2079-2083.
3. Holmquist, G. (1972) Chromosoma 36, 413-452.
4. Zhimulev, I.F. and Belyaeva, E.S. (1975) Chromosoma 49, 219-231.
5. Comings, D.E. (1972) in Advances in Human in Human Genetics, Harris, H. and Hirschhorn, K. eds., Plenum Press, New York 3, 237-431.
6. Comings, D.E. (1974) in The Cell Nucleus, Busch, H. Ed., Academic Press, 1, 537-563.
7. Hoehn, H. (1975) Am. J. Human Genet. 27, 676-686.
8. Yunis, J.J., Juo, M.T. and Saunders, G.F. (1977) Chromosoma 61, 335-344.
9. Comings, D.E. and Avelino, E. (1975) Chromosoma 51, 365-379.
10. Yunis, J.J. (1976) Science 191, 1268-1270.
11. Shafer, D.A. (personal communication).
12. Holmquist, G. and Steffensen, D.M. (personal communication)
13. Hannah, A. (1951) Advances in Genetics 4, 87-125.

14. Kaufmann, B.P. and Iddles, M.L. (1963) Port. Acta. Biol. 7, 225-248.
15. Arcos-Teran, L. (1972) Chromosoma Berl. 37, 233-296.
16. Carlson, E.A. (1959) Quart. Rev. Biol. 34, 33-67.
17. Serebrovsky, A.S. (1938) Acad. Sci. U.S.S.R. 19, 77-81.
18. Comings, D.E. and Okada, T.A. (1974) Cold Spring Harbor Symp. Quant. Biol. 37, 145-153.
19. Sorsa, V. (1972) Hereditas 72, 169-172.
20. Haapala, O.K. (1973) Hereditas 75, 61-66.
21. Ide, T., Nakane, M., Anzai, K. and Andoh, T. (1975) Nature 258, 445-447.
22. Cook, P.R. and Brazell, I.A. (1975) J. Cell Sci. 19, 261-279.
23. Benyajati, C. and Worcel, A. (1976) Cell 9, 393-407.
24. Berezney, R. and Cofffey, D.S. (1974) Biochem. Biophys. Res. Comm. 60, 1410-1417.
25. Berezney, R. and Coffey, D.S. (1976) Adv. Enzyme Regulation, Weber, G. Ed., 14, 63-100.
26. Comings, D.E. and Okada, T.A. (1976) Exp. Cell Res. 103, 341-360.
27. Comings, D.E. (1978) The Cell Nucleus, Busch, H., Ed., Vol. 5, Academic Press, New York (In press).
28. Comings, D.E. and Wallack, A.S. (unpublished data).
29. Comings, D.E. and Drets, M.E. (1976) Chromosoma 56, 199-211.
29a. Berezney, R. and Coffey, D.S. (1975) Science 189, 29-293.
30. DuPraw, E.J. (1968) Academic Press, New York, p. 560.
31. Stubblefield, E. and Wray, W. (1971) Chromosoma 32, 262-294.
32. Bahr, G.F. (1975) Fed. Proc. 34, 2209-2217.
33. Golomb, H.M. and Bahr, G.F. (1974) Exp. Cell Res. 84, 79-87.
34. Comings, D.E. (1974) in Chromosomes and Cancer, German, James, Ed., John Wiley and Sons, New York, p95-133.
35. Bak, A.L., Zeuthen, J. and Crick, F.H.C. (1977) Proc. Nat. Acad. Sci. 74, 1595-1599.
36. Daskal, Y., Mace, M.L., Jr., Wray, W., and Busch, H. (1976) Exp. Cell Res. 100, 204-212.
37. Brooke, J.H., Jenkins, D.P., Lawson, R.K. and Osgood, E.E. (1962) Ann. Hum. Genet. 26, 139-143.
38. Comings, D.E. and Okada, T.A. (1970) Exp. Cell Res. 63, 62-68.
39. Comings, D.E. and Okada, T.A. (1970) Cytogenetics 9, 436-449.
40. Engelhardt, P. and Pusa, K. (1972) Nature New Biology 240, 163-166.
41. Maul, G.G. (1977) J. Cell Biol. 74, 492-500.
42. Dẹscal, Y., Mace, M. and Busch, H. (Personal communication).
43. Carlson, R.D. and Olins, D.E. (1976) Nucleic Acid Research 3, 89-100.
44. Finch, J.T. and Klug, A. (1976) Proc. Nat. Acad. Sci. 73, 1897-1901.
45. Ohnuki, Y. (1968) Chromosoma 25, 402-428.

Chromosomes Today Volume 6, A. de la Chapelle and M. Sorsa eds.

USES OF FLUORESCENT DYES TO STUDY CHROMOSOME STRUCTURE AND REPLICATION

SAMUEL A. LATT, STEPHEN H. MUNROE, CHRISTINE DISTECHE, WILLIAM E. ROGERS, AND DOUGLAS M. CASSELL
Children's Hospital Medical Center, Boston, Massachusetts, 02115

The differential staining of metaphase chromosomes has been invaluable for chromosome identification, but it still awaits a molecular explanation. Since the work of Caspersson and coworkers pioneering the use of base-specific fluorochromes[1], a number of fluorescent dyes have been employed to highlight various combinations of chromosome regions [2-5]. However, the staining patterns produced are only partially accounted for by the spectroscopic properties of soluble dye-DNA or chromatin complexes [6,7]. Additional insight is provided by BrdU-dye techniques for detecting DNA synthesis [8-10], since these procedures can be employed to investigate the interrelation of chromosome structure and replication [11].

The chromosome banding pattern revealed by quinacrine and derivatives reflects a basic underlying partitioning of metaphase chromosomes. A similar pattern is produced by a variety of modified Giemsa techniques [12,13] and has been observed in the chromomere arrangement of meiotic chromosomes [14], while a reverse pattern can be elicited by selected Giemsa procedures and a few fluorescent dyes. In all cases, the junctions between differentially stained regions remain invariant.

The fluorescence of quinacrine increases upon interaction with A-T rich DNA [15]. Quinacrine binds with comparable affinity to a variety of natural DNA species, but exhibits a quantum yield which increases markedly with DNA A+T content [16,17]. The anthracycline dye daunomycin mimics the binding specificity, A-T dependent quantum yield [18] (Table 1), and staining properties[4] of quinacrine. Chromosome staining by quinacrine or daunomycin might also be influenced by other factors, such as uneven chromosome condensation, exclusion of dyes by chromosomal proteins, and perhaps energy transfer between bound dyes. Non-intercalating dyes with A-T binding specificity, such as 33258 Hoechst [2,19-22] and 2,7-di-t-butylproflavine[3], exhibit staining patterns resembling Q bands, but with relatively less contrast.

If chromosome banding were due solely to regional differences in DNA base composition, then dyes with opposite binding or fluorescence specificity might produce complementary staining patterns. Non-intercalating, chromomycinone dyes exhibit G-C binding specificity and produce a

TABLE 1

FLUORESCENCE QUANTUM YIELDS OF DYE-DNA COMPLEXES

DNA	(% A-T)	Quantum Yield*	
		Quinacrine	Daunomycin
Free Dye	---	0.08	0.031
Poly(dA)-Poly(dT)	100	0.49	0.037
Poly (dA-dT)	100	0.48	0.043
C. perfringens	70	0.13	0.006
Calf Thymus	56	0.06	0.002
E. coli	51	0.05	0.002
M. lysodeikticus	28	0.005	0.0004
Poly (dG-dC)	0	0.05	0.003

* 0.01 M NaCl, 0.005 M Hepes, pH 7; values determined relative to quinine sulphate in 0.1 N H_2SO_4 (quantum yield = 0.51). Phosphate:dye $\geq$ 87 (quinacrine), = 40 (daunomycin).

chromosome staining pattern which is described as the reverse of Q banding[5]. However, the fluorescent, 7-amino derivative of actinomycin D, an intercalating dye with G-C binding specificity[23], exhibits a DNA binding selectivity comparable to that of the chromomycinones [18,24], but a much less detailed chromosome staining pattern. For example, 7-aminoactinomycin D highlights the centromeric region of acrocentric cow chromosomes (Fig. 1), known to contain G-C rich satellite DNA[25], but stains only faintly the (presumptively A-T rich) long arm of the human Y. Thus, concentrations of DNA of extreme base

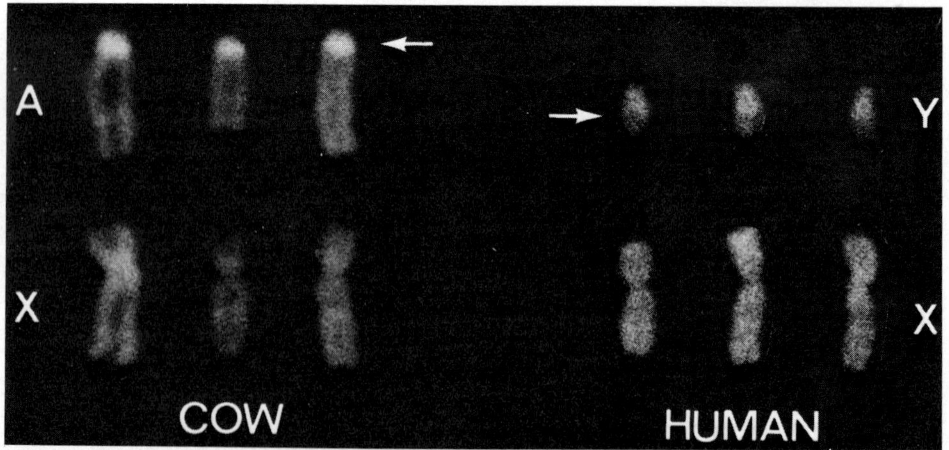

Fig.1 Staining of chromosomes with 7-aminoactinomycin D. Arrows point to bright centromeres (cow autosomes) and the dull region in the long arm of the human Y. Second row: X chromosomes.

composition might influence dye fluorescence to an extent depending on whether dye specificity was due to binding or quantum yield; the basis for the remainder of the staining patterns may be even more subtle.

Energy transfer measurements between an A-T specific dye such as 33258 Hoechst and a G-C specific dye such as 7-aminoactinomycin D[18] do not support the hypothesis that most of the fluorescence of dye-DNA complexes originates from large clusters of A-T or G-C base pairs. Negligible energy transfer occurs if these dyes are added to a mixture of poly (dA-dT) and poly (dG-dC) (Fig. 2), probably because the two dyes bind to separate molecules. However, with all natural DNA species examined, at least a 4 fold quenching of 33258 Hoechst fluorescence occurs upon addition of sufficient 7-aminoactinomycin D. The theoretical[26] separation for this donor-acceptor pair, below which energy transfer sharply becomes more than 50% efficient, is approximately 35A. Thus, little of the 33258 Hoechst is more than 10 base pairs away from sites ((G+C) base pairs) for 7-aminoactinomycin D. Similarly, the fluorescent lifetime of quinacrine bound to different types of DNA exhibits a slight dependence on overall base composition [17,27], suggesting that all of the fluorescence from these complexes does not result from dye molecules bound to long, uninterrupted runs of A-T pairs.

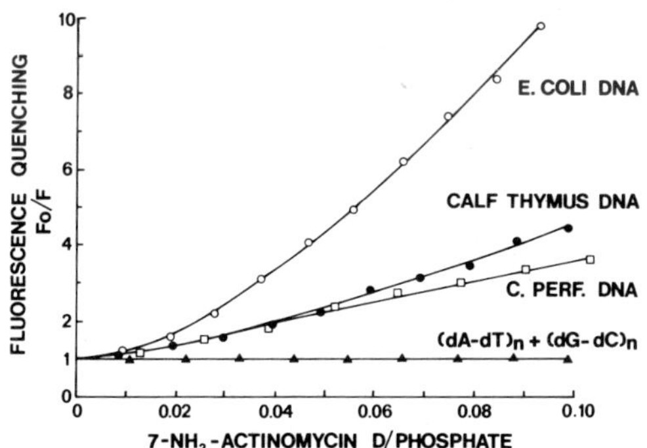

Fig. 2. Quenching of the fluorescence of 33258 Hoechst-DNA complexes by 7-aminoactinomycin D. Aliquots of 7-aminoactinomycin D were added to solutions of 33258 Hoechst ($5x10^{-7}$M) and 10^{-4} M DNA (calf thymus (●), E. coli (○), C. perfringens (□), 1:1 mixture of poly (dA-dT): poly (dG-dC) (▲)). F is the fluorescence amplitude of the 33258 Hoechst-DNA complex; F_o is the value of F in the absence of 7-aminoactinomycin D.

BrdU-dye techniques constitute a potentially useful approach for examining chromosome structure and afford a convenient, high resolution means of studying replication kinetics and sister chromatid exchange formation in cytological chromosome preparations. Biosynthetic substitution of BrdU for dT results in reduced fluorescence after chromosomes are stained with bisbenzimidazole dyes such as 33258 Hoechst. This effect is due to a BrdU-dependent reduction in dye fluorescence quantum yield[19]. Importantly, chromosomes from cells which have incorporated BrdU for one replication cycle show lateral differentiation in regions containing DNA with an asymmetric distribution of thymine [28-29]; however, they do not exhibit a reverse banding pattern, which would be expected if the Q bands were much higher in A+T than the intervening regions.

Exposure of BrdU-substituted chromosomes to dye, light, and warm buffer leads to reduced Giemsa staining[10]; this has been employed extensively to detect sister chromatid exchanges. Regions exhibiting pale Giemsa staining show reduced feulgen-staining, consistent with the hypothesis that DNA has been removed preferentially from BrdU-substituted regions [30]. Trypsin treatment does not reverse the differential staining, indicating that protein-dependent occlusion of dye from pale chromatin is not the major factor influencing staining.

The mechanism of BrdU-Giemsa differentiation was investigated further by measuring both chromosome staining and DNA elution throughout the procedure. Coverslips contained chromosomes specifically labelled with ^3H or ^{14}C so that loss of BrdU-substituted and unsubstituted polynucleotide chains could be distinguished. The dye promoted subsequent DNA elution, and differential elution in the absence of 2xSSC was appreciable only at very high light exposure. Loss of approximately 50% of bifilarily substituted DNA (Fig. 3) was associated with sister chromatid differentiation; additional light exposure resulted in removal of more than 70% of this DNA. Comparable treatment induced somewhat less elution of unifilarily substituted DNA. Interestingly, only the BrdU-containing chain of this DNA experienced appreciable loss. These experiments support the hypothesis that dye-sensitized sister chromatid differentiation involves photodegradation of BrdU-substituted DNA, involving single strand breaks and subsequent fragment elution during the SSC incubation.

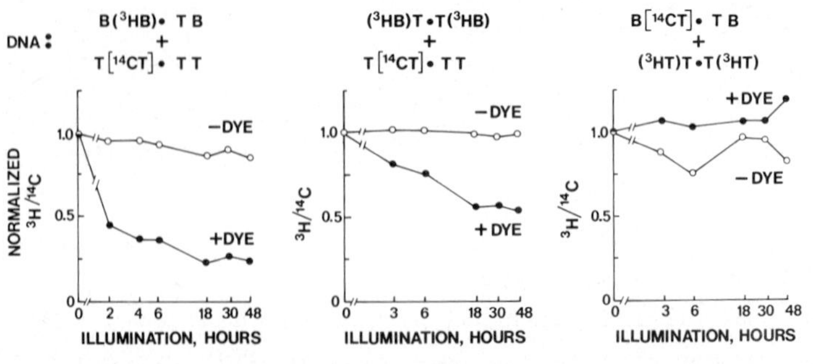

Fig.3. DNA elution during a BrdU-dye-Giemsa procedure. Synchronized CHO cells were cultured to produce DNA substituted as shown. Mixtures of colcemid-treated cells (average mitotic index approximately 40%) were applied to coverslips, mounted at pH 7 with or without prior staining with 33258 Hoechst, exposed 6 cm. below a 20 watt cool white lamp, and incubated 15 min. in 2xSSC at 65°. Relative elution of DNA species was estimated from the residual $^3H/^{14}C$ ratio.

BrdU-dye techniques have proved useful for the detection of sister chromatid exchanges induced by exposure of cells to clastogens [31-36]. Most of the agents capable of inducing SCE's appear to be mutagens and/or carcinogens[32]. S.C.E.'s, initially detected by autoradiography[37], are typically measured at the second metaphase following BrdU incorporation. After a subsequent cycle of BrdU incorporation, these events will be evident as non-reciprocal segmental staining, while approximately half of the exchanges produced during the third cycle will be evident as reciprocal alterations in chromosome staining.[38] The combination of 8-methoxypsoralen plus light is highly effective in inducing SCE's[33,39] (Fig. 4). Both reciprocal and non-reciprocal events are increased in third division cells, indicating that some damaged cells can survive at least three cycles and that some of the damage persists until (and perhaps beyond) the third cycle. Consistent with this idea, preliminary measurements of the interaction of ^{3}H-8-methoxypsoralen with cells indicate that perhaps 10^2 or more molecules of psoralen react with DNA for every sister chromatid exchange scored at the second metaphase.

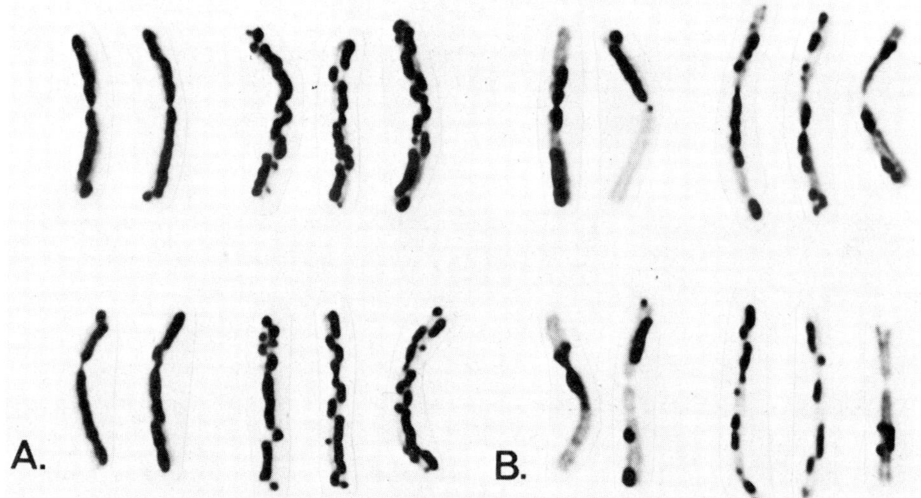

Fig. 4. Induction of sister chromatid exchanges in CHO chromosomes by 8-methoxypsoralen (6×10^{-6} M) plus light (10^{4} ergs/mm^2, mainly 365 nm). Synchronized CHO cells allowed to replicate twice (A) or three times (B) in the presence of 3×10^{-5} M BrdU. Slides were processed by a modified fluorescence plus Giemsa[10] procedure. Each set of 5 chromosomes consists of 2 controls (at the left) and 3 chromosomes from cells treated before release (at the right).

A particular advantage of the psoralen plus light system is the precision with which DNA substitution can be timed. Alkylation does not occur until the psoralen-DNA complex is irradiated [40-42]. In synchronized[43] CHO cells, SCE induction by a given dose of 8-methoxypsoralen plus light decreases progressively as cells traverse S [39], being disproportionately low very late in S. Late replicating DNA, which is primarily in G or Q bands, appears to be relatively resistant to SCE induction (Fig. 5). This effect may account in part for the observation that these bands are rarely the sites of exchange formation. Those SCE's which are induced late in S occur principally in regions which complete replication after alkylation. Preliminary experiments with ^{3}H 8-methoxypsoralen suggest that comparable alkylation occurs to both early and late replicating DNA. However, consistent with a previous report[44], it appears that damaged regions must replicate in order for sister chromatid exchanges to be produced.

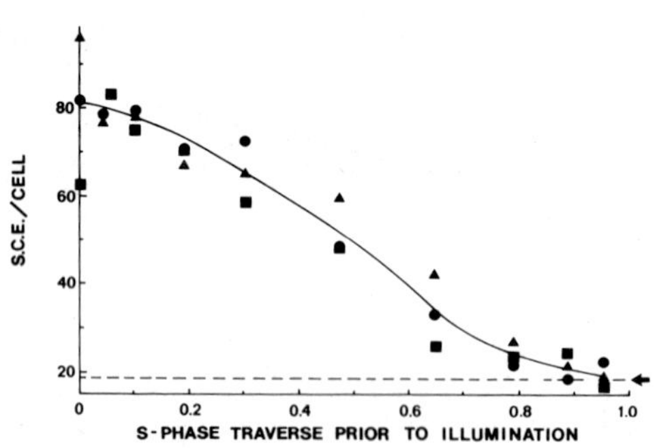

Fig. 5. Relative inducibility of SCE's in CHO cells during S phase. CHO cells which had undergone approximately one cycle of BrdU incorporation were synchronized at G1/S and then released in the presence of BrdU. At intervals, aliquots were tested for DNA synthesis rates (^{3}H dT incorporation experiment ●), which were summed to compute S-phase traverse, or SCE induction by 8-methoxypsoralen plus light (experiments ●, ■, ▲). Treatment conditions as in Fig. 4. Arrow: control SCE level.

The high resolution of BrdU-dye techniques substantiated an observation, based on autoradiography[45], that late replication is more pronounced in G or Q bands than in the intervening regions[8]. In fact, individual bands appear to replicate over faintly narrow intervals, and the bands can be differentiated according to the time that they complete replication. Fluorescence analysis of late replication has been especially useful in the differentiation of early and late replicating X chromosomes[46,47] and in the characterization of structurally abnormal X chromosomes[48].

Differentiation of G or Q bands from intervening regions by replication kinetics (Fig. 6) permits a biochemical characterization of the structural partitioning of chromosomes[11]. For example, in CHO cells examined in interphase at the end of S, the early replicating regions are more readily solubilized by nuclease than are late replicating regions (Fig. 7). Similarly, if chromatin is fractionated by DNA'se digestion and Mg++ precipitation [49], the soluble fractions are relatively enriched for early replicating material (C. Disteche, unpub-

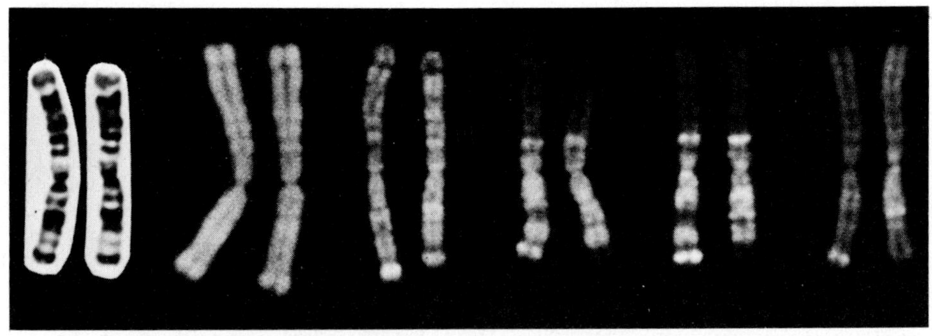

Fig. 6. Terminal replication patterns in pairs of #1 chromosomes from CHO cells. Synchronized cells released in BrdU were exposed to terminal pulses of dT (3,5,7,8, or 10 hours duration, from right to left). Bright 33258 Hoechst fluorescence indicates late replication. A pair of Giemsa banded chromosomes from a control cell is shown at the left.

lished observations). However, neither effect is very large, and the use of crude chromatin fractionation techniques, aimed at isolating material putatively localized to bands and intervening regions, may not be as effective as presumed. Selective BrdU-labelling might ultimately provide a better way to accomplish this purpose. Interestingly, while early replicating DNA in interphase chromatin is more readily solubilized by nuclease, the size of the resistant segments (presumably reflecting nucleosome[50] size) in metaphase chromatin is slightly greater in early than in late replicating regions (Fig. 8). Thus, it is unlikely that chromosome banding reflects gross differences in chromatin structure at the nucleosome or subnucleosome level. The importance of variations in condensation involving the higher order organization of nucleosomes remains to be determined.

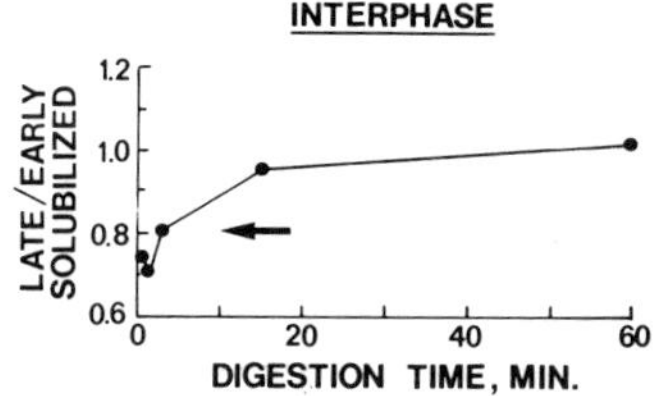

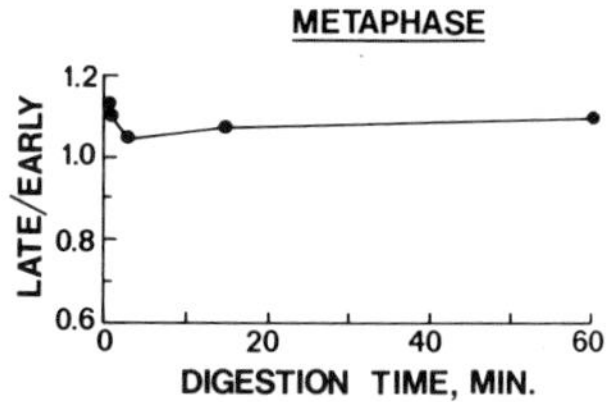

Fig. 7. Solubilization of early and late replicating chromatin from synchronized CHO cells. Mixtures of chromatin labelled with ^{14}C dT (first half of S) or ^{3}H dT (last half of S) were digested with micrococcal nuclease. The ratio of ^{3}H dT to ^{14}C dT solubilized is plotted versus digestion time for (A) interphase and (B) metaphase chromatin. Acid solubilization after 60 minutes digestion was 49-56%.

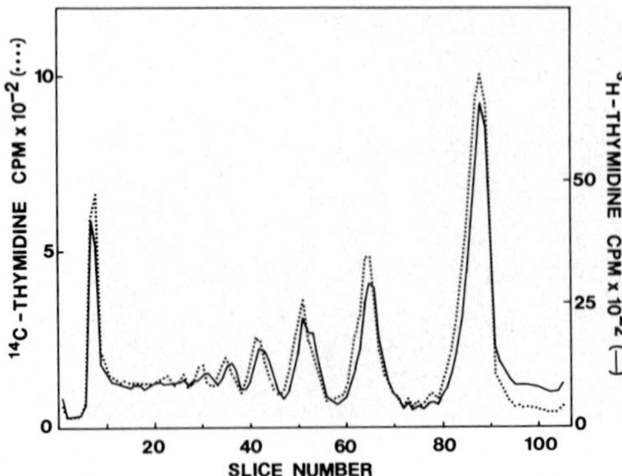

Fig. 8. Electrophoresis of early and late labelled DNA isolated from CHO metaphase chromatin after digestion with micrococcal nuclease. Chromatin from mixtures of cells labelled with ^{14}C dT during the first half of S or ^{3}H dT during the second half of S was digested with nuclease to nearly 10% acid solubilization and the DNA was electrophoresed in 2.3% acrylamide-agarose gels. Gels were sliced prior to radioactivity measurement. Mobility increases from left to right.

SUMMARY

Spectroscopic and biochemical studies have placed constraints on the possible mechanisms of chromosome banding. Additional information is derived from an increased number of fluorescent probes and from BrdU-dye techniques for correlating, via replication timing, the cytological, biochemical and functional properties of different chromosomal regions.

ACKNOWLEDGMENTS

The expert technical assistance of Ms. Lois Juergens and Mr. Michael Eisenhard is greatly appreciated, as are the gifts of cow chromosomes from Ms. Deborah Matthews, 33258 Hoechst from Dr. H. Loewe, and 8-methoxypsoralen from the Paul B. Elder Co. This research was supported by grants from the National Institutes of Health (GM 21121), the National Foundation March of Dimes (1-353) and the American Cancer Society, a Research Career Development Award (GM 00122) from the National Institute of General Medical Sciences (S.L.), and a National Foundation March of Dimes Summer Fellowship (8-13-17)(D.C.).

REFERENCES

1. Caspersson, T., Lindsten, J., Lomakka, G. Moller, A. and Zech, L. (1972) Int. Rev. Cytol., 11, pp 1-72.
2. Hilwig, I. and Gropp, A. (1972) Exper. Cell Res., 75, pp 122-126.
3. Disteche, C. and Bontemps, J. (1974) Chromosoma, 47, pp 263-281.
4. Lin, C.C. and Van de Sande, J. (1975) Science, 190, pp 61-63.
5. Van de Sande, J., Lin, C.C. and Jorgenson, K.F. (1977) Science, 195, pp 400-402.
6. Comings, D.E. and Drets, M.E. (1976) Chromosoma, 56, pp 199-211.

7. Latt, S.A. (1976) Ann. Rev. Biophys. Bioeng., 5, pp 1-37.
8. Latt, S.A. (1973) Proc. Nat. Acad. Sci. (Wash.), 70, pp 3395-3399.
9. Dutrillaux, B., Laurent, C., Couturier, J. and Lejeune, J. (1973) C.R. Acad. Sci., 276, pp 3179-3181.
10. Perry, P. and Wolff, S. (1974) Nature (London), 251, pp 156-158.
11. Munroe, S.H. and Latt, S.A. (1977) Exper. Cell Res. (In press).
12. Hsu, T.C. (1973) Ann. Rev. Genet., 7, pp 153-176.
13. Miller, O.J., Miller, D.A., and Warburton, D. (1973) Prog. Med. Genet., 9, pp 1-47.
14. Okada, T.A. and Comings, D.E. (1974) Chromosoma, 48, pp 65-71.
15. Weisblum, B. and de Haseth, P.L. (1972) Proc. Natl. Acad. Sci. USA 69, pp 629-632.
16. Pachmann, U. and Rigler, R. (1972) Exper. Cell Res., 72, pp 602-608.
17. Latt, S.A., Brodie, S. and Munroe, S.H. (1974) Chromosoma, 49, pp 17-40.
18. Latt, S.A., (1977) Canad, J. Genet. Cytol. (In press).
19. Latt, S.A., and Wohlleb, J.C. (1975) Chromosoma, 52 pp 297-316.
20. Weisblum, B. and Haenssler, E. (1974) Chromosoma, 46, pp 255-260.
21. Comings, D.E. (1975) Chromosoma, 52, pp 225-243.
22. Mueller, W. and Gautier, F. (1975) Eur. J. Biochm., 54, pp 385-394.
23. Mueller, W. and Crothers, D.M. (1968) J. Mol. Biol, 35, pp 251-290.
24. Modest, E.J. and Sengupta, S.K. (1974) Cancer Chemother. Rep., 58, pp 35-48.
25. Kurnit, D.M., Shafit, B.R. and Maio, J.J. (1973) J. Molec. Biol. 81, pp 274-284.
26. Forster, T. (1965), In Modern Quantum Chemistry, O. Sinanoglu, Ed. 3, Academic, N.Y., pp 43-137.
27. Duportail, G., Mauss, Y. and Chambron, J. (1977) Biopolymers, 16, pp 1397-1413.
28. Lin, M.S., Latt, S.A. and Davidson, R.L. (1974) Exper. Cell Res. 86, pp 392-395.
29. Latt, S.A., Davidson, R.L., Lin, M.S. and Gerald, P.S. (1974) Exper. Cell Res., 87, pp 425-429.
30. Goto, K., Akematsu, T., Shimazu, H. and Sugiyama, T. (1975) Chromosoma, 53, pp 223-230.
31. Latt, S.A. (1974) Proc. Natl. Acad. Sci. USA, 71, pp 3162-3166.
32. Perry, P. and Evans, H.J. (1975) Nature (London), 258, pp 121-125.
33. Latt, S.A., Allen, J.W., Rogers, W.E. and Juergens, L.A. (1977) Mutat. Res. (In press).
34. Vogel, W., Bauknecht, T. (1976) Nature (London), 260, pp 448-449.
35. Allen, J.W. and Latt, S.A. (1976) Nature (London), 260, pp 449-451.
36. Stetka, D. and Wolff, S. (1976) Mutat. Res. 41, pp 343-350.
37. Taylor, J.H., Woods, P.S. and Hughes, W.L. (1957) Proc. Nat. Acad. Sci. USA, 43, pp 122-128.

38. Tice, R., Chaillet, J., and Schneider, E.L. (1975) Nature, 251, pp 70-72.

39. Latt, S.A., Allen, J.W., Shuler, C., Loveday, K.S. and Munroe, S.H. (1977) ICN Symposium on Molecular Human Cytogenetics (In press).

40. Cole, R.S. (1971) Biochem. Biophys. Acta., 254, pp 30-39.

41. Isaacs, S.T., Shen, C.J., Hearst, J.E., and Rapoport, H. (1977) Biochemistry, 16, pp 1058-1064.

42. Cech, T., and Pardue, M.L. (1977) Cell, 11, pp 631-640.

43. Hamlin, J.L. and Pardee, A.B. (1976) Exper. Cell Res., 100, pp 265-275.

44. Wolff, S., Bodycote, J. and Painter, R.B. (1974) Mutat. Res., 25 pp 73-81.

45. Ganner, E. and Evans, H.J. (1971) Chromosoma, 35, pp 326-341.

46. Willard, H.F. and Latt, S.A. (1976) Amer. J. Hum. Genet., 28, pp 213-227.

47. Willard, H.F. (1977) Chromosoma, 61, pp 61-73.

48. Latt, S.A., Willard, H.F. and Gerald, P.S. (1976) Chromosoma, 57, pp 135-153.

49. Gottesfeld, J.M., Garrard, W.T., Bagi, G., Wilson, R.F. and Bonner, J. (1974) Proc. Nat. Acad. Sci. USA, 71, pp 2193-2197.

50. Kornberg, R.D. (1977) Ann. Rev. Biochem. 46, pp 931-954.

Chromosomes Today Volume 6, A. de la Chapelle and M. Sorsa eds.
© 1977 Elsevier/North-Holland Biomedical Press, Amsterdam, The Netherlands

MULTIPLE TELOMERIC FUSIONS AND CHAIN CONFIGURATIONS IN HUMAN SOMATIC CHROMOSOMES

B. DUTRILLAUX, A. AURIAS, J. COUTURIER, M.F. CROQUETTE, E. VIEGAS-PEQUIGNOT
Institut de Progenèse, 15 rue de l'Ecole de Médecine F-75006 PARIS

ABSTRACT

Two observations are presented, showing multiple cases of chromosome end-to-end fusions or associations. In the first one, more than 100 abnormal cells were analysed, in a patient with a Thiberge-Weissenbach syndrome. The disposition of the chromosomes, which is apparently anarchical, seems to show that it does not reflect a physiological distribution in the interphasic nucleus. In the second observation, a single cell shows a clear organization, with chromosome pairing.

The results obtained are analysed and compared, with reference to the hypothesis of the telomere constituted by a palindromic base sequence of DNA.

INTRODUCTION

The telomere hypothesis, first formulated by Muller[1], is generally accepted, although no direct proof has ever been provided. Originally, the presumed function of the telomere was to preserve the individuality of chromosomes, preventing their fusion. Nevertheless, some exceptional observations, as in Ascaris[2], showed that chromosomes may temporarily fuse, or inversely that interchromatidic structures may behave as telomere does.

Furthermore, electron microscopy has often suggested the existence of interchromosomal connections, and some have postulated that all the chromosomes could be associated during the interphase, forming a giant ring[3].

In 1974, Cavalier-Smith[4] proposed a molecular model for telomeres which may be summarized as follows :

- telomeres consist of a short DNA segment,
- this DNA segment has a palindromic base sequence, and is common to all chromosome ends,
- according to the cell phase, each palindrome could be self-paired, or paired with an other one.

The telomeres would then permit both the individualization of the chromosomes and their attachment to each other.

Observations are reported here, indicating that in Man, where the chromosomes are quite generally seen as independent units, it is possible to observe multiple end-to-end fusions. The whole karyotype appears then like a giant ring or chain.

MATERIAL AND METHODS

The major part of this work was done on a patient affected by a Thiberge-Weissenbach syndrome (observation N°1) which associates sclerodermia, calcinosis and telangiectasia. Her lymphocytes were cultured according to our standard technique, for periods ranging from 48 H to 96 H. In some cultures, there was no addition of colchicine and no hypotonic shock. In other cultures, BrdU was added in various concentrations, during 72 H or during the last 7 H.

The majority of the abnormal cells were analysed with three consecutive stainings: Giemsa, QFQ[5] and RFA[5]. Some slides were stained by Feulgen, C-banding and acridine orange.

The second observation concerns a single cell detected in a culture using blood from a normal individual to which a high dose of BrdU had been added.

RESULTS

Observation N°1. The anomalies all consist in end-to-end fusions leading to formation of dicentrics, tricentrics, chain multicentrics and various rings. In the abnormal cells, the number of fusions showed a wide range of variations; from one dicentric to a giant chain or ring including nearly all the chromosomes (fig.1).

The frequency of the abnormal cells, and the degree of ploidy is given in table I :

TABLE I
FREQUENCY OF ABNORMAL CELLS, AND OF POLYPLOID CELLS, AT VARIOUS TIMES OF CULTURE

Time of culture (H)	Percentage of abnormal cells	Frequency of polyploids among abnormal cells	Frequency of polyploids among normal cells
48	0.6	0.25	0.00
56	0.5	0.20	0.01
72	0.2	0.60	0.03
96	0.2	I.00	0.03

The aspects of the BrdU-treated cells indicate that the abnormal diploid cells have undergone a single division in culture, and the abnormal tetraploid ones two divisions (fig.2). This indicates that

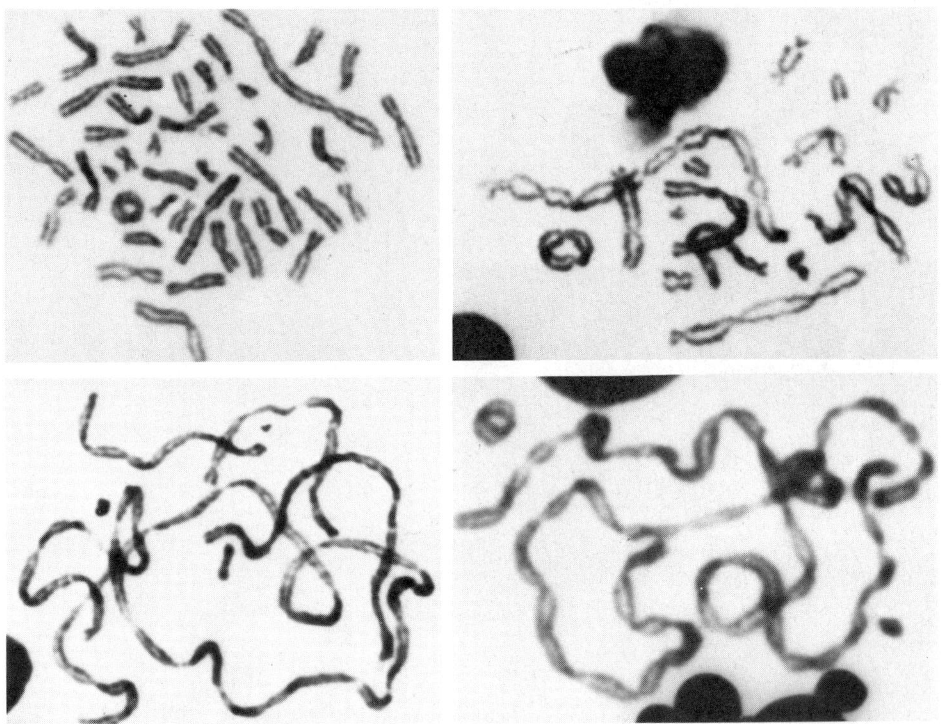

Fig. 1. Cells with dicentrics, rings and multicentric chains

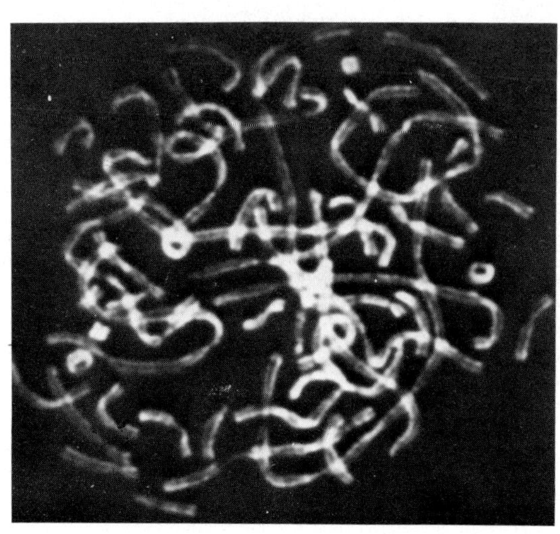

Fig. 2. Tetraploid cell cultured with BrdU during two cell cycles. No obvious relationship between fusion points and sister chromatid exchanges was seen.

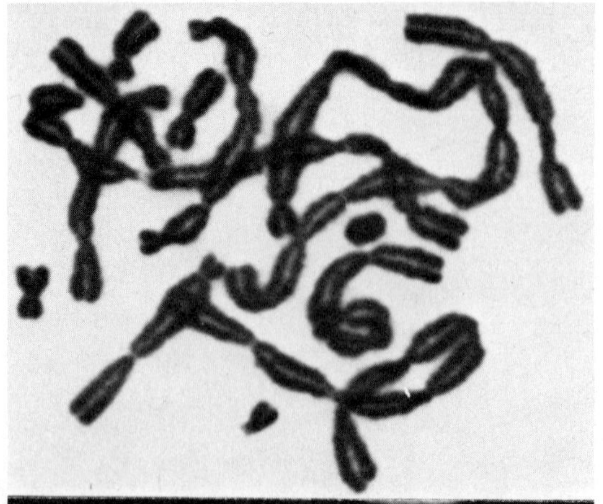

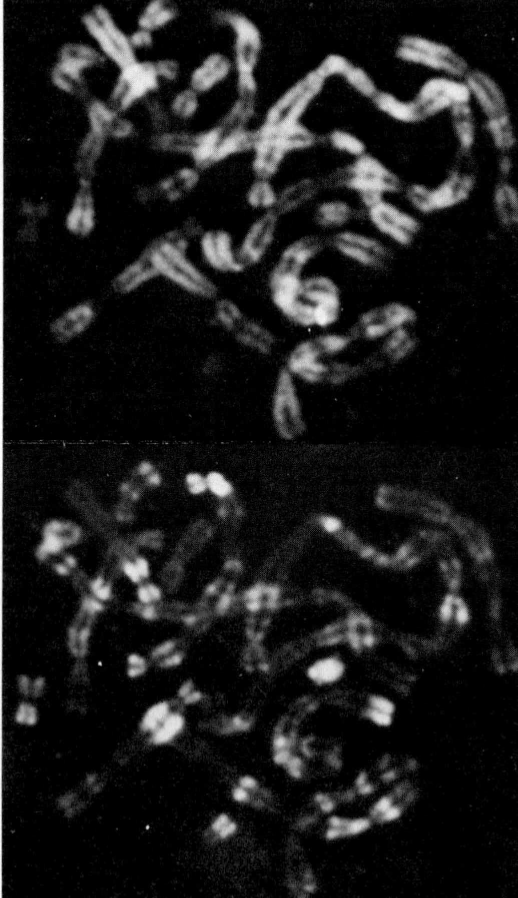

Fig. 3. Giemsa staining,Q-banding and R-banding of the same mitosis. The chromosomes included in the chains seem entire.

Fig. 4. Nucleus with clear pairing of the chains. The chromosomes have a single chromatid and are linked by thin filaments.

the anomalies which always affect the two sister chromatids occurred either in G_0-phase or in G_1-phase of the first cell cycle, or else preexisted in vivo.

The analysis of the cells not treated by colchicine nor by hypotonic shock seems to indicate the incapacity of the cells carrying chains to form equatorial plates and to undergo a normal anaphase.

Among more than 100 abnormal cells observed, the ring and chain configurations were never associated with acentrics. This obviously indicates that the anomalies do not result from a breakage-reunion mechanism, as is the case after irradiation, but effectively from end-to-end fusions. This is confirmed by the integrity of the chromosomes included in the chains (fig.3).

Table II gives the distribution of more than 500 chromosomal fusions observed. This distribution seems random.

TABLE II

DISTRIBUTION OF MORE THAN 500 CHROMOSOME FUSIONS

For each chromosome, the short arm and the long arm are respectively indicated by p and q.

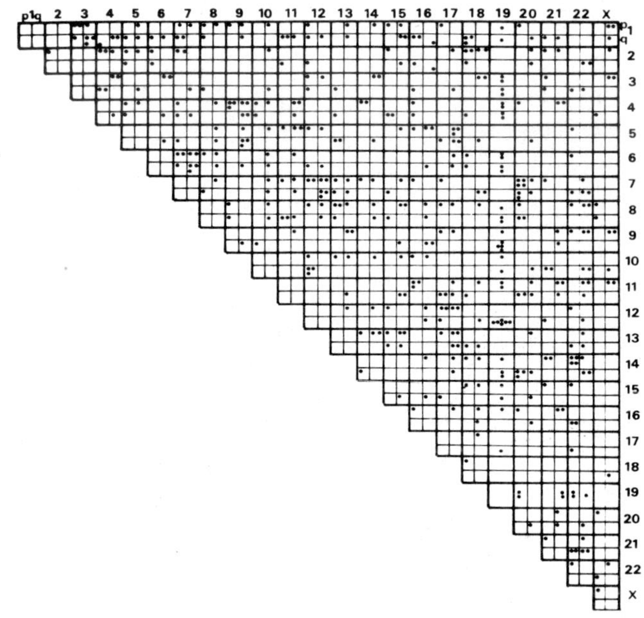

In a given diploid cell, no tendency to chromosome pairing was detected, and there are never two identical sequences corresponding to the homologues.

A clear tendency to pairing was often seen in tetraploid cells, but it was obviously between the brother chromosomes included in the duplicated chains[6]. The pairing was complete in a few cells, interpreted as endoreduplications.

The C-banding and NOR-staining did not bring complementary information. The Feulgen showed continuous staining intensity all along the chains, and the terminal pulse of BrdU did not reveal any drastic change in DNA replication time.

Observation N°2. A single nucleus was observed with chromosome chainings. It is comprised of approximately 46 chromosomes, with a single chromatid. Very near, another nucleus has a normal metaphasic appearance. This was interpreted as the result of a cell-fusion having induced a premature chromosome condensation of the nucleus in the G_0-phase.

In this last nucleus, the chromosomes have a brilliant green fluorescence, after staining with acridine orange, and they are associated in chains, linked by a red filament.

Furthermore, there is a loose, but clear, pairing of the homologues and all the chains seem to converge towards the centre of the nucleus where they are linked by a long filament.

DISCUSSION

Observation N°1 provides some information which is not a formal proof, but constitutes a strong argument in favour of the theory of the existence of a palindromic base sequence of DNA at the chromosome ends.

First, it demonstrates that all the chromosomes of a given cell may fuse end-to-end. This fusion never links more than two ends at the same time, except sometimes for the short arm of acrocentrics.

This fusion persists during two or more cell generations, and always involves the two sister chromatids in the same way. It probably results from a strong chemical binding and, in so far as no gap is detectable, it may be concluded that the DNA molecules of each fused chromosome finally constitute a single giant molecule.

On the other hand, each chromosome end seems able to fuse with any other one. For instance, Table II shows that the following connections exist : 1p-3p, 3p-6q, 6q-1q, 1q-3q, 3q-4p, 4p-9p, 9p-1p etc...

If we admit that the DNA of each chromosome terminates in a single 5'-end strand, each must be able to pair with any other one.

This implies that all these strands are complementary, and by this fact, self-complementary. The only base sequence with such a property is a palindrome. Consequently, in this hypothesis, all the telomeres would be constituted by a palindromic base sequence of DNA which would be identical for all the chromosome ends. This correspond exactly to the postulate formulated by Cavalier-Smith[4].

A second piece of information concerns the hypothetic existence of chains in the normal interphasic nucleus. Obviously, if it had been possible to detect a logical and consistent sequence of the chaining of the chromosomes, this would have strongly favoured such a hypothesis. As this is not the case, it must be concluded that : either chaining does not exist in normal cells; or it does and it is apparently random; or else it exists, but the type of chaining observed here does not correspond to the normal situation.

Although it is not possible to choose definitively between these three alternatives, the anarchical distribution of the chromosomes, and for instance the presence of monocentric rings, as in human pathology, makes it more probable that the chaining observed here does not correspond to a physiological situation.

Observation N°2 is based on a single cell, and must be considered with caution. Like Observation N°1, it shows that interchromosomic linkages are possible. They could exist here in "interphasic" chromosomes observed after premature condensation.

The interchromosomic connections have a red fluorescence (RBA) whereas the chromosomes are green. If these connections are constituted by DNA, the BrdU should already have been incorporated in the latter. The PCC should thus have occurred at the very beginning of the S-phase, when only the telomere zones should have started their replication.

On the other hand, the 16 chains are clearly disposed in 8 pairs. This indicates that each homologue is paired, and consequently the same sequence exists in each homologous chain. A rational order should thus exist in this cell.

The comparison of these two observations seems to show that, if chromosomic chainings normally exist during interphase, it is more likely that they would correspond to those of Observation N°2.

In this eventuality, the following scheme could be proposed:

- the associations of many telomeres would occur at the beginning of the G_1-phase, when the uncoiling of the chromosomes is just achieved.
- the specificity of the associations would depend on juxta-palin-

dromic structures (all palindromes being identical), which would be recognized, two by two, by specific enzymes of the ligase type.
- some other telomeres would not be recognized and would constitute the ends of the chains.
- the associations would normally be suppressed before chromosome condensation, by the action of another group of enzymes, of the endonuclease type, which could also be specific.

The rational organization of the cell would then depend on the telomere-ligase specificity, which would impose a precise order for the chromosome associations.

The abnormal character of Observation N°1 could result from the loss of specificity, either by the modification of the telomeric sites (abnormal condensation, for instance) or by the defectiveness of the enzymatic function. The persistence of the chains would be a direct consequence of this trouble, the "endonucleases" becoming unable to cleave the abnormal connections. Finally, the anomaly of a single enzymatic function could be in cause, and it may be interesting to recall that the patient suffers from an affection where the presence of anti-nucleoprotein antibodies has been noticed.

The accumulation of data is needed to test such hypothesis.

Finally, the study of this case may be a good approach for the understanding of the formation of rings and terminal rearrangements, as they are frequently observed in pathology.

ACKNOWLEDGEMENTS

The technical assistance of Misses Anne Marie Fosse and Martine Lombard was greatly appreciated. This research was supported by grants of CNRS (ERA n° 47), CEA (BC n° 1127) and CNPq (Bresil).

REFERENCES

1. Muller, H.J. (1938) Collecting Net, Woods Hole, 13, 181-195 and 198.
2. Bonnevie, K. (1902) Jena Z. Naturwiss, 36, 275-288.
3. DuPraw, E.J. (1970) DNA and Chromosomes. New York, Holt, Rinehart and Winston.
4. Cavalier-Smith, T. (1974) Nature, 250, 467-470.
5. Paris Conference (1972) Birth Defects, Original Article Series, 8.
6. Dutrillaux, B., Croquette, M.F., Viegas-Péquignot, E., Aurias, A., Coget, J., Couturier, J. and Lejeune, J. (1977) Cytogenet. Cell Genet. (in press).

Chromosomes Today Volume 6, A. de la Chapelle and M. Sorsa eds.

SILVER-STAINING OF NUCLEOLUS ORGANIZER REGIONS IN MAN: LIGHT - AND ELECTRONMICROSCOPIC OBSERVATIONS

H.G.SCHWARZACHER, A-V.MIKELSAAR*, W. SCHNEDL
Histologisch-Embryologisches Institut der Universität Wien
Schwarzspanierstraße 17, A 1090 Vienna, Austria

ABSTRACT

The Ag-stainability of the NORs during interphase and mitosis is correlated with their transcription activity. Electron micrographs of metaphase chromosomes reveal the Ag-positive material to be located around the NORs but not in the chromosomes themselves. In interphase nuclei the fibrillar material around the chromosomal components are strongest stained. Extraction experiments indicate that the Ag-stainable substance is an acidic protein, probably part of the RN-protein. Individual differences in NOR-Ag-stainability may be due to different amounts of rDNA or to different capacities of NOR activation.

Introduction

The nucleolus organizer regions (NORs) can be specifically stained in chromosome preparations by a silver (Ag) staining method[1,2]. Only those NORs are stained by this method which were functionally active during the preceeding interphase[3,4]. In man NORs can be found in the secondary constrictions of the short arms of all 5 acrocentric chromosomes[5,6]. There are, however, individual and heritable differences in the patterns of the Ag-stainability of the NORs, and also a smaller degree of intraindividual variations as we have shown previously[7,8].

Until now, it was assumed by most authors that a chromosomal component ist stained with the Ag-method. In the present paper we shall present evidences that the Ag-stainable substance is rather a protein component of the RN-protein surrounding the NOR. A more detailed description of our findings will be given elsewhere[9].

---

* AV. Mikelsaar is visiting exchange scientist of the Österreichische Bundesministerium für Wissenschaft und Forschung. His permanent address is: Central Research Laboratory, Tartu State University, Tartu, U.S.S.R.

MATERIAL AND METHODS

Fibroblast cultures and short term lymphocyte cultures from 3 karyotypically normal persons were grown after standard procedures. Chromosome preparations as well as directly fixed (alcohol-acetic acid) cells grown on coverglasses were used. Ag-staining was carried out similar to the methods of Goodpasture and Bloom[1] and Denton et al.[10]. Part of the preparations were pretreated in different ways: 1.: 0.1% RNase at 37°C for 2^h; 2.: 4% trichloracetic acid (TCA) at 90°C for 30 min; 3.: Trypsin(Wellcome) 0.05% at 37°C for 10 min. Total preparations from Ag-stained chromosome preparations as well as ultrathin sections from glutaraldehyd fixed and epon embedded fibroblast cultures were used for electron microscopy.

RESULTS

In Interphase the nucleoli show an intensive Ag-staining. In air dried preparations of lymphocyte cultures made 48^h after onset and the addition of phytohaemagglutinine non activated lymphocytes as well as all stages of activated lymphocytes are found. The size of the Ag-stainable masses of the nucleoli is according to the stage of activation, being absent in the small non-activated nuclei and large in fully activated blast cells (figure 1).

In early prophase the amount of Ag-positive material is about the same as in activated G2-interphase nuclei (figure 2). In late prophase and metaphase, however, only small Ag-positive dots are seen at the sites of the NORs (figure 3). Also in anaphase the Ag-dots are very small (figure 4). In late telo- and reconstruction-phase again larger Ag-positive masses are seen (figure 5).

In electron microscopic pictures of Ag-stained metaphase chromosomes the Ag-deposit is seen only on the outside of the NORs in cases of weak reaction (figure 6). When a particular NOR is strongly stained the Ag-positive substance is seen around the NOR but the chromatids themselves remain unstained (figure 7). The differences in the degree of staining, seen in the light microscope (see our previous papers[7,8]) are very obvious in electron microscopic preparations. The Ag-staining is sometimes very weak and completely negative NORs are also seen. In so called "satellite-association"figures, the Ag-positive material is mainly found between the chromosomes (figure 8).

Electron micrographs from total preparations of Ag-stained interphase nuclei reveal that the Ag-deposit is most probably localized around the NORs within the nucleolus. This would correspond to the dense fibrillar mass of the nucleolus (figures 9 and 10).

After extraction of RNA by RNase, as well as after treatment with TCA which extracts all nucleic acids and solves basic proteins, the Ag-stainability of nucleoli remains the same as without extractions. After trypsin treatment an Ag-staining is not possible. This is in accordance with earlier findings on metaphase chromosomes[2,1] and indicates that the stainable material is an acidic protein.

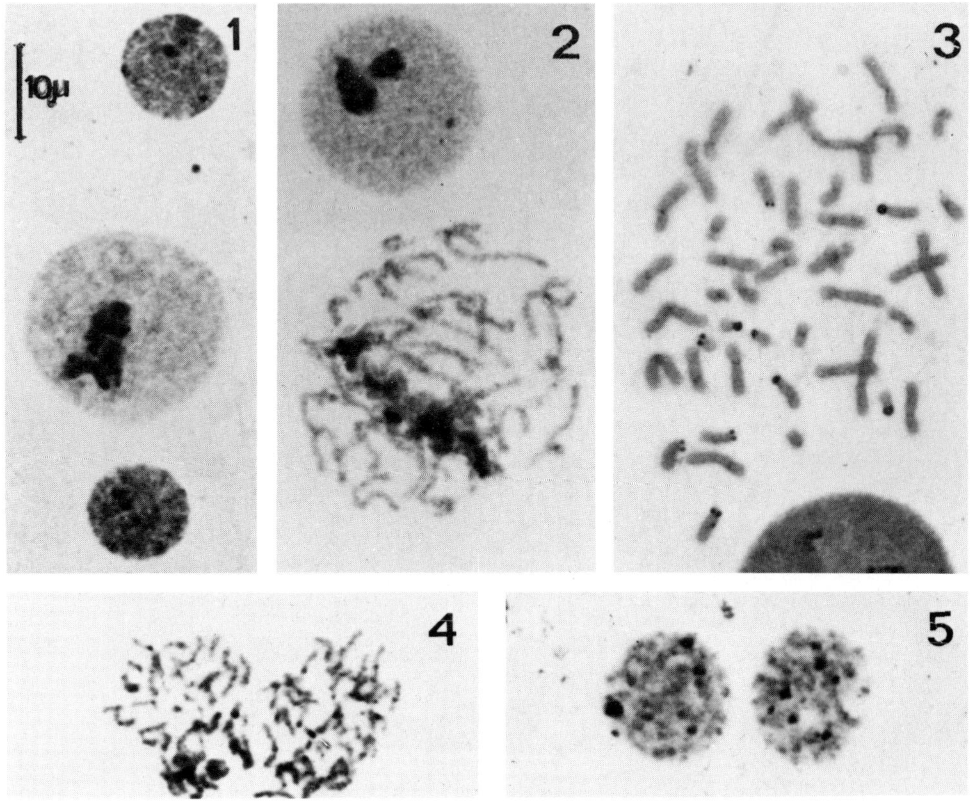

Figs. 1 - 5 Different stages of interphase and mitosis from a lymphocyte culture. Ag-staining of the NORs and nucleoli corresponding to activity

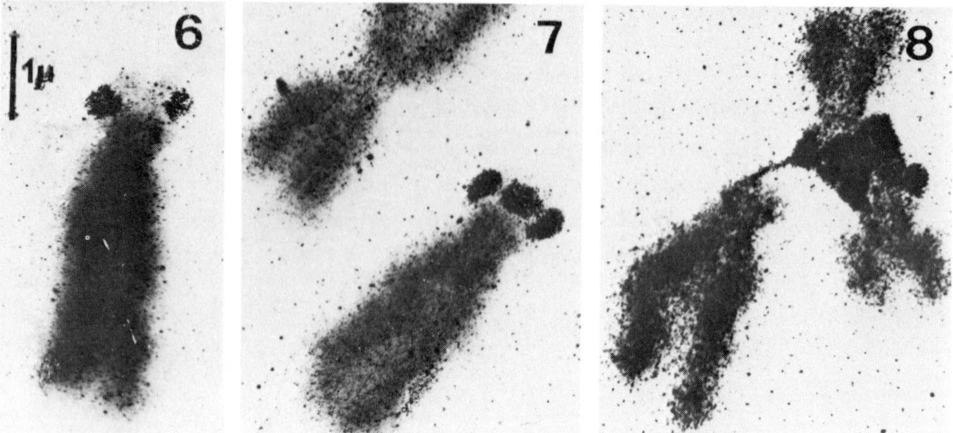

Figs. 6 - 8 Electron micrographs of Ag-stained acrocentric metaphase chromosomes. Total preparations.

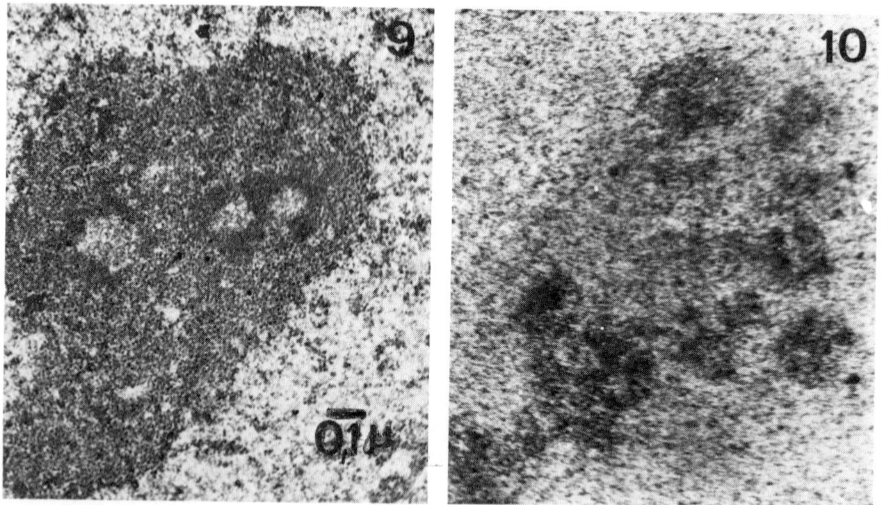

Figs. 9 and 10 Electron micrographs of nucleoli of interphase nuclei. Standard thin section (fig. 9) and Ag-staining of total preparation (fig. 10).

DISCUSSION

The Ag-staining method used consists of two steps: in the first the Ag is reduced and precipitates at the specific stainable sites; in the second more Ag is precipitating at the original Ag granules and forms larger masses. The size of the Ag-masses depends therefore of course on technical factors such as duration of impregnation and concentration of Ag-solutions. It depends, however, also on the size of the area where the first precipitation occurs and is therefore under standardized conditions a measure of the amount of stainable substance present.

Our electron microscopic findings that the chromatids themselves remain unstained contractict a direct Ag-stainability of the chromosomal DNA-protein.

The obvious conclusion from our observations is therefore that the Ag-method stains an acidic protein component of the material surrounding active NORs. The localization within nucleoli suggests, that it is the RN-protein in the form of the fibrillar component of nucleolus. Another possibility is, that an acidic protein of unknown function accumulates at the sites of the NORs in correlation to their genetic activity. In either case, the more active a NOR is, the more of this material is accumulated and the stronger is the Ag-staining. In prophase, when the activity of the NOR ceases[11,12] most of the Ag-positive material is removed from the site of its production but some remains around the NORs through mitosis. In telophase the Ag-positive material increases again parallel to the anew activation of the NORs. It should be pointed out that in metaphase the Ag-positive substance is seen to a greater amount on the outside of the chromatids. This may reflect the direction of the gene product after replication, when the two chromatids are very close together.

The individual and heritable differences in the Ag-stainability of NORs found in our previous studies[7,8] may be due to differences in the amount of rDNA[6] but also to differences in the capacity of NOR activation. A maximal activation of the NORs may be assumed to take place in dividing lymphocytes stimulated by phytohaemagglutine. The observed intraindividual differences between cells[7,8] may, however, reflect differences in the degree of activation of the NORs under these conditions.

ACKNOWLEDGMENTS

This work was partly supported by the Österreichische Fonds zur Förderung der Wissenschaft (Projekt Nr. 2514) and the Jubiläumsfonds der Österreichischen Nationalbank.

REFERENCES

1. Goodpasture, C. and Bloom, S.E. (1975) Chromosoma 53, 37-50.
2. Howell, W.M., Denton, T.E. and Diamond, J.R. (1975) Experientia 31, 260-262.
3. Miller, D.A., Dev, V.G., Tantravahi, R. and Miller, O.J. (1976) Expl. Cell Res. 101, 235-243.
4. Miller, O.J., Miller, D.A., Dev, V.G. and Tantravahi, R. (1976) Proc. nat. Acad. Sci. USA 73, 4531-4535.
5. Henderson, A.S., Warburton, D. and Atwood, K.C. (1972) Proc. nat. Acad. Sci. USA 69, 3394-3398.
6. Evans, H.J., Buckland, R.A. and Pardue, M.L. (1974) Chromosoma 48, 405-426.
7. Mikelsaar, A.-V., Schmid, M., Krone W. and Schwarzacher, H.G. (1977) Hum. Genet. 37, 73-77.
8. Mikelsaar, A.-V., Schwarzacher, H.G., Schnedl, W. and Wagenbichler, P. (1977) Hum. Genet., in press.
9. Schwarzacher, H.G., Mikelsaar, A.-V. and Schnedl, W. (1977) Cytogenet. Cell Genet., in press.
10. Denton, T.E., Howell, W.M. and Barret, J.K. (1976) Chromosoma 55, 81-84.
11. Das, N.K. and Alfert, M. (1966) Natl. Canc. Inst. Monograph. 23, 337-351.
12. Arrighi, F.E. (1967) J. Cell Physiol. 69, 45-52.

Chromosomes Today Volume 6, A. de la Chapelle and M. Sorsa eds.

THE USE OF X-RAY MICROANALYSIS TO STUDY CHROMOSOME BANDING.

A.T. SUMNER
M.R.C. Clinical and Population Cytogenetics Unit, Western General Hospital,
Edinburgh EH4 2XU, Scotland, U.K.

ABSTRACT

X-ray microanalysis in the scanning electron microscope has been used to study changes in elemental composition and distribution in chromosomes during banding procedures. Fixed, untreated chromosomes contain a high concentration of calcium, as well as phosphorus (from DNA) and sulphur. The calcium is lost during G-, C- and Q-banding techniques. In the BSG (C-banding) technique, barium becomes attached to chromosomes during barium hydroxide treatment, but is subsequently lost. Giemsa staining introduces large amounts of sulphur (from thiazine dyes) and bromine (from eosin). Quinacrine staining introduces chlorine, the distribution of which appears to match that of the Q-bands.

THE PRINCIPLES OF X-RAY MICROANALYSIS

When electrons strike any object, a number of interactions occur which can be used microscopically (figure 1). The electrons, after interaction with the object, can be used to produce either a transmission or scanning electron microscope image; the electrons also excite visible radiation (cathodoluminescence) and X-rays. The X-rays are of two types (see figure 2): a continuum ("white radiation" or Bremsstrahlung) which forms a background to the spectrum at all energy levels; and specific peaks, characteristic of particular elements. The specific peaks permit identification of the elements present, and their size is related to the quantity of the elements; the size of the continuum is related to the total mass of material present (in a thin specimen). Thus for a thin specimen in constant conditions the concentration of a particular element is proportional to the ratio of the peak size to the continuum. Comparison of concentrations of different elements, and absolute quantitation, both require the use of standards, and have not been done in the work described here. It should be noted that, as normally carried out, X-ray microanalysis does not detect elements lighter than sodium, thus excluding the organic matrix of biological materials. For a full account of the principles and practice of X-ray microanalysis, see Chandler[1].

APPLICATIONS TO THE STUDY OF CHROMOSOME BANDING

X-ray microanalysis can be profitably applied to two types of problems which cannot easily be tackled by conventional histochemical techniques. Firstly, it is

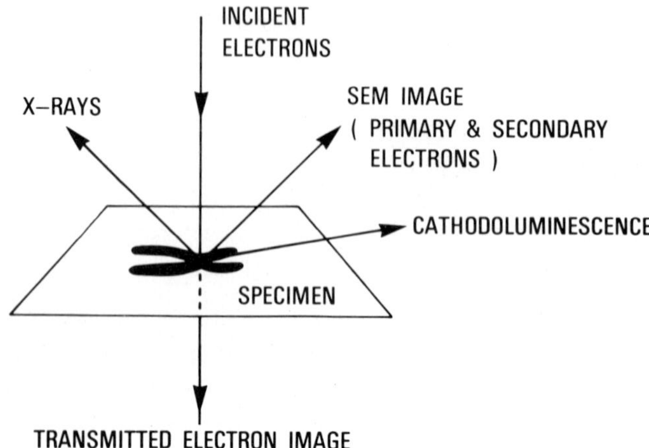

Fig. 1. Interactions of electrons with matter which have been used in microscopy. Transmitted electrons are used to produce a transmission electron micrograph, while backscattered (primary) and secondary electrons are used to form a SEM image. Electron-induced visible radiation produces a cathodoluminescence image, and electron-induced X-rays can be used for specific elemental identification.

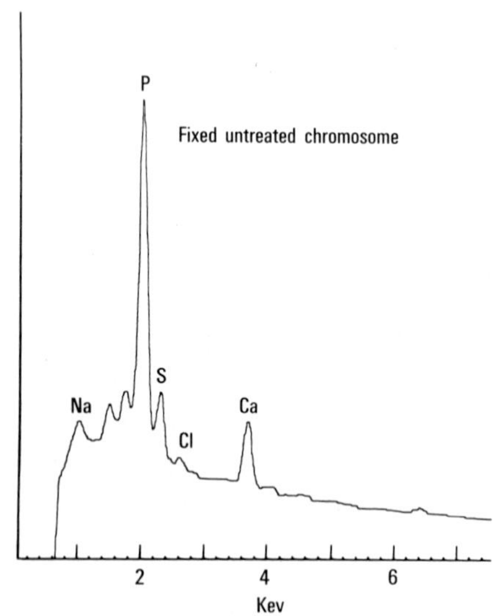

Fig. 2. X-ray spectrum of a human chromosome, fixed in methanol-acetic acid but otherwise untreated. Specific elemental peaks are referred to by their chemical symbols.

particularly useful for studies on inorganic ions and their distribution in biological materials. An example of this type of problem which has been studied here is the changes in elemental composition of chromosomes during G- and C-banding procedures. It has been found that only certain ionic solutions can produce various types of banding[15,16], and it has been suggested that removal of ions from chromosomes might be involved in banding[17]. Knowledge of the binding of various ions to chromosomes should help in understanding this problem.

Secondly, X-ray microanalysis can be used for the detection of low concentrations of substances which are not revealed by other techniques. The fluorescence of chromosomes stained with quinacrine is readily detected, but the absorption of the dye is so low that it is impossible to tell whether or not the distribution of fluorescence reflects the distribution of the dye. Such information is necessary to understand the mechanism of Q-banding of chromosomes. As we shall see, the chlorine atom in the quinacrine molecule permits its assay by X-ray microanalysis.

INSTRUMENTATION

In the work described here, X-ray microanalysis was carried out in a Cambridge Stereoscan S180 scanning electron microscope. The microscope was used in the scanning transmission mode, using a specially modified stage and specimen holder[2]. X-rays were detected and the signals processed with a Link Systems model 290 energy dispersive X-ray microanalyser system. The output was obtained in one of three forms: plots on an XY recorder of the X-ray spectrum (e.g. figures 2,4); X-ray counts of selected elemental peaks and of a selected portion of continuum, for relative quantitation; and line scans for a selected element, to show variations in amount of that element along the direction of scanning (e.g. figure 5).

RESULTS AND DISCUSSION

G- and C-banding. Human metaphase chromosomes fixed in methanol-acetic acid and mounted on formvar films[2] were analysed after the various stages of the ASG G-banding technique[3] and the BSG C-banding technique[4]. An X-ray spectrum for a fixed, untreated chromosome is shown in figure 2, and most of the changes found during these banding procedures in figure 3. Untreated chromosomes contain substantial amounts of phosphorus (presumably mainly in DNA, but possibly also in RNA and phosphorylated proteins), sulphur, and calcium.

In the ASG technique, 2 x SSC treatment at 60°C produces a reduction in phosphorus, and a complete loss of calcium. Similar changes also occur in the early stages of the BSG technique. During barium hydroxide treatment in the BSG technique, barium becomes attached to the chromosomes, but is lost again during the subsequent 2 x SSC treatment. With both techniques, Giemsa staining results in an increased

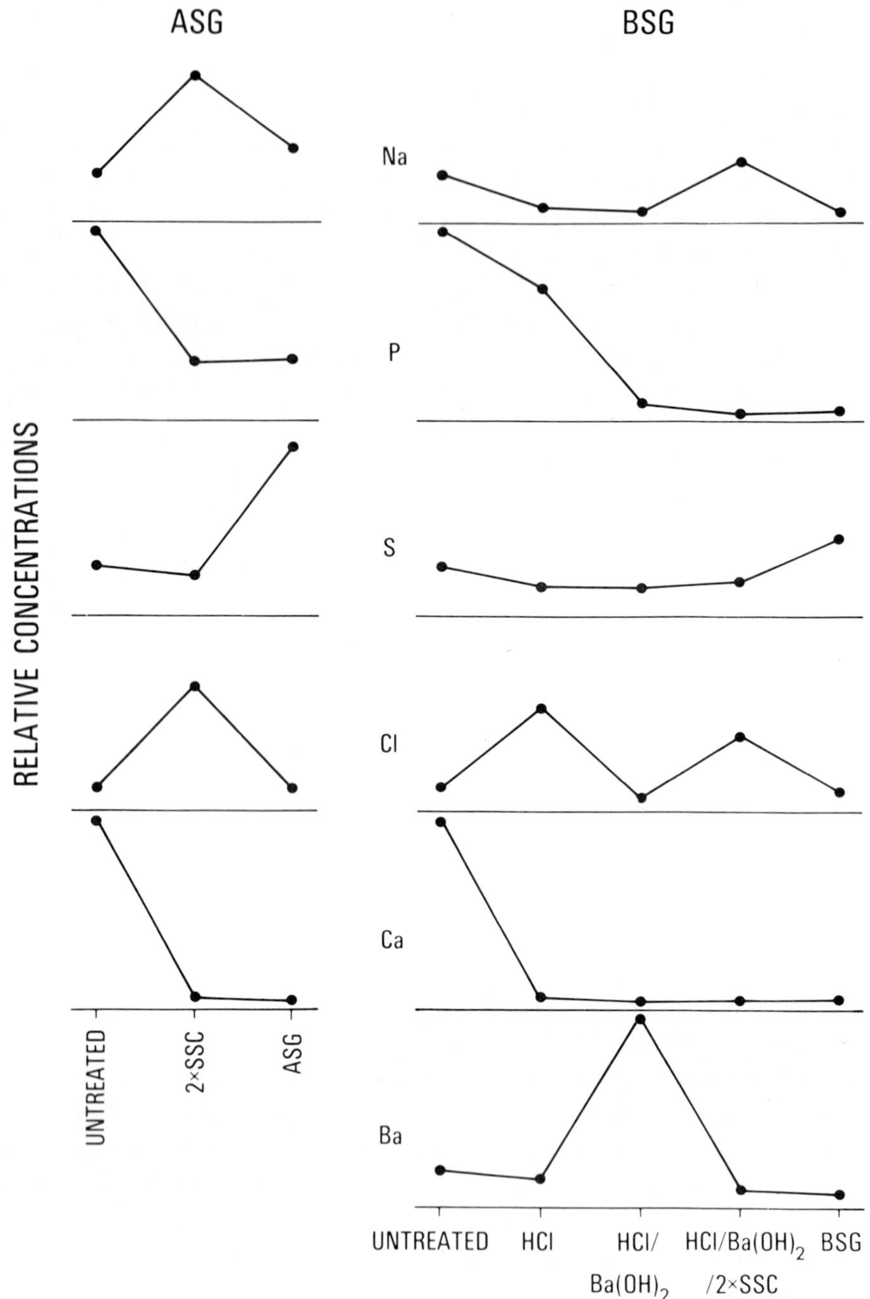

Fig. 3. Changes in elemental concentrations (of sodium, phosphorus, sulphur, chlorine, calcium and barium) during the ASG (G-banding) and BSG (C-banding) procedures. Note that concentrations can only be compared within, and not between, elements.

concentration of sulphur in the chromosomes (from thiazine dyes), and the introduction of bromine (from the eosin in the Giemsa - not shown in figure 3). Freely diffusible ions, such as sodium and chlorine, are readily taken up by the chromosomes and lost again, when the preparations are placed in appropriate solutions.

The presence of substantial amounts of calcium in acid-fixed chromosomes is surprising. Divalent cations such as calcium appear to be involved in aggregation of chromatin fibres[5]; thus the loss of calcium might be related to the collapse of chromosome structure in 2 x SSC[6], and this loss might be necessary so as to permit the swelling thought to be required for differential staining of banded chromosomes[7].

The loss of phosphorus during the ASG procedure is unexpected, since the Feulgen reaction indicates that no DNA is lost[8]. The amount of RNA and phosphorylated protein in the chromosomes is not known, but is probably too small to account for the large decrease observed. On the other hand, the occurrence of a decrease in DNA during C-banding procedures is well established[9]. The amount of phosphorus lost during the BSG technique is, however, unexpectedly large. The occurrence of large amounts of sulphur and bromine in chromosomes after staining provides further evidence that both thiazines and eosin are involved in G- and C-banding of chromosomes with Giemsa[7].

Q-banding. Metaphase chromosomes from CHO cells were fixed in methanol-acetic acid and spread on a thin film of "Optilux"[10]. Quinacrine staining was carried out according to standard procedures[11].

An X-ray spectrum of a quinacrine stained chromosome is shown in figure 4. Compared with an untreated chromosome, the calcium has been lost, and there is a large chlorine peak. The chlorine peak is no doubt due to the presence of quinacrine, which has a covalently bound chlorine atom in its molecule[12].

Figure 5 shows three line scans of the largest acrocentric chromosome in the CHO karyotype. While the phosphorus distribution along the chromosome is essentially uniform apart from a dip at the centromere, the line scan for chlorine shows a number of small but definite and reproducible peaks. These peaks correspond reasonably well in position with Q-bands as shown on the line scan of the density of a photographic negative of a quinacrine stained chromosome of this type.

These experiments are in a preliminary phase, and most chromosomes analysed have not shown such clear and consistent differences between their phosphorus and chlorine distribution. This is no doubt largely due to technical difficulties. Nevertheless, it should be emphasized that the phosphorus and chlorine line scans for a particular chromosome have not been found so far to have identical shapes; in other words, the distributions of these two elements, and therefore presumably of DNA and quinacrine, are not identical. It may therefore be concluded that the

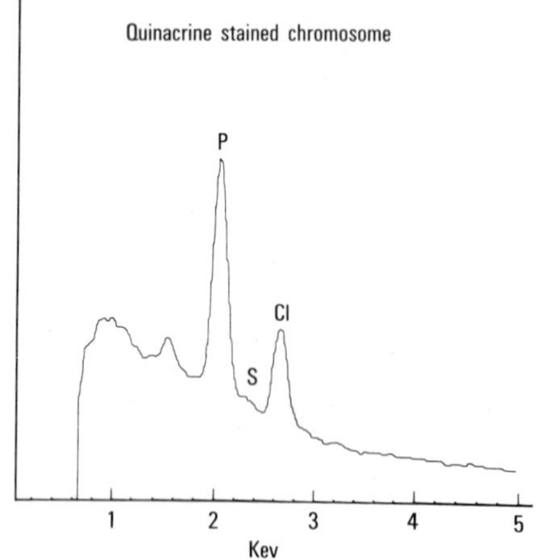

Fig. 4. X-ray spectrum of a CHO chromosome, fixed in methanol-acetic acid, and stained with quinacrine. Note the large chlorine peak due to quinacrine (cf. fig. 2)

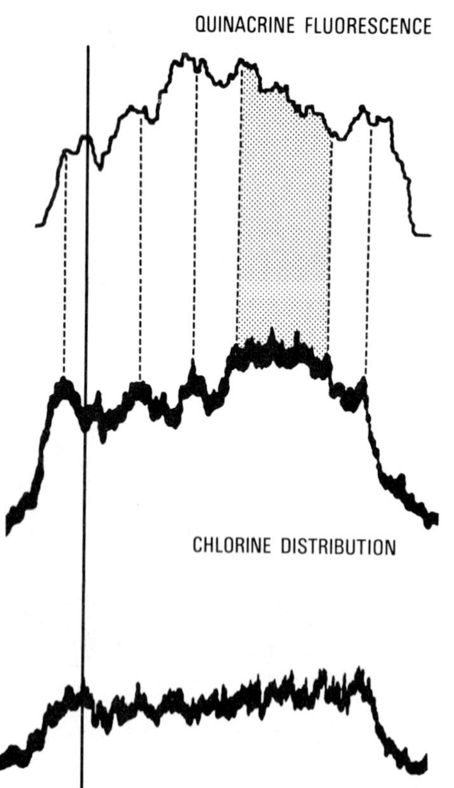

Fig. 5. Line scans of a certain CHO chromosome, stained with quinacrine. Top to bottom: density of a photographic negative of a Q-banded preparation; chlorine (quinacrine) distribution by X-ray microanalysis; phosphorus distribution. Note the similarity in position of the Q-band and chlorine peaks. The continuous vertical line indicates the position of the centromere.

ratio of quinacrine to DNA varies along the chromosomes. It has been supposed that Q-banding patterns are a consequence of differential excitation and quenching of fluorescence along the chromosome[13,14], implying that the distribution of the dye itself is uniform. The results described here, however, suggest that the distribution of the dye itself could be important in the production of Q-bands. If the correspondence of quinacrine distribution to the fluorescence patterns could be confirmed for other chromosomes, it would suggest that the distribution of quinacrine along the chromosome is the major determinant of Q-banding patterns, rather than differential excitation and quenching of fluorescence.

CONCLUSIONS

The work described here is preliminary, but has already raised some interesting problems. The presence of calcium in fixed chromosomes was unexpected, although its occurrence in native chromatin is established[18]. It still has to be established whether the loss of calcium during G-banding and C-banding procedures is essential to the production of these types of bands, and indeed whether it occurs with all such procedures. One anomaly found during the G-banding is the loss of phosphorus, in contrast to the established retention of DNA[8]. Do other phosphorus compounds occur in substantial quantities in chromosomes, and is their loss important for the production of bands? Examination of chromosomes during other G- and C-banding, as well as R-banding, techniques are needed to assess the degree of universality of these findings. The implications of the observations on quinacrine-stained chromosomes have already been discussed.

X-ray microanalysis of chromosomes has much greater potential than the two examples described here. It could be used to study the distribution along chromosomes of a wide variety of substances, either already in the chromosome or introduced in a histochemical procedure. Possible examples of the former are inorganic ions; of the latter specific nucleic acid bases or amino acid residues in proteins identified by their binding of heavy atoms. It is not necessary to restrict such studies to fixed chromosomes, and although there are great technical problems (see, for example, Echlin[19]), the study of the elemental composition of chromosomes as close as possible to living state would be particularly interesting. It is hoped that the examples described here will have given some idea of the potentialities of X-ray microanalysis in the study of chromosomes. In particular, the technique can easily provide information not readily obtainable in any other way, and can provide a novel viewpoint for our understanding of chromosome banding.

REFERENCES

1. Chandler, J.A. (1977) X-ray Microanalysis in the Electron Microscope. North-Holland, Amsterdam.
2. Sumner, A.T. (1977) Histochem. J. (in press).
3. Sumner, A.T., Evans, H.J. and Buckland, R.A. (1971) Nature New Biol. 232, 31-32.
4. Sumner, A.T. (1972) Exptl. Cell Res. 75, 304-306.
5. Ris, H. and Kubai, D.F. (1970) Ann. Rev. Genet. 4, 263-294.
6. Gormley, I.P. and Ross, A. (1972) Exptl. Cell Res. 74, 585-587.
7. Sumner, A.T. and Evans, H.J. (1973) Exptl. Cell Res. 81, 223-236.
8. Sumner, A.T., Evans, H.J. and Buckland, R.A. (1973) Exptl. Cell Res. 81, 214-222.
9. Pathak, S. and Arrighi, F.E. (1973) Cytogenet. Cell Genet. 12, 414-422.
10. Felluga, B. and Martinucci, G.B. (1976) J. submicr. Cytol. 8, 347-352.
11. Evans, H.J., Buckton, K.E. and Sumner, A.T. (1971) Chromosoma 35, 310-325.
12. Lillie, R.D. (1969) Biological Stains, 8th edition. Williams and Wilkins, Baltimore.
13. Pachmann, V. and Rigler, R. (1972) Exptl. Cell Res. 72, 602-608.
14. Weisblum, B. and de Haseth, P.L. (1972) Proc. nat. Acad. Sci. 69, 629-632.
15. Kato, H. and Moriwaki, K. (1972) Chromosoma 38, 105-120.
16. Scheres, J.M.J.C. (1977) Cytogenet. Cell Genet. 18, 2-12.
17. Dev, V.G., Warburton, D. and Miller, O.J. (1972) Lancet i, 1285.
18. Naora, H., Naora, H., Mirsky, A.E. and Allfrey, V.G. (1961) J. gen. Physiol. 44, 713-742.
19. Echlin, P. (1975) J. Microsc. Biol. Cell. 22, 215-226.

Chromosomes Today Volume 6, A. de la Chapelle and M. Sorsa eds.

ON THE STRUCTURE OF CHROMATIN

ALEXANDER J. VARSHAVSKY
Institute of Molecular Biology, USSR Academy of Sciences,
Vavilov street, 32, Moscow B-312, USSR *

ABSTRACT

1) A mild treatment of the eukaryotic chromatin with staphylococcal nuclease or DNase I results in formation of mononucleosomes and of smaller DNP particles (subnucleosomes) of several discrete kinds differing from each other by both protein composition and size of DNA fragments.

2) Evidence for histone H1-H1 interactions in solution (as revealed by formaldehyde cross-linking) is presented.

3) Analysis of *Escherichia coli* chromosomes isolated under conditions similar to those used for isolation of the eukaryotic chromatin has shown that the proteins of highly purified *E.coli* deoxyribonucleoprotein are mainly in addition to RNA polymerase two specific proteins of apparent molecular weight 17,000 and 9,000. Some properties of these two proteins suggest that they may be prokaryotic "counterparts" of the eukaryotic histones.

4) We have found that the isolated SV40 viral minichromosome which contains all five histones[1] exists in solution under approximately physiological ionic conditions not as a circular beaded fiber but as a compact roughly spherical particle approximately 300 Å in diameter. The compact form of the minichromosome in contrast to its extended (circular beaded) form is virtually completely *resistant* to staphylococcal nuclease, strongly suggesting that in particular, nuclease-sensitive parts of the internucleosomal DNA regions are not exposed on the outside of the compact SV40 minichromosome.

INTRODUCTION

In this report I shall briefly summarize our recent experimental findings obtained in three different but mutually interdependent lines of research namely, in studies on the structure of eukaryotic[2-4], prokaryotic[4,5] and viral[1,4] chromatin. This work was carried out in the laboratory of Professor G.P.Georgiev in collaboration with V.V.Bakayev, T.G.Bakayeva, S.A.Nedospasov, V.V.Shmatchenko and P.M.Chumackov.

RESULTS

Heterogeneity and composition of eukaryotic di-, mono- and subnucleosomes. Chromatin was prepared from mouse Ehrlich tumor cells, treated with staphylococcal nuclease[6-8] and then fractionated by low ionic strength gel electrophoresis (Fig. 1a). As was shown previously[2,3], mononucleosomes[9-13] are fractionated into

---

*Present address: Department of Biology, Massachusetts Institute of Technology, Cambridge, Massachusetts 02139, USA.

two major discrete bands MN_1 and MN_2 (Fig. 1a). Mononucleosome MN_1 lacks histone H1 and contains a 140-bp DNA fragment, whereas the MN_2 contains all five histones and a 170-bp DNA fragment[3]. Three discrete bands DN_1-DN_3 (Fig. 1a) correspond to three dinucleosomes differing from each other by both the number of H1 molecules per particle and the size of DNA fragment[2,3]. More detailed analysis described in ref. 3 revealed a third type of mononucleosomes (MN_3) with unusual properties (see ref.3 for detail).

The electrophoretic pattern shown in Figure 1a was virtually one and the same for several different chromatins, in particular, for chromatins from mouse Ehrlich tumor cells, green monkey kidney cells (CV-1 line) and calf thymus. However, when the above-described method of analysis was applied to the hen erythrocyte chromatin in which histone H1 is partially replaced by histone H5[11], the patterns obtained differed from those in the case of mouse Ehrlich tumor chromatin (Fig. 1a; cf. Fig. 1b-f). Specifically, it was found that the mononucleosomes MN_2 from nuclease digests of the erythrocyte chromatin migrate in two discrete bands with close but nonidentical mobilities (Fig. 1c-f). Analysis of the histone composition of MN_2 by two-dimensional gel electrophoresis has shown that the faster MN_2 band (Fig. 1c-f) contains histone H5 plus all other histones except H1, whereas the other MN_2 band contains all five histones but no histone H5 (see ref.4 for detail).

It was found previously that a small but significant proportion of the DNP in a mild nuclease digest of the chromatin sedimented more slowly than mononucleosomes[2,14]. Electrophoresis of these particles (subnucleosomes) through low-ionic-strength polyacrylamide gels resolved them into eight major discrete species some of which were found to be complexes of discrete DNA fragments (from 20 to 50 bp long) with specific nonhistone proteins[3,4]. These specific basic nonhistone proteins belong to a group of so-called HMG (high mobility group) nonhistone proteins[15] (see refs. 3 and 4 for a detailed description of these findings).

Oligomers of H1 histone in solution as revealed by formaldehyde cross-linking. Histone H1 is generally assumed to be a "cross-linking" histone, that is, a histone which is at least partially responsible for maintenance of the compact state of chromatin under approximately physiological ionic conditions[16,17]. However, there are a number of conflicting reports on H1-H1 interactions in the current literature. Histone H1 within isolated chromatin is not cross-linked to other histones by treatment with diimidoesters at pH 8.5-9.0 but apparently forms long chains of covalently joined H1 molecules[18].

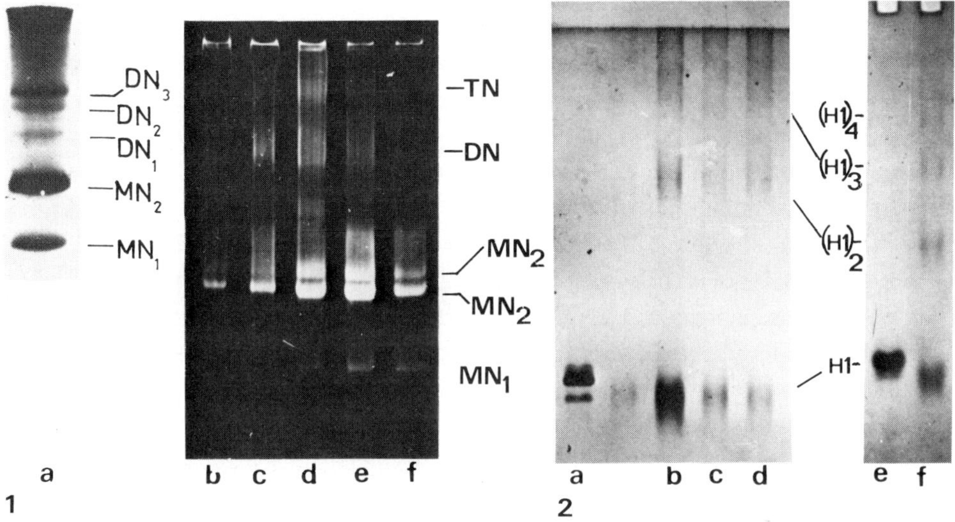

Fig. 1. Polyacrylamide gel electrophoresis of deoxyribonucleoproteins[2-4].

(a) Total 5% acid-soluble staphylococcal nuclease digest of the mouse Ehrlich tumor chromatin; (b-f) Total nuclease digests of the histone H5-containing hen erythrocyte chromatin at various extents of digestion[4]. Electrophoresis in a 5% gel. Staining with Coomassie in a and with ethidium bromide in b-f.

Fig. 2. Cross-linking of H1 histone in solution with formaldehyde.

Purified histone H1 (100 μg/ml) was treated with 1% HCHO for 6 hr at ~4°C followed by addition of SDS to a final SDS concentration of 1%, dialysis against 1% SDS, 1 mM triethanolamine (TEA--HCl), pH 7.5 and electrophoresis through a 5/15 percent polyacrylamide gel in a discontinious buffer system (a-d) or through a 5 per cent gel in a continious buffer system (e,f). The gels were stained with Coomassie. Untreated (control) H1 was processed identically except that the HCHO was omitted.

(a) and (e) Control (untreated) H1 histone; (b) and (f) H1 treated with HCHO in 0.1 M NaCl, 2 mM Na-EDTA, 20 mM TEA-HCl, pH 7.5 (pH of buffers was measured at 20°C); (c) The same as b but the HCHO treatment at pH 9.0; (d) The same as b but 2 M NaCl instead of 0.1 M NaCl.

On the other hand, results of cross-linking of H1 at pH 8.5--9.0 may not correspond to a "native" conformation of H1 molecules as was suggested by the recent work of Irwning and Isenberg[19]. Their data suggest that there are no significant H1-H1 interactions in solution, in contrast to strong interactions of some H1 subfractions with nonhistone proteins HMG1 and HMG2[18]. As concerning the use of aldehydes for cross-linking of H1 molecules, an early work of Olins and Wright[20] suggested formation of H1- and/or H5-containing oligomers within chicken erythrocyte chromatin upon treatment with formaldehyde or glutaraldehyde. However, to our knowledge no comparable study has been carried out so far on the purified H1 in solution. To partially clarify this point we treated the purified H1 histone at a concentration of 0.1 mg/ml with 1% formaldehyde in a solution with approximately physiological ionic strength and pH and also under some other solvent conditions (see Fig. 2 and the legend to it for details). The results clearly show that formation of H1 oligomers does occur in solution upon treatment with formaldehyde, and hence, suggest the existence of relatively strong H1-H1 interactions even in the absence of DNA (Fig. 2). Notice that in contrast to sharp electrophoretic bands of untreated (monomer) H1 histone the bands corresponding to HCHO-treated H1 histone are much more diffuse in appearance and in addition the monomer band of the HCHO-treated H1 migrates slightly faster than that of the untreated H1 (Fig. 2a,e; cf. Fig. 2 b-d,f). Both effects may be due to a general decrease of a positive charge of H1 molecules upon chemical modification with formaldehyde which in turn, leads to an increase of an anomalously low mobility of H1 in SDS-gels[21]. In addition, the HCHO treatment may result in intramolecular cross-links in H1 which also can lead to an increase of H1 mobility in SDS-gels.

Formation of oligomers of H1 (Fig. 2) and probably even of long polymers of H1 strongly suggests the existence of multiple affinity sites (most probably of two sites) within each H1 molecule. Attractive candidates for these sites are firstly, the central weakly charged "globular" region of the H1 molecule[17,22] and secondly, the N-terminal region of the molecule, namely, the first 10-12 residues among which there are no basic amino acids[22]. Moreover, similarly to the central "globular" region of H1 this short N-terminal stretch is also highly conserved in different subfractions[17] of H1 from a given tissue or even between H1 histones from different species[22]. Thus formation of a regular "lattice"of interacting H1 molecules may well form the basis of its "cross-linking" role in the chromatin.

Histone-like proteins of the Escherichia coli chromosome. In the last few years methods have been developed for the isolation of the bacterial chromosomes[23,24].However, until now there was practically no direct data on the possible existence of specific histone-like proteins in association with the folded E.coli DNA. On the other hand, a variety of low-molecular weight DNA-binding proteins of apparently nonribosomal origin have been isolated from DNase-treated extracts of E.coli cells[25,26]. The isolation of these proteins generally started from DNase-treated extracts of E.coli cells and therefore could not permit unambiguous identification of a particular basic protein as a chromosome-bound protein in vivo. Nevertheless, some of these proteins have properties sufficiently in common with eukaryotic histones to suggest similar functions in E.coli cells. These and related data[27] have led us to a search for histone-like proteins in the isolated E.coli chromosome which could organize bacterial DNA into nucleosome-like structures.

Folded E.coli chromosomes were prepared by a modification of previously developed methods[28]. The E.coli deoxyribonucleoprotein was extensively purified from RNP particles and free proteins which were present in the initial E.coli chromosome preparation by Sepharose 2B gel chromatography of the mildly DNase-treated E.coli chromosomes (see ref. 5 for a detailed description of methods and results). Figure 3 shows that the purification of the E.coli DNP greatly reduces the complexity of the protein pattern of the preparation as revealed by SDS-gel electrophoresis (Fig. 3j; cf. Fig. 3a-c). Two specific protein bands (bacterial histones BH1 and BH2; see below) corresponding to apparent molecular weights of 17,000 and 9,000 respectively, constitute a major protein component of the purified E.coli DNP (Fig. 3j). The SDS-gel electrophoretic pattern of the total protein complement of the purified E.coli DNP contains in addition to the major bands of BH1 and BH2 the bands corresponding to subunits of RNA polymerase and also a few other minor protein bands (Fig. 3j)[5]. However, a 0.25 N HCl-extract of the purified E.coli DNP contains almost exclusively the BH1 and BH2 proteins (Fig. 3i; cf. Fig. 3j) thus indicating both the basic nature of these proteins and the absence of other acid-soluble proteins in the purified E.coli DNP.

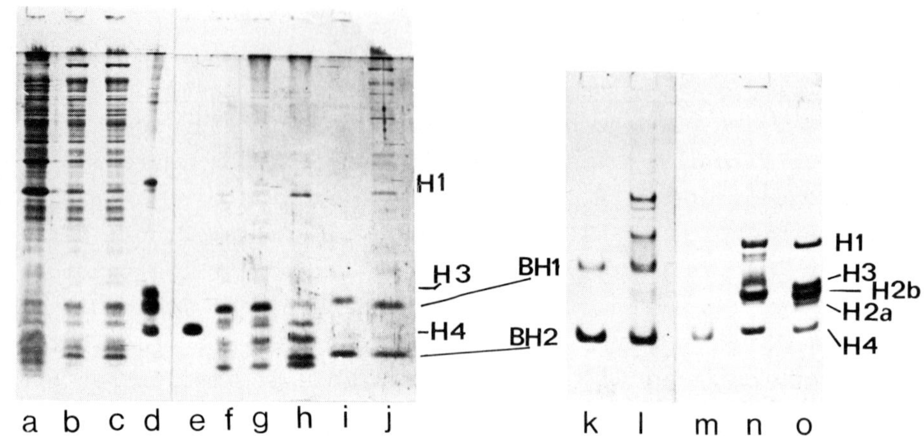

Fig. 3. Polyacrylamide gel electrophoresis of E.coli chromosomal proteins .

Electrophoresis was carried out in the presence of SDS (slots a-j) and in the acetic-urea system (slots k-o). (a) Proteins of the initial unfractionated E.coli chromosome preparation. (b) The same as a but after purification of the DNP from a major part of membraneous material[5]. (c) The same as b but from another experiment. (d) Total protein of the mouse Ehrlich tumor chromatin. (e) Egg white lysozyme. (f)-(h) Heat-stable proteins from extracts of E.coli cells (see ref. 5 for details). (i) Proteins of the 0.25 N HCl-extract of the purified E.coli DNP. (j) Total protein of the purified E.coli DNP. (k) The same as i but low-pH electrophoretic system. (l) A 0.25 N HCl-extract of the DNP preparation shown in b,c. (m) HU protein[25]. (n) Total protein of the mononucleosomes isolated from a nuclease digest of the mouse Ehrlich tumor chromatin[2] (a doublet of bands below the eukaryotic histone H1 are specific HMG nonhistone proteins). (o) Total mouse histone.

Figure 3 k,l shows the results of analysis of the BH1 and BH2 proteins in the acetic acid-urea gel electrophoretic system[29]. Although future sequencing of BH1 and BH2 may reveal significant similarities between primary structures of the E.coli BH1 and BH2 proteins and the eukaryotic histones, it is already clear from the comparison of these proteins in two different electrophoretic systems (Fig. 3; see also ref. 5) that neither BH1 nor BH2 are strictly identical to any of the five eukaryotic histones. It should be noted that while the BH2 protein is readily and quantitatively extracted from the E.coli DNP with 0.25 N HCl, the BH1 protein is extracted much less efficiently and is therefore underrepresented in the electrophoregrams of the acid extracts (Fig. 3k; cf. Fig. 3i). Finally, we have found[5] that the BH2 is apparently identical to the DNA-binding protein HU which was isolated previously from DNase-treated extracts of E.coli cells[25,26].

Summarizing the above-considered data on the BH1 and BH2 proteins we conclude that by several criteria, in particular histone-like molecular weights and electrophoretic properties, presence in the purified E.coli DNP in approximately equal molar amounts and in a large excess over any other protein in the DNP (see ref.5 for details), these two proteins closely resemble the eukaryotic histones. Since in addition, the BH1 and BH2 were found to afford a localized protection to DNA in the E.coli DNP upon nuclease digestion[5] we propose to name these two proteins bacterial histones (BH1 and BH2).

Compact form of SV40 viral minichromosome is resistant to nuclease: possible implications for chromatin structure. Simian Virus 40 appears to be a particularly attractive experimental object for studies on the structure and functioning of nucleosomes since SV40 DNA and cellular histones are associated in infected cells and in SV40 virions in a chromatin like structure called a minichromosome[30]. The SV40 and polyoma minichromosomes were visualized in the electron microscope as circular beaded fibers which consisted of 20-22 nucleosomes joined by short DNA-like threads[30-33]. We have previously found that the SV40 minichromosomes extracted with 0.15 M NaCl from nuclei of lytically infected cells contain not only "nucleosomal" histones but histone H1 as well[1].

Sedimentational and electrophoretic analysis of the SV40 minichromosomes has shown that under "physiological" salt conditions ($\mu \approx 0.15$) minichromosomes exist in a much more compact conformation than at a low ionic strength ($\mu \approx 0.005$) (see ref.4 for detail). In another line of experiments the SV40 minichromosomes were examined by electron microscopy. Most of the material on the metal-shadowed grids were roughly spherical particles 350-400 Å in diameter (Fig. 4a,c). A striking difference between the previously observed extended (circular beaded) form of the histone H1-containing SV40 minichromosome and its compact configuration which is capable of fitting within the virus capsid is illustrated by Fig. 4a,b. In contrast to metal-shadowed compact minichromosomes (Fig. 4a,c) the stained particles (Fig. 4d-g) show a distinct internal structure. On the basis of staining properties of the mononucleosomes (see e.g., refs. 34, 35) we interpret the particles in Fig. 4d-g as consisting of a tightly packed circular oligonucleosomal filament in which individual mononucleosomes with their characteristically stained central area are clearly seen.

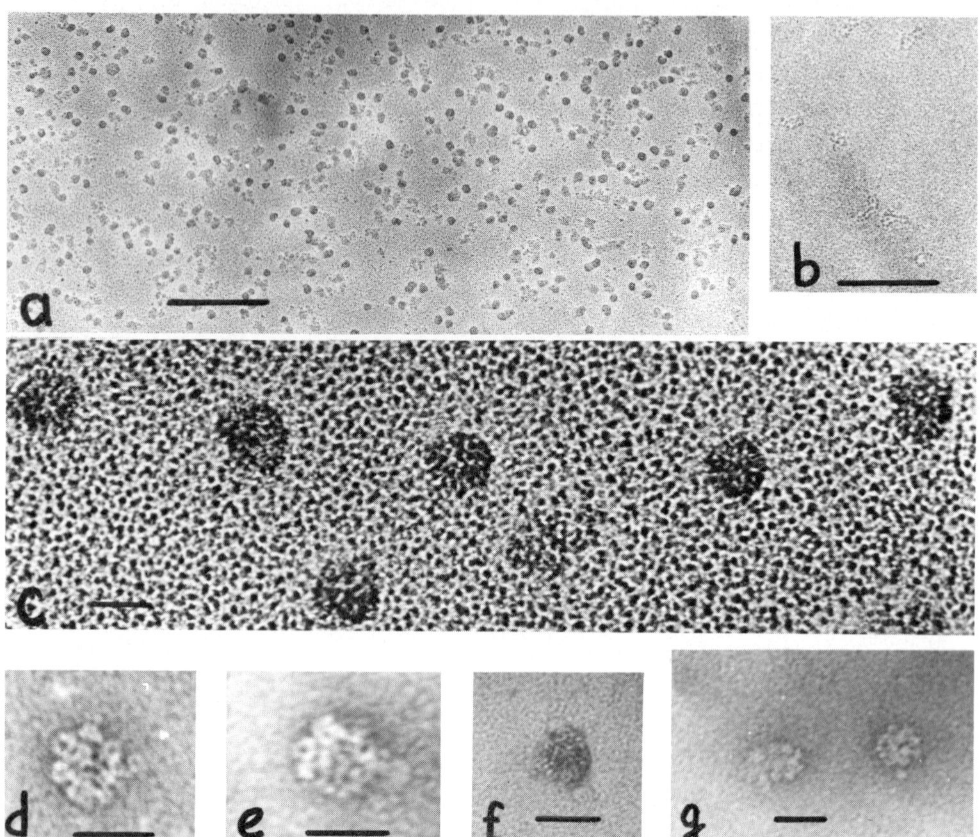

Fig. 4. Electron microscopy of compact and extended SV40 viral minichromosomes.

(A) A low-magnification micrograph of the metal-shadowed compact minichromosomes. (B) Extended (circular beaded) minichromosomes in a low-ionic-strength buffer. (C) The same as A but a higher magnification. (D,E) Compact minichromosomes stained with uranyl formate. (F) The same but stained with uranyl acetate. (G) The same but stained with sodium phosphotungstate. The bars correspond to 0.5 μ in A,B and to 300 Å in C-G.

Figure 5 shows that the treatment of the compact SV40 minichromosomes in 10% sucrose, 0.13 M NaCl, 3 mM $CaCl_2$, 1 mM TEA-HCl, pH 7.5 with staphylococcal nuclease does not lead to any significant degradation of the covalently closed supercoiled SV40 DNA I in the minichromosomes even in the presence of relatively high concentrations of the nuclease both at a low (4°C) and high (37°C) temperature (see ref. 4 for details). Appropriate controls have shown that staphylococcal nuclease was enzymatically active under the above-mentioned solvent conditions[4].

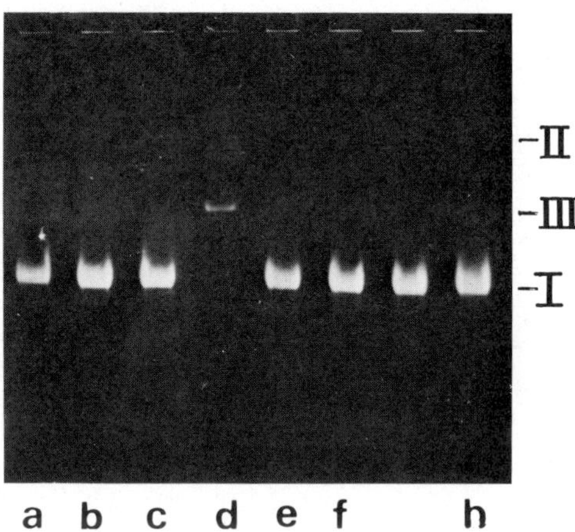

Fig. 5. Compact SV40 minichromosome is resistant to staphylococcal nuclease.

Isolated SV40 minichromosomes were treated with nuclease in the presence of 0.13 M NaCl at 4°C for 0 (no enzyme), 3, 10, 30 and 60 min, respectively (slots a-c, e-f) followed by incubation at 37°C for 10 and 60 min, respectively (slots g and h) and agarose gel electrophoresis of DNA. Slot d contains the EcoRI-produced linear SV40 DNA III marker[4].

Since staphylococcal nuclease apparently preferentially attacks internucleosomal DNA stretches (linkers) in the chromatin[2,3,6,11,12] the observed resistance of the compact minichromosome to the nuclease strongly suggests that in particular nuclease-sensitive parts of the linker region are not exposed on the outside of the compact minichromosomes. On the other hand, we have found that in contrast to staphylococcal nuclease pancreatic DNase (DNase I) readily attacks the compact SV40 minichromosomes (unpublished data). This result correlates with the previously established mode of action of DNase I on chromatin, according to which DNase I attacks both inter- and intranucleosomal DNA stretches with comparable efficiences[36,37].

Figure 6 shows some of the possible models of the compact SV40 minichromosome (see ref. 4 for discussion of these models).

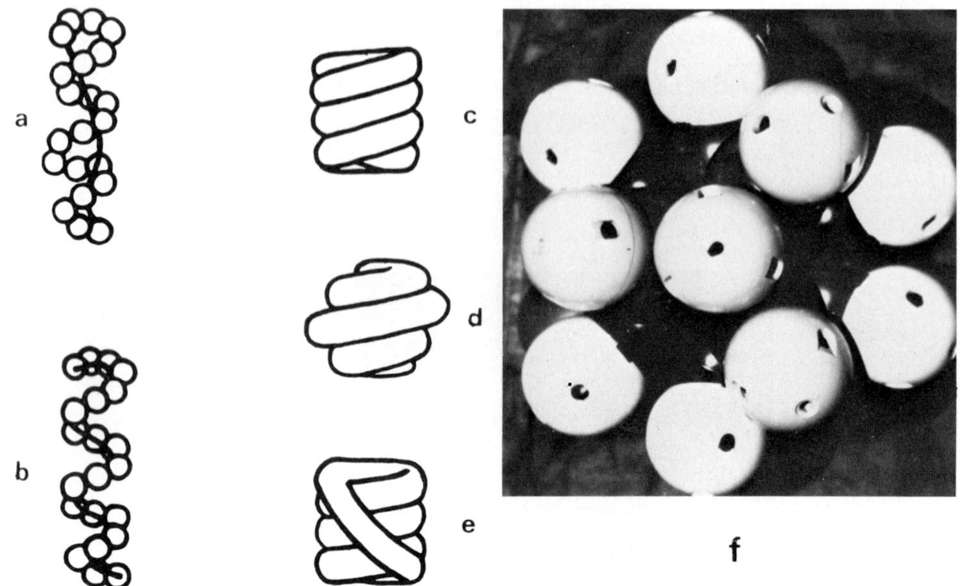

Fig. 6. Some of possible models of the compact SV40 minichromosome[4].

(a) and (b) are highly schematic drawings of a partially relaxed solenoidally coiled SV40 nucleosomal filament with a relatively short (a) and long (b) "return" fiber (see ref. 4 for a terminology). Both a and b types of structure may correspond to general side views depicted roughly to-scale in (c) and (d). The model (c) is based on a constant number of nucleosomes per solenoidal turn (approximately six[38]). In the dodecahedral model (d), (f) the number of nucleosomes per turn varies from approximately ten in the central turn to approximately five in the two other turns of the nucleosomal helix. The dodecahedral arrangement of nucleosomes permits also a "nonsolenoidal" mode of packing[4]. The model (e) is an unlikely but formally not completely excluded structure in which the two ends of the nucleosomal helix are joined together outside the helix.

REFERENCES

1. Varshavsky, A.J., Bakayev, V.V., Chumackov, P.M. and Georgiev, G.P. (1976) Nucl.Acids Res. 3, 2101-2118.
2. Varshavsky, A.J., Bakayev, V.V. and Georgiev, G.P. (1976) Nucl.Acids Res. 3, 477-496.
3. Bakayev, V.V., Bakayeva, T.G. and Varshavsky, A.J. (1977) Cell 11, 619-629.
4. Varshavsky, A.J., Bakayev, V.V., Nedospasov, S.A. and Georgiev, G.P. (1977) Cold Spring Harb. Symp. Quant. Biol.,in press.
5. Varshavsky,A.J., Nedospasov, S.A., Bakayev, V.V., Bakayeva, T.G. and Georgiev, G.P. (1977) Nucl.Acids Res., in August issue.
6. Noll, M. (1974) Nature 251, 249-250.

7. Axel, R., Melchior, W., Sollner-Webb, B. and Felsenfeld, G. (1974) Proc. Nat.Acad.Sci. USA 71, 4101.
8. Greil, W., Igo-Kemenez, T. and Zachau, H.G. (1976) Nucl.Acids Res. 3, 2633-2645.
9. Olins, A.L. and Olins, D.E. (1974) Science 184, 330-331.
10. Kornberg, R.D. (1974) Science 184, 868-871.
11. Shaw, B.R., Herman, T.M., Kovacic, R.T., Beadreau, G.S. and Van Holde, K.E. (1976) Proc.Nat.Acad.Sci.USA 73, 505-509.
12. Weintraub, H. (1975) Proc.Natl.Acad.Sci.USA 72, 1212-1216.
13. Oudet, P., Gross-Bellard, M. and Chambon, P. (1975) Cell 4, 281.
14. Bakayev, V.V., Melnickov, A.A., Osicka, V.D. and Varshavsky, A.J. (1975) Nucl.Acids Res. 2, 1401-1419.
15. Goodwin, G.H., Sanders, C. and Johns, E.W. (1973) Eur.J.Biochem. 38, 14-20.
16. Bradbury, E.M., Carpenter, B.G. and Rattle, H.W. (1973) Nature 241,123-126.
17. Cole, R.D. (1977) in Molecular Biology of the Mammalian Gene Apparatus (P.Ts'o, ed.), Elsevier, Amsterdam, pp.93-104.
18. Thomas, J.O. and Kornberg, R.O. (1975) Proc.Natl.Acad.Sci. USA 72, 2626-2630.
19. Smerdon, M.J. and Isenberg I. (1976) Biochemistry 15, 4242-4247
20. Olins, D.E. and Wright, E.B. (1973) J.Cell Biol. 59, 304-311.
21. Weber, K. and Osborn, M. (1975) in The Proteins (H.Neurath and R.L.Hill, eds.). Acad.Press, N.Y. pp.180-221.
22. Yaguchi, M., Roy, C., Dove, M. and Seligy, V. (1977) Biochem. Biophys.Res.Commun. 76, 100-106.
23. Stonington, G.O. and Petijohn, D.E. (1971) Proc.Nat.Acad.Sci. USA 68, 6-10.
24. Worcel, A. and Burgi, E. (1972) J.Mol.Biol. 71, 127-137.
25. Rouviere-Yaniv, J. and Gross, F. (1975) Proc.Nat.Acad.Sci.USA 72, 3428-3431.
26. Haselkorn,R. and Rouviere-Yaniv, J. (1976) Proc.Nat.Acad.Sci. USA 73, 1917-1920.
27. Griffith, J.D. (1976) Proc.Nat.Acad.Sci.USA 73,563-567.
28. Portalier, R. and Worcel, A. (1976) Cell 8, 245-252.
29. Panyim, S. and Chalkley, R. (1969) Biochemistry 8, 3972-3976.
30. Griffith, J.D. (1975) Science 187, 1202-1203.
31. Bellard, M., Oudet, P., Germond, J. and Chambon, P. (1976) Eur. J. Biochem. 70, 543-551.
32. Cremisi, C., Pignatti, P.G., Groissant, O. and Yaniv, M. (1976) J.Virol. 17, 204-210.
33. Christiansen, G. and Griffith, J. (1977) Nucl.Acids Res. 4, 1831-1851.
34. Varshavsky, A.J. and Bakayev, V.V. (1975) Mol.Biol.Reports 2, 247-251.
35. Olins, A.L., Senior, M.B. and Olins, D.E. (1976) J.Cell Biol. 68, 787-788.
36. Noll, M. (1974) Nucl. Acids Res. 1, 1573-1579.
37. Sollner-Webb, B. and Felsenfeld, G. (1977) Cell 10, 537-546.
38. Klug, A. and Finch, J.T. (1976) Proc.Nat.Acad.Sci.USA 73, 1897-2000.

Chromosomes Today Volume 6, A. de la Chapelle and M. Sorsa eds

MICROSPREADING AND THE SYNAPTONEMAL COMPLEX IN CYTOGENETIC STUDIES

MONTROSE J. MOSES
Department of Anatomy, Duke University Medical Center, Durham, North Carolina 27710

ABSTRACT

A simple and rapid surface microspreading technique for preparing complete complements of synaptonemal complexes (SC) in spermatocytes and oocytes has been used to study meiosis in a number of mammals. Pachytene SC karyotypes have been constructed, autosomal and XY morphology and pairing behavior have been followed through meiotic prophase, and multivalents and mouse chromosome rearrangements, including translocations, a duplication and an inversion analyzed. It is concluded that the chromosomal axes, which form the lateral elements of the SC, provide faithful representations of chromosome behavior. SC formation at zygotene is observed to depend on homology, while subsequent SC accommodations in pachytene are independent of it.

INTRODUCTION

The synaptonemal complex (SC) regularly forms the axis of meiotic bivalents in animals and plants. It is itself composed of the synapsed proteinaceous axes of homologous chromosomes, joined in parallel by transverse filaments which meet in a linear central element. These essential architectural facts, which were deduced from early studies on thin sections[1] (review[2]) have since become established and extended by studies on whole meiocyte nuclei, either through reconstructing electron micrographs of serial sections (reviews[3,4]), or through whole mounts obtained by spreading cells by surface tension on an aqueous surface (see discussion[5]). It is now clear that the single axis constitutes a backbone of the chromosome which in turn consists of two chromatids. The axis, to which chromatin fibrils are attached, probably as loops, appears prior to or at synapsis, and disappears at or following desynapsis.

Following the first demonstrations by serial section reconstructions that axes synapse to form SC's[6,7] and separate again at desynapsis[8], it became evident that the axes provide simple, linear representations of chromosome behavior in meiosis, the SC that they form being a characteristic of the synaptic state. The biological information possible from quantitative analysis of the SC in normal cells[9] and in cells carrying chromosomal rearrangements[10] was recognized, and has been pursued in several laboratories by serial section reconstruction techniques.

SURFACE MICROSPREADING

The introduction of a simple and rapid surface microspreading technique has provided new information about the synaptonemal complex as well as a substantial improvement in its visualization[11]. Subsequent applications of the method to mammal-

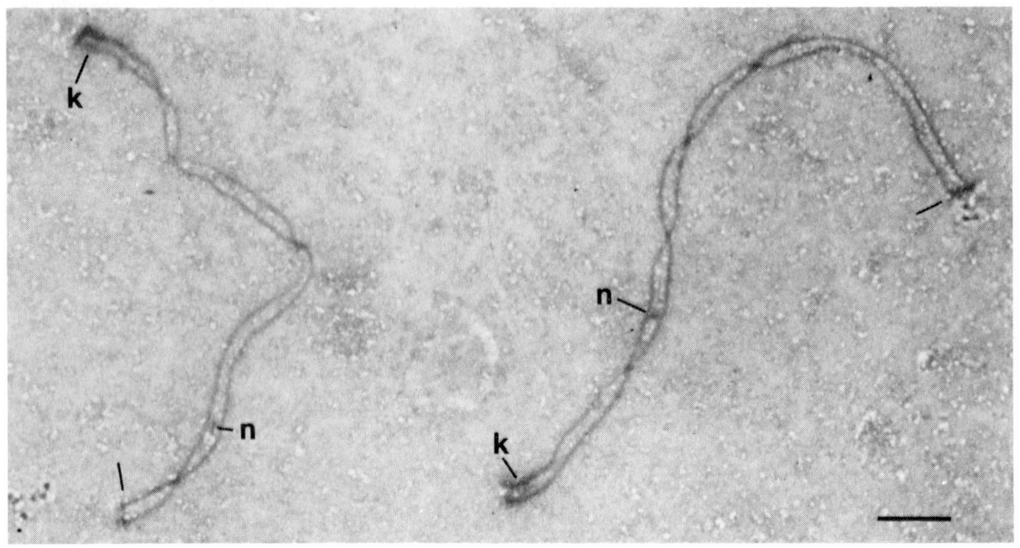

Fig. 1. Electron micrograph of mouse spermatocyte SC's prepared by the Counce-Meyer microspreading procedure. (—) = attachment plaques to the nuclear envelope; k = kinetochore (terminal); n = "recombination nodule" of Carpenter[24] (other SC's of the complement also show nodules). Scale bar = 1.0 μm.

ian spermatocytes in which full complements of SC's are preserved[5,12,13,14,15,16] has further extended this knowledge and has demonstrated the potential of this method as a means of analyzing chromosome structure and behavior during meiotic prophase, a period of particular cytogenetic significance.

The method (description[5,11,13]) provides flattened nuclei in which full SC complements, selectively stained, are displayed in two dimensions. Terminal plaques for attachment to the nuclear envelope, and differentiations of the SC that represent kinetochores, are also stained (Fig. 1), together, in some cells, with nucleoli, annuli of the nuclear envelope, and centrioles. The morphological details in such preparations correspond to those seen in thin sections with respect to axial and SC structure, attachment to the nuclear envelope, etc. In addition, some details, such as the kinetochore, are often not visible in thin sections, but are distinct in the microspread preparations. While precise spatial interrelationships are disrupted during the spreading process, such relationships as end-to-end associations, association with nucleoli, and peripheral position of the sex body are maintained. The method has now been used in our laboratory to investigate spermatocytes and oocytes in a number of mammals, including mouse, man, Syrian, Chinese and Armenian hamsters, the Egyptian sand rat (Psammomys obesus), dog, macaque, lemur, and deer mouse.

OBSERVATIONS ON MEIOTIC PROPHASE IN MICROSPREAD PREPARATIONS

The behavior of the axes during synapsis and desynapsis, as typified by the

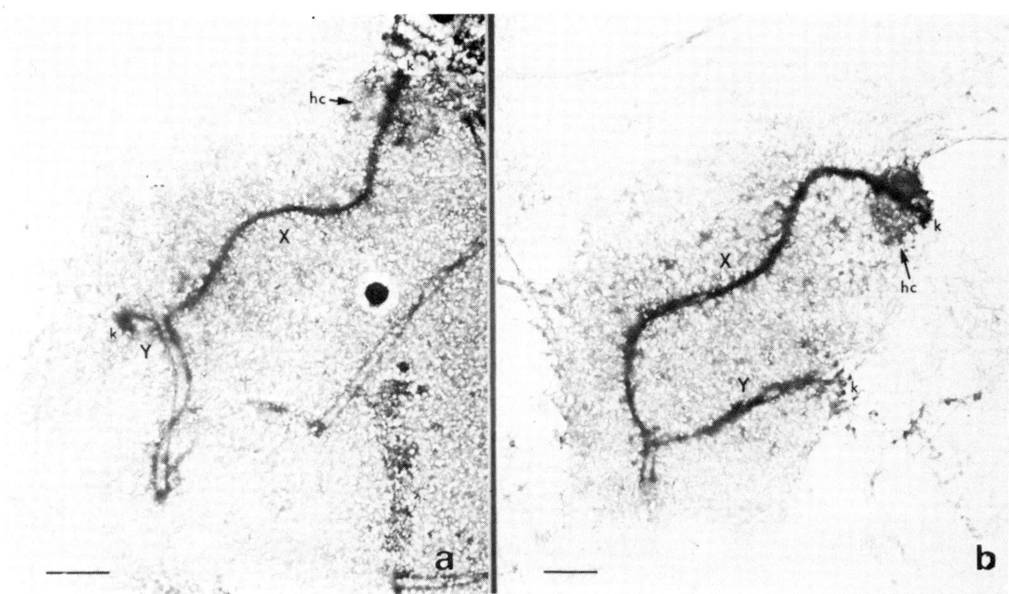

Fig. 2. Mouse XY pair. a) Early pachytene: most of the Y axis has paired distally with the distal end of the X to form an SC. b) At later pachytene the X and Y axes have desynapsed precociously, leaving a short length of SC at their distal ends. Unpaired axes are thicker and denser, and in places appear double. k = kinetochore region; hc = heterochromatic knob. (a. from Moses, 1977[5]; b. from Moses et al., 1977[15]).

Chinese hamster[5], has certain common features among the mammals studied. Incomplete unpaired axes appear at leptotene, initiating from terminal attachment points on the nuclear envelope. Synapsis (SC formation) follows quickly, often before the axes are completely formed, again with initiation usually occurring at the nuclear envelope. Ends of unpaired axes are not associated with each other prior to synapsis and therefore must move together over the surface of the nuclear envelope. In chromosomes with interstitial kinetochores, this region is the last to synapse. In the mouse, where kinetochores are terminal (Fig. 1), pairing may occur via the kinetochore associated heterochromatin of the terminal knob, while the SC forms interstitially, or at the distal end, joining the kinetochore region later. At pachytene, SC's are continuous and complete and in spreads are often connected with each other at their ends by packed arrays of annuli.

In the Chinese hamster, the axes disassemble prior to autosomal desynapsis[5], while in the mouse, as in most of the other mammals studied, axes are observed to separate at diplotene and then to disassemble[16]. As reported by Solari[8] from thin section reconstructions, residual segments of SC occur at diplotene. That these may represent chiasma sites, as postulated, is supported by observations of such a persistent segment in the known site of a localized chiasma in the XY pair of the Armenian hamster (Solari and Moses, in preparation).

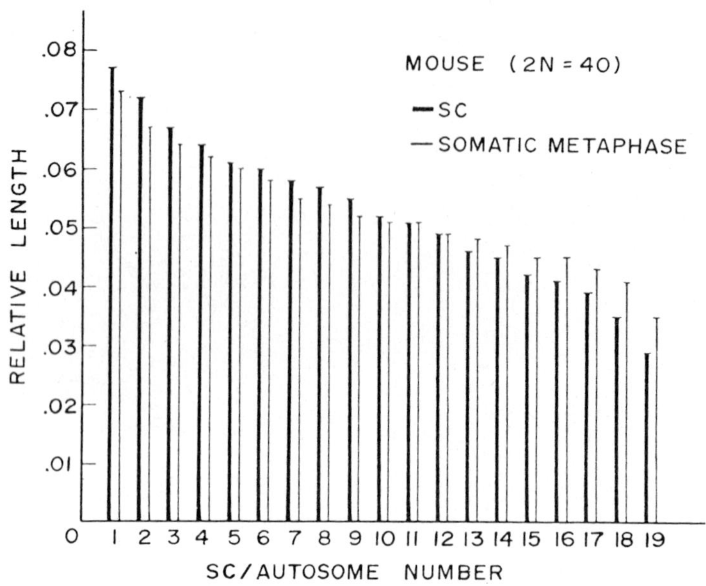

Fig. 3. Idiogram of mouse autosomal karyotype constructed from relative length measurements of eight full SC complements (dark bars), compared with somatic metaphase karyotype.

The axes of the X and Y chromosomes are easily distinguished from those of the autosomes (Fig. 2). They are generally thicker where they are not synapsed, often with distinctive differentiations such as bulges, hollows, multiple strands, branches, excrescences, loops, and association of external material. They are segregated from the autosomes during pachytene within a chromatin body ("sex vesicle"). Progressive changes in the XY pair during pachytene are species-characteristic, and, in some cases, may be taken as a timing index of progress through pachytene. However, as the synapsis and desynapsis of the XY pair (Fig. 2) are generally out of phase with the autosomes, and because there is variation in this asynchrony from cell to cell, use of the XY pair to define substages in anything but a general way, must be made with caution[14, cf., 17].

Quantitative measurements of autosomal bivalents in the Chinese hamster[15] show that they may be characterized by SC length and kinetochore position (arm ratio). From such data, pachytene SC karyotypes may be constructed which agree with those of somatic metaphase (Fig. 3). Relative autosomal SC lengths and arm ratios are constant for each bivalent despite SC shortening and lengthening through pachytene[15,18] indicating that all autosomes change length synchronously. The evidence that relative lengths and arm ratios are constant at once rules out physical distortions in preparation that might affect the linear measurement, and indicates as well that bivalent length is under regular biological control. SC relative lengths are equal to those of somatic metaphases (Fig. 3), as are arm ratios, implying that meiotic prophase and mitotic metaphase chromosomes are under common length controls.

KINETOCHORE AND CHROMOSOME AXIS

The finding of a transient axis continuous with the kinetochore in somatic cells[16,19] suggests that the axis is not exclusively a meiotic feature. It well may be a structure involved in the regulation of chromosome length, and thus could be regarded as a structural expression of a system that provides linear order at the chromosomal level and determines the chromosome to be a rod, rather than a blob.

The important discovery that the kinetochore, or at least a component of it, is itself a differentiation of the axis[11] means that, in a sense, the axis is an extension of the kinetochore. That the latter does in fact represent the kinetochore has been verified by its position on the SC from arm ratios in the Chinese hamster[15]. Moreover, a comparable structure in preparations of mitotic chromosomes has been shown to be the site of _in vitro_ microtubule assembly[20].

The behavior of the two chromatids as a unit at meiotic prophase has long been attributed to the singleness of the kinetochore, and it has been assumed that doubling of the kinetochore is an expression of the individualization of the chromatids. A contemporary restatement of this thesis would have the axis holding the chromatids as a unit along their length; a corollary would be that the chromatids do not behave individually until either the axis becomes double, or disassembles and releases its hold. In mammals generally, neither autosomal axes nor interstitial kinetochores are seen as double structures, except occasionally at late pachytene or diplotene when the ends adjacent to the attachment plaques may appear double. In telocentric chromosomes (as in mouse) this phenomenon represents doubling of the terminal kinetochore region. In the grasshopper (_Melanoplus_) kinetochores are double at late diplotene[21], by which time the axes have disassembled.

From such evidence it may now be concluded that the doubleness sometimes observed in the unpaired axes of mammalian X and Y chromosomes[7, 14, 22, 17] represents a precocious individualization of the chromatids, another expression of the allocycly of the sex pair.

The axes do not appear to participate directly in chromatid exchanges during crossing over. It has been argued that twists, often seen in SC's (Fig. 1) represent intersection points at which recombinational exchange may occur[23]. However, the frequency and position of twists are highly variable[15] and they seem more likely to be the results of internal stresses than specific recombinational events involving exchange between the axes. Further, both the "recombination nodules"[24,16] (Fig. 1) and the persistent segments of SC that apparently mark chiasma sites at diplotene[8] are independent of twists (unpublished observations).

CHROMOSOME ABNORMALITIES

If the axes are indeed faithful representations of chromosomal behavior, as indicated by the foregoing, then they should also express irregularities in chromosome

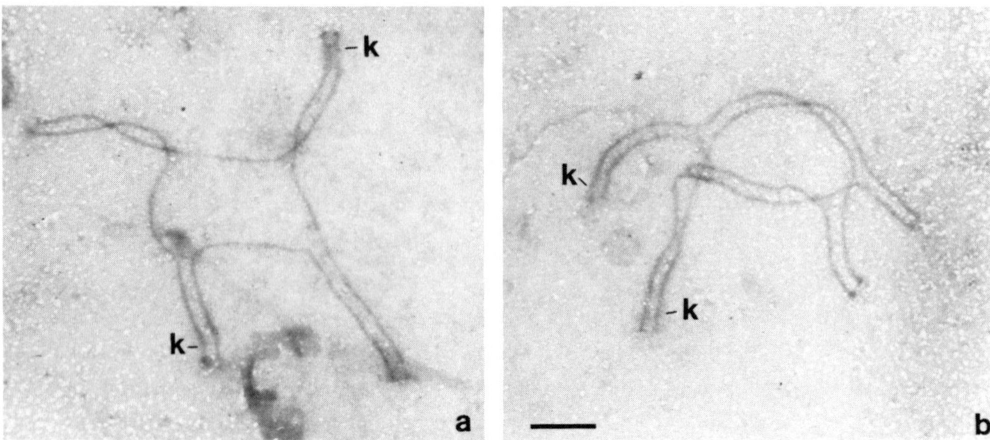

Fig. 4. Two tetravalents from a tetraploid mouse spermatocyte; a) shows one switch-over of pairing partners, and b) shows two. (From Solari and Moses, 1977[27]). Scale bar = 1 μm.

behavior. Indication that this is the case has come from serial section reconstructions of a translocation in the mouse[25] and an inversion in Maize[10]. Further studies of chromosome irregularities, taking advantage of the larger sample sizes provided by the microspreading technique, should lead to new insights into how chromosomes handle abnormalities during synapsis, and into the basis of the genetic consequences, such as irregular disjunction, recombination inhibition, etc.

Multivalents

The axes reflect the pairing behavior of the chromosomes in miltivalent, as well as bivalent formation. The switching of pairing partners in a triploid Lily has been followed in serial sections[26]. A tetraploid mouse spermatocyte has recently been observed[27] in which quadrivalents show both single and double switch-overs (Fig. 4). The multiple SC configuration reported for chicken triploids[28] has not been observed in these studies. The formation of a trivalent SC between a metacentric and two acrocentrics in a lemur hybrid has been taken as evidence of homology[29], corroborating that deduced from the similarity of G-banding patterns between the acrocentrics and the metacentric arms.

Rearrangements

A study of chromosome rearrangements in the mouse has been undertaken in order to compare the precision of information obtainable from microspreading analysis with that known from genetics and light microscopy. In each case, heterozygotes of aberration-bearing stocks were analyzed as unknowns in blind labeled experiments. The results were later compared with genetic and cytological data.

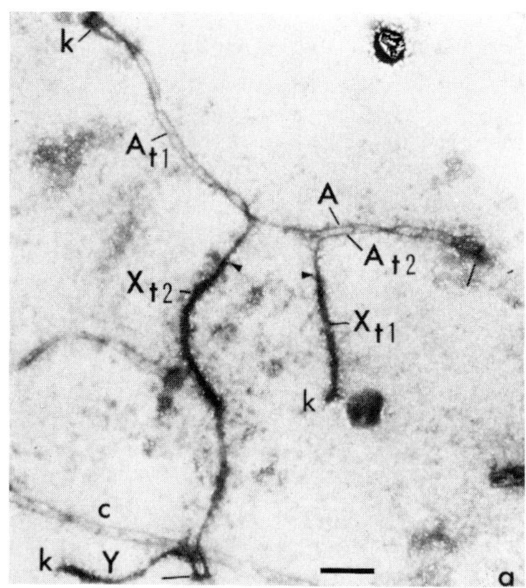

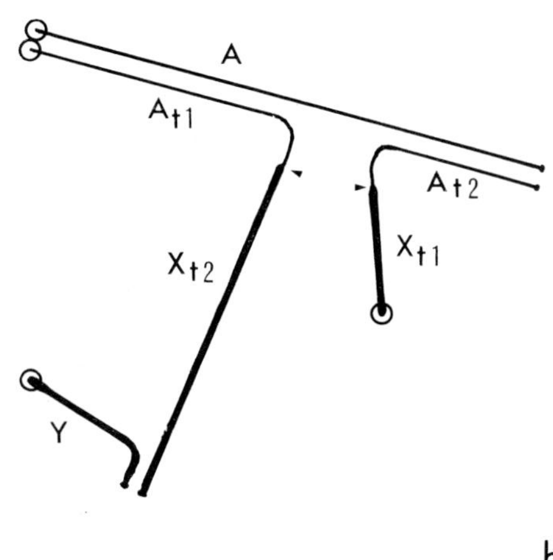

Fig. 5. a) Pairing figure of mouse X-autosome translocation heterozygote. A = non-translocated autosomal axis; A_{t1} and A_{t2} = axes of translocated portions of autosome; X_{t1} and X_{t2} = translocated portions of X axis; C = indifferent autosomal SC; k = kinetochore; Y = Y axis; (-) = distal attachment plaque; arrowheads = translocation breakpoints. b) Schematic diagram of (a). Scale bar = 1 μm. (From Moses, Russell and Cacheiro, 1977[22]).

<u>Translocations</u>. Three translocations were studied in collaboration with Drs. L. B. Russell and N. L. Cacheiro, Oak Ridge National Laboratory[22]. Two were translocations between the X and autosome 7, and one between autosomes 10 and 18. The X-autosome translocations were most easily analyzed; the thickened X axes are readily distinguished from autosomal axes, and breakpoints are sharply marked as transitions from thick to thin axes (Fig. 5). The synaptic figures observed were exactly those to be predicted from genetic and light microscopic cytological information, the axes being representations of the chromosomes. In one X-autosome translocation the EM analysis provided new information not known from light microscopic cytology; where the breakpoint was within the pairing region of the X, there was no synapsis with the Y. By contrast, in the translocations in which the breakpoint in the X was outside of the pairing region, XY synapsis was observed (Fig. 5). Quantitative data showed good agreement between measured breakpoints on the SC's and those estimated from banded chromosome analyses.

<u>Tandem duplication</u>. Another rearrangement, a tandem duplication[30], illustrates the potential of this method for quantitative cytogenetic analysis, and the way in which it can lead to new knowledge about chromosome behavior in meiosis. An Ω-like loop was observed in one lateral element of an SC in early pachytene complements (Fig. 6a). Measurement of 20 SC's bearing such loops showed a 16% difference in the two axes (lateral elements), the center of the loop being 59% of the length of the SC

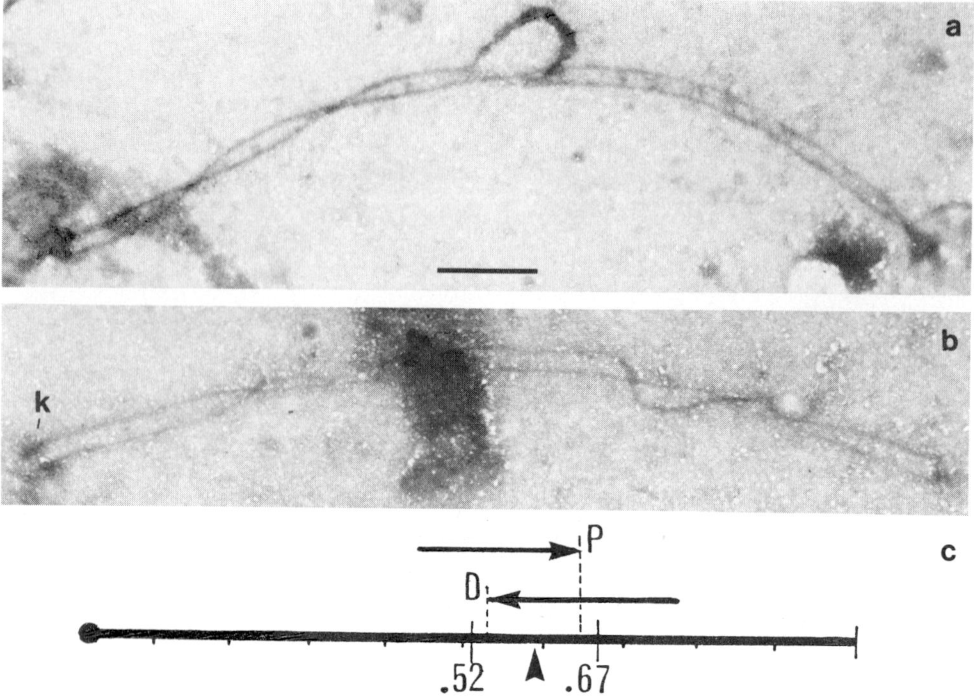

Fig. 6. Tandem duplication[31] in a mouse heterozygote. a) Duplication loop at early pachytene. Axis lengths differ by 13%, reflecting the length of the duplicated segment. b) Reduced duplication loop at mid-pachytene. Axis lengths now differ by only 4%. k = kinetochore. c) Length map of normal chromosome 7 showing the limits (.52 and .67) and center (arrowhead = .59) of the segment that is duplicated. The most proximal position (.53) of the distal end of the loop (D) produced by synapsis proceeding from the distal end, and the most distal position (.65) of the proximal end of the loop (P) produced by synapsis proceeding from the proximal end, are indicated. Data from a sample of 20 early pachytene bivalents. Scale bar = 1 μm.

distal to the kinetochore. Karyotypes were constructed using alternatively the long and short axes; these were then compared with control karyotypes. The karyotypes constructed with the short axes showed a slightly better agreement with the normal, indicating that the rearrangement was a tandem duplication and not a deletion. The tandem duplication was verified as one that included the Hbb, c, and possibly the Mod-2 loci[31]. From somatic chromosome banding, the abnormal chromosome (7) was estimated to be approximately 20% longer than the normal, and the calculated center of the segment to be 55-60% of the length distal to the kinetochore. The agreement between the electron and light microscope data is good considering difficulties of measuring banded preparations in which wide bands are involved.

The presence of a duplicated segment in one axis provides an opportunity to test the dependence of SC formation on chromosome homology. Assuming that synapsis may proceed from the ends in either direction (observations on zygotene in the mouse indicate that this is so; unpublished observations), either the proximal or the distal duplicated segments would synapse with the normal, depending on which reached it first. Completion of synapsis of this segment would leave non-homologous regions next in register, and SC formation, if it is dependent on homology, would have to cease in that direction. Synapsis from the opposite direction could proceed, however, leaving a single loop containing the duplicated segment. From these assumptions, one would expect the proximal end of the loop, resulting from synapsis proceeding away from the kinetochore, never to exceed the distal limit of the segment on the normal axis. Similarly, the distal end of the loop, resulting from synapsis proceeding toward the kinetochore, would never be more proximal than the proximal limit of the segment itself. In the 20 cases analyzed, the proximal and distal ends of the loops were observed to fall within these limits, the extreme positions being within 2 or 3% of the limits (Fig. 6c).

The above is taken as conclusive evidence that SC formation at zygotene is closely dependent upon genetic homology, and is unable to proceed in its absence.

Inversion. In collaboration with Drs. T. Roderick and M. Davisson, Jackson Laboratory, a mouse paracentric inversion heterozygote was analyzed. Inversion loops were clearly visible at early pachytene (Fig. 7). From preliminary measurements of a few such loops, the inversion appears to involve approximately 48% of the chromosome, with breakpoints 40% and 87% of the length distal to the kinetochore. The inversion was verified from genetic and banded chromosome data[32], its length being 49%, and the breakpoint positions 23 and 72% distal to the kinetochore. The regular formation of an inversion loop is further evidence for the respect that SC formation has for homology.

Accommodation at late pachytene

A surprising observation emerged from the tandem duplication and inversion studies that was only possible by virtue of a large sample size in which early, mid, and late pachytene stages could be distinguished (by the extent to which the X and Y were synapsed[7,22,17]; Fig. 2). In the tandem duplication, loops were observed mainly at early pachytene stages. The largest loops occurred at late zygotene, as judged by one or two SC's in which synapsis was incomplete. Unexpectedly, at late pachytene, no loops could be found. Reduced loops observed at intermediate stages indicated that they must open out and form twists (Fig. 6b). Measurements showed that the long axis shortens while the short axis maintains its relative length with respect to the rest of the complement. At late pachytene, there was no detectable length difference between the two axes of any of the SC's in the size range of that carrying the duplication, and the SC's appeared normal. The inversion apparently follows a pattern similar to the duplication: while inversion loops are present at early pachytene, they are no longer present at late pachytene and only normal SC's are to be seen. At intermediate stages there is again evidence to suggest that the loop opens out,

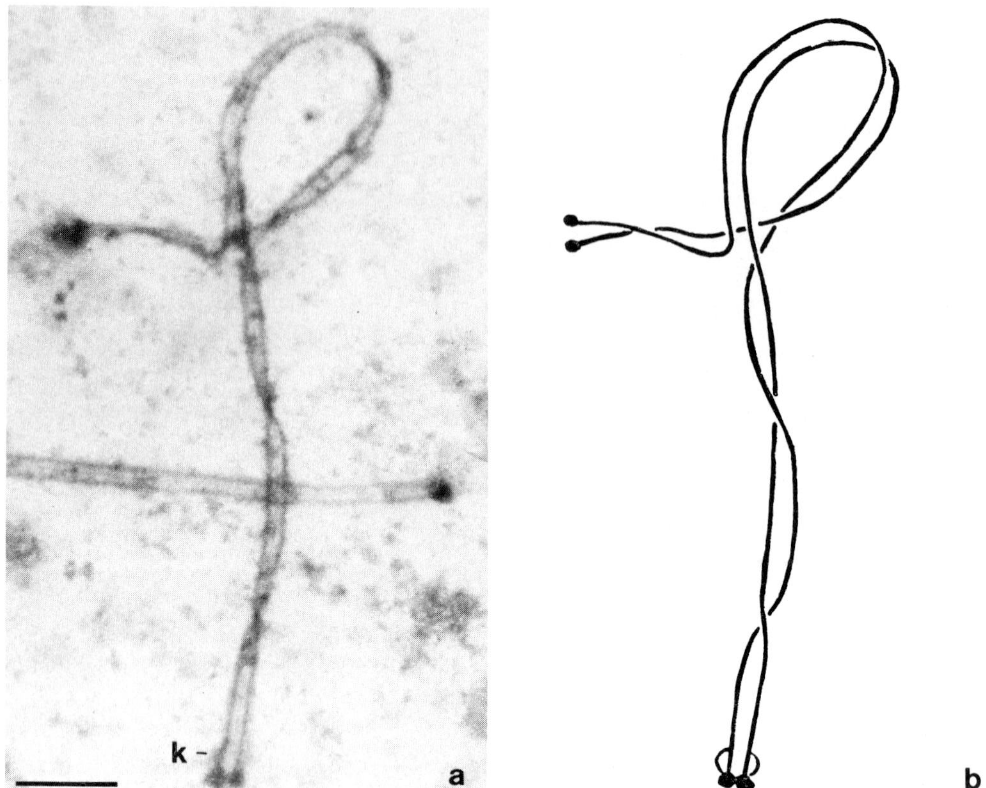

Fig. 7. a) Early pachytene SC inversion loop from a mouse heterozygote (In(2)5Rk). b) Interpretive tracing of (a). k = kinetochore. Scale bar = 1 µm.

leaving a normal SC, but with non-homologous regions now synapsed.

In our studies on the lemur metacentric/acrocentric heterozygote[29], the two non-homologous short arms were found to be unpaired at late zygotene-early pachytene, while the long arms formed regular SC's with the arms of the metacentric. At later pachytene, however, the short arms had synapsed and formed typical SC's, another instance of non-homologous synapsis during pachytene.

These examples indicate that an unsuspected accommodation mechanism operates during pachytene to correct structural irregularities that result from the requirements of homology operating between unlike axial partners. The rearrangement figure evidently represents an unstable state. The linear SC to which it is reduced by differential alteration in axial length, and by the synapsis of partners irrespective of homology, may be a more stable condition that precedes desynapsis (diplotene). Evidently, initial SC formation depends closely on homology, while subsequent compensations leading to apparently normal SC's are independent of such homology. Thus, any model of the SC in which a mechanism for homologue recognition is an integral part of the SC[e.g.,33] is insufficient to account for subsequent conversion to an SC that disregards homology. Recognizing that both homologous and non-homologous SC formation may occur sequentially within the same bivalent, it is now possible to reconcile the evidence that SC formation is a prerequisite for crossing over, with that for SC formation in non-homologous pairs. One could conceive, for example, that in homoeologous pairing[e.g.,34], SC formation might initiate between homologous regions, and be completed at later pachytene when the requirement for homology does not operate.

The homology-independent SC accommodation mechanism could have significant implications to crossing over. For example, if the recombinational event that requires both homology and the presence of SC occurs at mid to late pachytene[35], then crossing over within the duplicated and inverted regions would be reduced, depending on how far the non-homologous synapsis had progressed at the time of the recombination event. Reduced crossing over within these rearranged regions is in fact observed[32,36].

SUMMARY

From the observations summarized here, the lateral elements of the SC, as axes of the homologues, provide simplified but faithful representation of the chromosomes themselves during synapsis and desynapsis. Both qualitative and quantitative analyses of microspread preparations provide a powerful, practical approach to the routine study of chromosomes at those meiotic prophase stages (leptotene, zygotene, pachytene and early diplotene) that in many organisms have so far been intractable to detailed examination with the light microscope. The speed and ease with which such preparations can be made and observed in the electron microscope are quite comparable to those for light microscopy. We have examined grids within an hour after taking the biological specimen.

The advantages of SC analysis at pachytene are enormous. Synaptic irregularities resulting from chromosomal alterations are easily detected at early pachytene. Light microscopic analysis at diakinesis and metaphase I reflect post-synaptic events from which synaptic behavior can only be inferred. By comparison, SC analysis reveals synaptic behavior directly. It is possible to observe all cells in synapsis, including those that never progress further and escape detection at later stages, such as the X-autosome translocations which seldom progress beyond pachytene. Further, it is possible to detect first generation chromosome irregularities produced in the germ line, for example following exposure to mutagenic agents, thus providing an opportunity to scan for induced chromosomal abnormalities. In addition, the examples presented here indicate how SC analysis in spread preparations can provide biological information about chromosome behavior in synapsis that has long lain beyond the limits of light microscopic resolution, hidden in the confused appearance of the early prophase meiotic chromosomes.

ACKNOWLEDGEMENT

It is a pleasure to acknowledge the collaboration at various times of Drs. S. J. Counce, A. J. Solari, T. Ashley, A. E. Hamilton, Ms. P. A. Poorman, Mr. G. Slatton, and Ms. P. A. Karatsis in the studies reported, and the excellent technical assistance of Mr. T. Gambling and Mss. N. Staddon and M. Johnson.

Research was supported by grants from the NSF (PCM-76-00440) and NIH (GM-23047 to M. J. Moses and 5-S01-RR-05404 and CA-14236 to Duke University).

REFERENCES

1. Moses, M. J. (1958) J. Biophys. Biochem. Cytol, 4, 633-638.
2. Moses, M. J. (1968) Ann. Rev. Genet., 2, 363-412.
3. Westergaard, M. and von Wettstein, D. (1972) Ann. Rev. Genet., 6, 71-118.
4. Gillies, C. B. (1975) Ann. Rev. Genet., 9, 91-109.
5. Moses, M. J. (1977) Chromosoma, 60, 99-125.
6. Moens, P. B. (1969) Chromosoma, 28, 1-25.
7. Solari, A. J. (1970) Chromosoma, 29, 217-236.
8. Solari, A. J. (1970) Chromosoma, 31, 217-230.
9. Moens, P. B. (1973) Cold Spring Harb. Symp. Quant. Biol., 38, 99-107.
10. Gillies, C. B. (1973) Chromosoma, 43, 145-176.
11. Counce, S. J. and Meyer, G. F. (1973) Chromosoma, 44, 231-253.
12. Moses, M. J. and Counce, S. J. (1974) in Mechanisms in Recombination, Grell, R. F. ed., Plenum, New York, pp. 385-390.
13. Moses, M. J., Counce, S. J. and Paulson, D. F. (1975) Science, 187, 363-365.
14. Moses, M. J. (1977) Chromosoma, 60, 127-137.
15. Moses, M. J. et al. (1977) Chromosoma, 60, 345-375.
16. Moses, M. J. (1977) in Molecular Human Cytogenetics; ICN-UCLA Symposia on Molecular and Cellular Biology, Vol. VII, Sparkes, R.S., Comings, D. and Fox, C. eds., Academic Press, New York, in press.
17. Tres, L. L. (1977) J. Cell Sci., 25, 1-15.
18. Carpenter, A.T.C. (1975) Chromosoma, 51, 157-182.
19. Moses, M. J., Counce, S. J. and Solari, A. J. (1974) J. Cell Biol., 63, 234a.
20. Telzer, B.R., Moses, M. J. and Rosenbaum, J. L. (1975) Proc. Nat. Acad. Sci. USA, 72, 4023-4027.
21. Solari, A. J. and Counce, S. J. (1977) J. Cell Sci., in press.
22. Moses, M. J., Russell, L. B. and Cacheiro, N. L. (1977) Science, 196, 892-894.
23. Moens, P. B. (1974) in Mechanisms in Recombination, Grell, R. F. ed., Plenum, New York, pp. 377-383.
24. Carpenter, A. T. C. (1975) Proc. Nat. Acad. Sci. USA, 72, 3186-3189.
25. Solari, A. J. (1971) Chromosoma, 34, 99-112.
26. Moens, P. B. (1969) J. Cell Biol., 40, 273-279.
27. Solari, A. J. and Moses, M. J. (1977) Exp. Cell Res., in press.
28. Comings, D. E. and Okada, T. A. (1971) Nature (Lond.), 231, 119-121.
29. Moses, M. J., Karatsis, P. A. and Hamilton, A. E. (1975) J. Cell Biol., 67, 297a.
30. Moses, M. J. et al. (1977) J. Cell Biol., in press.
31. Russell, L. B. et al. (1976) Proc. Nat. Acad. Sci. USA, 73, 2843-2846.
32. Davisson, M. T. and Roderick, T. H. (1973) Cytogenet. Cell Genet. 12, 398-403.
33. von Wettstein, D. (1971) Proc. Nat. Acad. Sci. USA 68, 851-855.
34. Menzel, M. R. and Price, J. M. (1966) Am. J. Bot. 53, 1079-1086.
35. Stern, H. and Hotta, Y. (1973) Ann. Rev. Genet. 7, 37-66.
36. Russell, L. B. and Cacheiro, N. L. (1976) Oak Ridge National Lab. Biol. Div. Ann. Report ORNL 5195, pp. 12-13.

Chromosomes Today Volume 6, A. de la Chapelle and M. Sorsa eds.

THREE-DIMENSIONAL RECONSTRUCTIONS OF MEIOTIC CHROMOSOMES IN HUMAN SPERMATOCYTES

PREBEN BACH HOLM & SØREN WILKEN RASMUSSEN
Department of Physiology, Carlsberg Laboratory, Gl. Carlsbergvej 10, DK-2500 Copenhagen Valby (Denmark)

ABSTRACT

Meiosis in human males has been analyzed by three-dimensional reconstruction of bivalents. The total mean length of the synaptonemal complexes of the autosomal bivalents at pachytene amounted to 231 μm. Differences in mean lengths among individuals or among different stages of pachytene were insignificant. Most bivalents can be classified by length and centromere indices. Pairing of the X and Y chromosomes by a short telomeric region was observed in 19 of 22 nuclei. All acrocentric bivalents can be associated with a nucleolus. In the chromatin of the metaphase bivalents 30-60 fragments of central region material of the synaptonemal complex have been observed. These may represent synaptonemal complex chiasmata.

INTRODUCTION

A detailed picture of chromosome pairing during the meiotic prophase can be obtained through the analysis of the synaptonemal complex - a tripartite structure which joins the homologous chromosomes at pachytene. (For general reviews on the structure and function of the synaptonemal complex, see[1,2,3]). This structure is easily identified in the electron microscope and can be followed by serial sectioning and reconstruction of whole nuclei. The method has successfully been used to establish the chromosome number and the absolute lengths of zygotene and/or pachytene bivalents in a variety of organisms[4,5,6,7,8,9,10]. Also the sequence of events ultimately resulting in regular disjunction of the homologous chromosomes at anaphase I has been characterized with this method in several organisms[9,11,12].

The ultrastructure of human spermatocyte nuclei has been analyzed recently by the three-dimensional reconstruction of 22 pachytene nuclei[10]. The present report summarizes these results and extends the information on meiosis in the human male by a preliminary analysis of the ultrastructure of bivalents at metaphase I.

MATERIALS AND METHODS

Testicular biopsies were obtained from five men with a normal spermatogenesis. A description of the material and the methods employed for three-dimensional reconstructions from electron micrographs are given in[10].

RESULTS

The pachytene nucleus

Synaptonemal complexes. In all 22 nuclei investigated, the 22 autosomal chromosome pairs are present as bivalents each with a continuous synaptonemal complex from telomere to telomere (a survey picture of a mid pachytene nucleus is shown in Figure 1). The fine structure of the synaptonemal complex and its dimensions are similar to those described for other eukaryotic organisms[3]. Each of the homologous chromosomes is flanked by a lateral component which is closely associated with the chromatin of the two sister chromatids. The lateral components are electron dense with a filamentous substructure and with a round to avoid shape in cross section. The diameter varies between 300 and 400 Å. The two lateral components are held in register at a distance of 1100 to 1200 Å by the central region which contains a medially located central component with a diameter of 100 to 300 Å. The central component is connected to the lateral components by irregularly spaced traversing filaments. The type of nodes in or close to the central component analyzed by Gillies[13], Carpenter[14] and Zickler[9] were not discernible in the spermatocyte nuclei. The telomeres of all bivalents are attached to the inner membrane of the nuclear envelope through local thickenings of the lateral components.

Centromeric heterochromatin and knobs. Each pachytene bivalent contains one large block of compacted chromatin associated with the lateral components of the synaptonemal complex. In addition, a number of bivalents possess up to three smaller chromatin condensations, knobs. The reconstructions have revealed that the relative position of the larger blocks of chromatin corresponds to the position[15] of the primary constriction of diakinesis bivalents and somatic metaphase chromosomes. Inside the centromeric heterochromatin of all meta- and sub-metacentric bivalents, material of low electron density is present marking the position of the centromere. In the acrocentric bivalents, a morphological distinguishable centromere could only be identified in one early pachytene nucleus and centromere indices have therefore not been calculated for this group of bivalents.

The centromeric heterochromatin of bivalents 1, 9, and 16 exhibits a bipartite organization in all nuclei except for one in early pachytene: The large block of compacted centromeric chromatin is associated with the diffuse heterochromatin of the long arms of these bivalents, consisting of a mixture of compacted and diffuse chromatin.

Random fusion between the centromeric heterochromatin of two or more bivalents was frequently observed, fusion being most prevalent between heterochromatin of the acrocentric bivalents. A fusion of the synaptonemal complexes from different bivalents does not occur; they always retain their individuality.

The location and frequency of the knobs are shown in Figure 2. The chromatin of the knobs is generally distributed asymmetrically around the synaptonemal complex similar to the centromeric heterochromatin. Whether the observed differences in

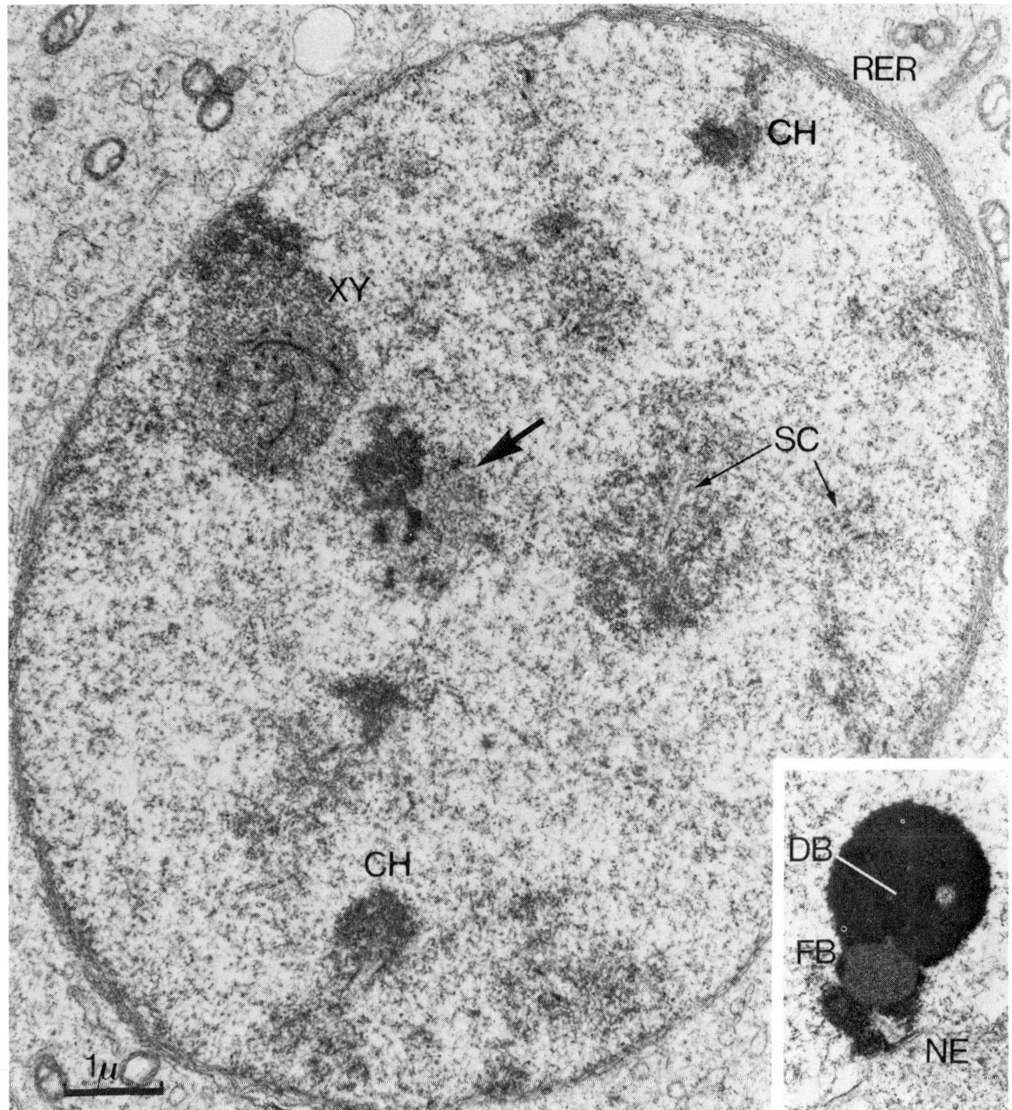

Fig. 1. Survey micrograph of a mid pachytene spermatocyte nucleus. The bipartite centromeric and diffuse heterochromatin of bivalent 9 is denoted by the arrow. Within the chromatin of the XY bivalent (XY) the lateral components of the sex chromosomes can be recognized. The inserted micrograph illustrates the association between an acrocentric bivalent and a nucleolus. A sphere of filamentous material (FB) marks the position of the nucleolus organizer region. The nucleolus contains a body of high electron density (DB). The chromatin of the short arm of the acrocentric bivalent is condensed. NE, nuclear envelope. Tl, telomere. SC, synaptonemal complex. CH, centromeric heterochromatin. RER, rough endoplasmic reticulum.

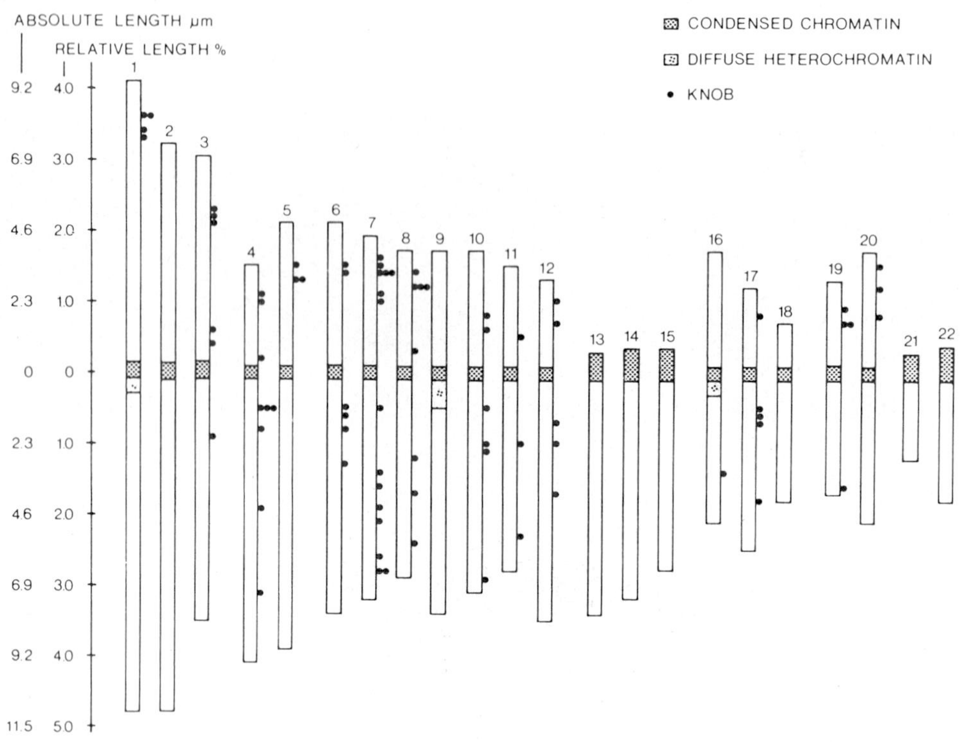

Fig. 2. Idiogram showing the absolute and relative lengths of the synaptonemal complexes of the 22 autosomal bivalents. The position of the centromeric heterochromatin is indicated by the dark hatched area and the position of the diffuse heterochromatin of bivalents 1, 9, and 16 by a lighter hatching. The filled circles denote the position and the number of heterochromatic knobs.

frequency of knobs on individual bivalents are dependent upon the stage of development or reflect natural polymorphism is not known.

The nucleolus. The investigation demonstrates that each of the five acrocentric bivalents is capable of organizing a nucleolus but that in most cases only one major nucleolus is present in each nucleus. In 19 of the 22 nuclei examined, one of the acrocentric bivalents was attached to a large nucleolus with a diameter of about 1.5 µm. In addition, smaller nucleoli with diameters ranging from 0.2 µm to 0.5 µm were frequently found attached to different acrocentric bivalents. In a few cases, major nucleoli were lying free in the nucleoplasm or were associated with the XY bivalent. In 3 nuclei, one or two major nucleoli were lying free in the nucleoplasm and none of the acrocentric bivalents were associated with a nucleolus.

The XY bivalent. At early pachytene the XY bivalent is condensed as a discrete body which is connected to the nuclear envelope through the attachment sites of the telomeres of the X and Y chromosomes. As shown in Figure 1, the chromatin of the XY bivalent is less compacted than the centromeric heterochromatin. One short piece of synaptonemal complex extending up to 1 μm from the attachment sites of one X and one Y telomere was revealed in 19 of the 22 nuclei analyzed. The length of the paired region varies between 0.1 μm and 1.0 μm and is not correlated with the developmental stage of the nuclei. In 12 nuclei, the central region of the synaptonemal complex had not attained its normal structure. The two lateral components were, however, positioned 1100 Å apart, a distance characteristic of completed complexes. In two nuclei, one telomere of the X and one of the Y chromosome had attachment sites on the nuclear envelope in proximity of each other but a central region could not be detected.

The synaptonemal complex in pachytene nuclei. The total length per nucleus of all autosomal synaptonemal complexes varies among the 22 nuclei between 199 and 258 μm with an average value of 231 ± 16 μm. As judged from t-test comparisons, the total lengths of autosomal synaptonemal complexes do not differ significantly among nuclei of different developmental stage or among nuclei from different individuals.

The criteria used in the classification of the 22 autosomal bivalents are with the exception of bivalents 1, 9, and 16 entirely based on relative lengths and centromere indices (CI = length of short arm x 100 / total synaptonemal complex length of a bivalent) and are comparable to those given by Hultén at the Paris Conference 1971[16].

It was possible to identify unequivocally 14 of the 22 autosomal bivalents. Five of the remaining bivalents (numbers 6-8, 10, and 11) belong to the C group and the last three (numbers 13-15) to the D group. The mean absolute and relative lengths and the centromere indices for the 22 autosomal bivalents from the 22 nuclei analyzed are given in Table I, an example of a serial reconstruction of a whole nucleus is presented in Figure 3.

A comparison between light microscopical data on bivalent lengths and centromere indices at diakinesis[15] and relative synaptonemal complex lengths and centromere indices at pachytene gives correlation coefficients of 0.97 for the relative lengths and of 0.91 for the centromere indices. This demonstrates that the classification of the pachytene bivalents corresponds very closely to that obtained from light microscopical data on relative lengths and centromere indices at diakinesis.

Variation in length of individual bivalents. The error in length determination by the reconstruction is estimated to be less than 10 per cent, primarily due to variation in section thickness and microscope magnification. Taking bivalents 1, 9, and 22 as examples, the variation of the absolute lengths is smallest in bivalent 1 with a standard deviation of 9.3% of the mean and largest in bivalent 9 and 22 with standard deviations amounting to 12.9% and 14.0% of the mean. The longest bivalent 9 measured 14.4 μm and the shortest 9.6 μm. Such differences are considered to exceed significantly the variation caused by reconstruction and measuring inaccuracies. This variation is not due to variations between individuals, as only minor differences are found between the mean absolute bivalent lengths of the nuclei from each of the five individuals. Thus, it is concluded that significant differences in absolute length of the synaptonemal complexes occur for a given bivalent among the meiocytes of a single individual.

TABLE 1

MEAN ABSOLUTE LENGTH, RELATIVE LENGTH, AND CENTROMERE INDEX FOR THE 22 AUTOSOMAL BIVALENTS FROM 22 PACHYTENE NUCLEI. (s.d., standard deviation)

Chromosome number	Absolute length in μm	s.d.	Relative length	s.d.	Centromere index	s.d.
1	20.5	1.9	8.9	0.4	46	2
2	18.5	2.0	8.0	0.5	40	3
3	14.9	1.7	6.5	0.6	47	3
4	13.0	1.3	5.6	0.4	26	3
5	13.7	1.5	6.0	0.4	35	6
6	12.8	1.1	5.5	0.3	38	4
7	11.7	1.0	5.1	0.2	37	4
8	10.6	0.9	4.6	0.2	36	5
9	11.6	1.5	5.1	0.4	33	4
10	11.2	1.0	4.8	0.2	35	5
11	10.0	1.0	4.3	0.3	35	5
12	11.1	1.2	4.8	0.4	27	3
13	8.5	0.5	3.7	0.2	8	-
14	7.9	0.6	3.5	0.2	10	-
15	7.2	0.8	3.1	0.3	11	-
16	8.9	1.0	3.8	0.4	46	4
17	8.6	1.8	3.7	0.8	32	5
18	5.8	0.8	2.5	0.2	29	5
19	6.8	1.3	3.0	0.5	43	4
20	8.8	1.2	3.8	0.4	44	4
21	3.5	0.4	1.5	0.2	20	-
22	5.0	0.7	2.2	0.3	18	-

Fig. 3. A complete reconstruction of the synaptonemal complexes of a mid pachytene spermatocyte nucleus. The continuous synaptonemal complexes of all autosomal bivalents are attached to the nuclear envelope at both ends. Each bivalent has a block of heterochromatin marking the position of the centromere. None of the acrocentric bivalents are associated with a nucleolus and the centromeric heterochromatin of bivalents 13, 14, 21, and 22 have fused. The bipartite organization of the centromeric heterochromatin of bivalents 1, 9, and 16 can be recognized, the heterochromatin of bivalents 1 and 16 being partly fused. The X and Y chromosomes are paired in the telomeric region with a synaptonemal complex. The remainder of the sex chromosomes have condensed into a typical XY body. The XY body is associated with a nucleolus which contains a dense body and a filamentous sphere. The latter has been found to mark the position of the nucleolus organizer in cases where a nucleolus is associated with an acrocentric bivalent. An assembly of Golgi apparatuses is present in the cytoplasm, surrounding the two centrioles. The latter are in the process of duplication. (The underlined numbers are bivalent numbers).
K, knob.

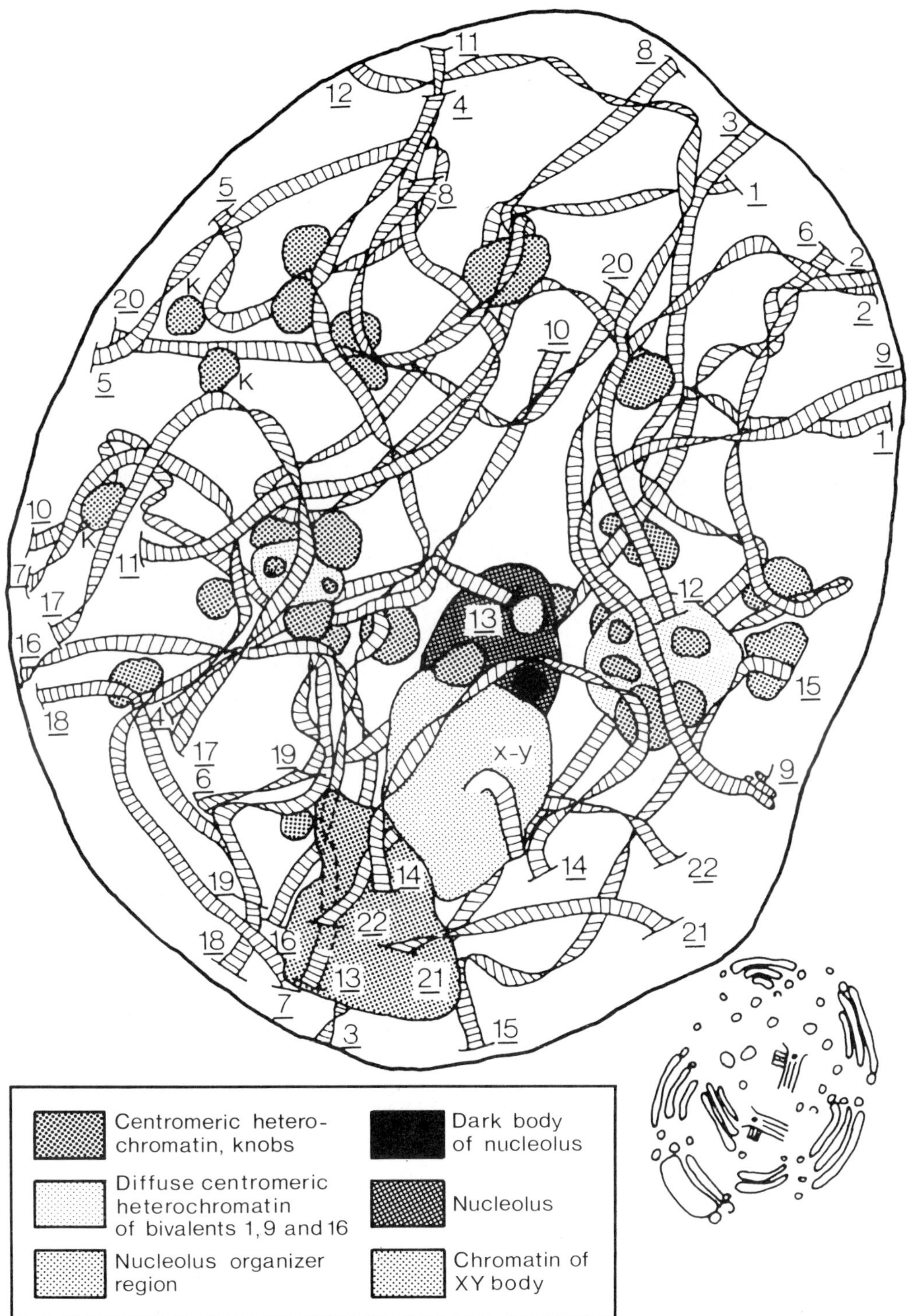
x-y
K
Centromeric hetero-chromatin, knobs
Dark body of nucleolus
Diffuse centromeric heterochromatin of bivalents 1,9 and 16
Nucleolus
Nucleolus organizer region
Chromatin of XY body

Differences in relative synaptonemal complex length of individual bivalents among different nuclei tend to be smaller than the differences in absolute length. Also the centromere indices of individual bivalents vary among different nuclei, the variation exceeding that attributable to errors in the reconstruction and measuring procedure.

Metaphase I

Three metaphase I cells from one individual have been serially sectioned and reconstructed (Figures 4-5). The bivalents are roughly organized into a metaphase plate. At each pole, a pair of centrioles is present. The chromatin of the bivalents

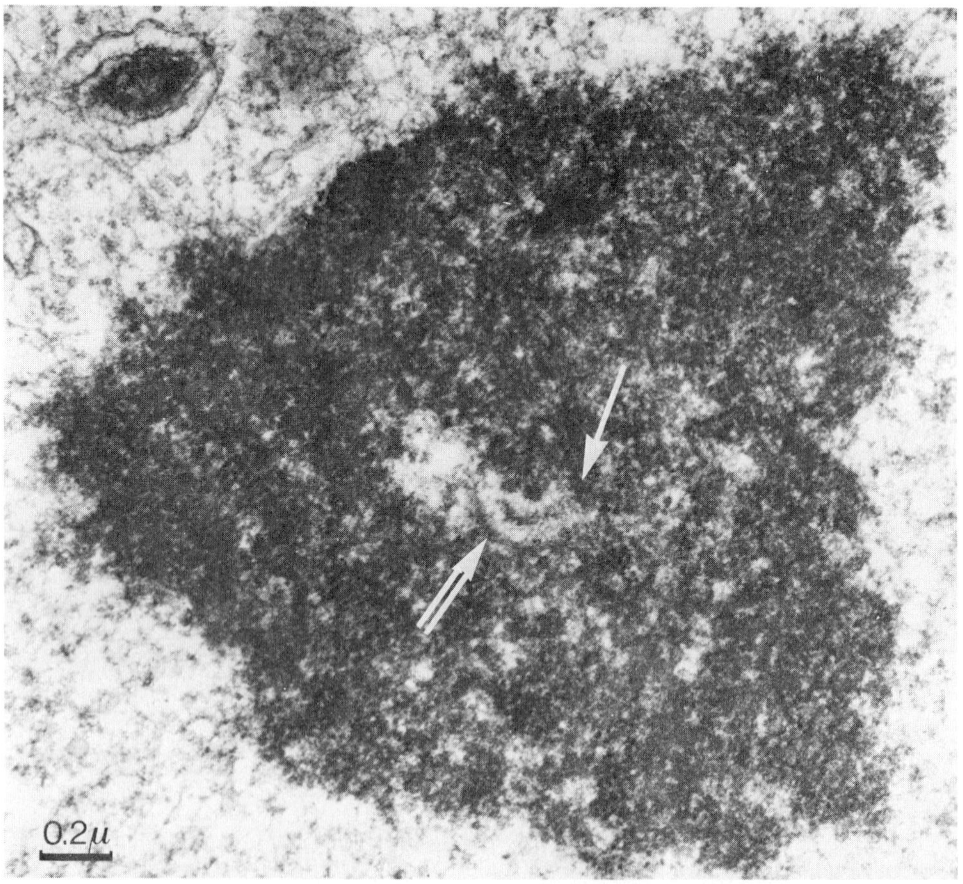

Fig. 4. Micrograph of an early metaphase I bivalent. A piece of central region of the synaptonemal complex is present in the chromatin of the bivalent (arrow). The double arrow denotes lateral component-like material.

is heavily compacted and the bivalents are often too closely associated to allow an unambigous tracing of the individual bivalent. The centromeric heterochromatin can be identified by its association with the kinetochore and by a higher degree of compaction as compared to the rest of the chromatin. The mixture of compacted and dif-

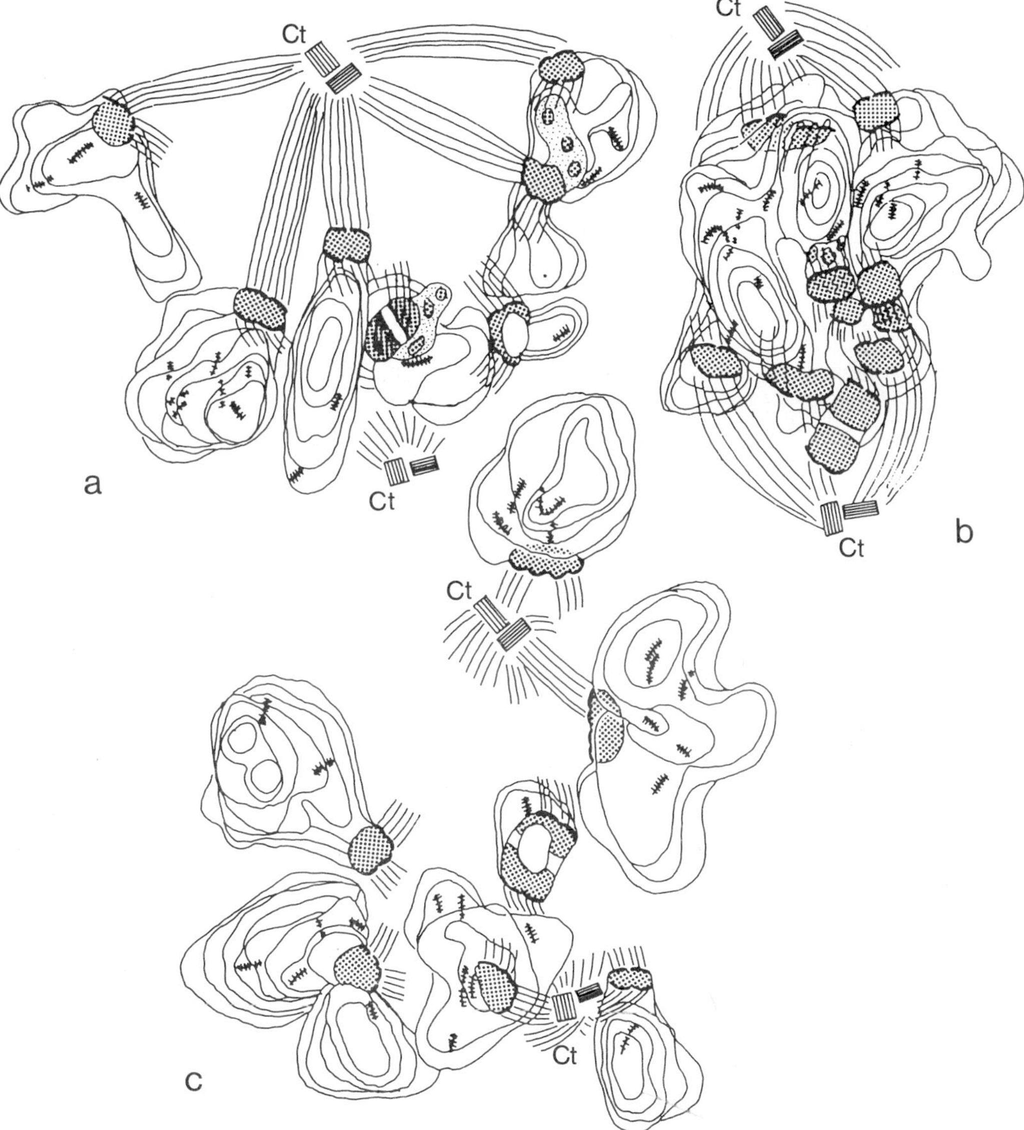

Fig. 5. Complete reconstruction of an early metaphase I. At each pole one pair of centrioles (Ct) is present. The centromeric heterochromatin is shown with a dark hatched signature and the diffuse heterochromatin of bivalents 1, 9, and 16 with a lighter hatching. The kinetochores are shown with solid lines. Inside the chromatin, fragments of the central region of the synaptonemal complex are illustrated. The nine bivalents depicted in Fig. 5b are too closely associated to allow an unambigous tracing of the individual bivalents.

fuse chromatin characteristic of the secondary constrictions of bivalents 1, 9, and 16 can still be recognized associated with the uniformly compacted portion of the centromeric heterochromatin. A total of 92 kinetochores can be identified, demonstrating that each chromatid carries its own kinetochore. As seen in Figure 5, the pair of kinetochores of the homologous chromosomes are not completely positioned in repulsion, suggesting that the investigated cells are in an early metaphase stage.

Fragments of the central region of the synaptonemal complex can be identified in the chromatin of the metaphase bivalents, often associated with material of an electron density comparable to that of lateral components. The number of fragments is given in Table 2.

TABLE 2

Nucleus number	Number of fragments
1	57
2	36
3	58

DISCUSSION

The 22 pachytene nuclei have been arranged into a temporal sequence by degree of chromatin condensation, assuming a progressive condensation between zygotene and diplotene. Differences in the total length of the synaptonemal complex complement among the nuclei of a single or several individuals are insignificant. The total complement length is also unaffected by the progressive chromatin condensation which occurs during pachytene. These results are in agreement with earlier reports on constancy of total synaptonemal complex length at pachytene in maize[4], Bombyx[7], and Sordaria[9].

The comparison of the relative synaptonemal complex lengths at pachytene and the relative lengths at diakinesis[15] have shown that the relative lengths of the individual chromosomes are similar in the two stages. Also the relative position of the centromere is very similar in pachytene and diakinesis bivalents. These observations demonstrate that differential shortening of bivalents during the meiotic prophase occurs only to a limited extent. It also implies that the classification of the bivalents on the basis of relative synaptonemal complex lengths and centromere indices only in rare cases is expected to be incorrect (with the exception of the bivalents in the C and D groups as previously mentioned). A good correlation between light and electron microscopical data on relative chromosome lengths and/or centromere indices has also been reported for other organisms[4,5,13,17].

In contrast to the variation encountered in bivalent length among nuclei, the lateral components of the two homologous chromosomes within a single nucleus have the same length, i.e., an unpaired lateral component extending beyond the central region has never been observed. These observations are in agreement with previous analyses of the synaptonemal complex and its constituents in leptotene, zygotene, and pachytene nuclei[4,5,7,8,9].

A positive correlation between synaptonemal complex length and crossing over frequencies has been reported in Zea mays[4,17]. It is possible that the differences in chiasma frequencies of individual bivalents among different nuclei reported by

Hultén[15] are related to the differences in synaptonemal complex length.

The metaphase I reconstructions demonstrate that the number of bivalents can be counted and the behaviour of the bivalents analyzed in detail. The presence of synaptonemal complex fragments in metaphase I bivalents has previously been reported by Moens[19] for grasshoppers and by Tres[20] for human spermatocytes. The fate and function of the fragments of synaptonemal complexes within the chromatin during co-orientation and disjunction of the bivalents is the subject of further investigation. Available evidence suggests that retained fragments of the synaptonemal complex between the homologous chromosomes at diplotene are the ultrastructural counterparts of the chiasmata seen in the light microscope[11,18]. The number of fragments found in this study equals the number of chiasmata at diakinesis[15]. It is thus likely that the fragments are synaptonemal complex chiasmata.

ACKNOWLEDGEMENTS

This work was supported by grant 202-76-1 BIO DK from the Commission of the European Communities.

REFERENCES

1. Gillies, C.B. (1975) Ann. Rev. Genet., 9, 91-109.
2. Moses, M.J. (1968) Ann. Rev. Genet., 2, 363-412.
3. Westergaard, M. and von Wettstein, D. (1972) Ann. Rev. Genet., 6, 71-110.
4. Gillies, C.B. (1973) Chromosoma (Berl.), 43, 145-176.
5. Holm, P.B. (1977) Carlsberg Res. Commun., 42, 103-151.
6. Moens, P.B. (1973) Cold Spr. Harb. Symp. Quant. Biol., 38, 99-107.
7. Rasmussen, S.W. (1976) Chromosoma (Berl.), 54, 245-293.
8. Rasmussen, S.W. (1977) Carlsberg Res. Commun., 42, 163-197.
9. Zickler, D. (1977) Chromosoma (Berl.), 61, 289-316.
10. Holm, P.B. and Rasmussen, S.W. (1977) Carlsberg Res. Commun., 42, 283-323.
11. Gillies, C.B. (1975) Compt. Rend. Trav. Lab. Carlsberg, 40, 135-161.
12. Rasmussen, S.W. (1977) Chromosoma (Berl.), 60, 205-221.
13. Gillies, C.B. (1972) Chromosoma (Berl.), 36, 119-130.
14. Carpenter, A.T.C. (1975) Proc. Nat. Acad. Sci. USA, 72, 3186-3189.
15. Hultén, M. (1974) Hereditas, 76, 55-78.
16. Paris Conference 1971 (1972) Cytogenetics, 11, 313-362.
17. Mogensen, L.L. (1977) Carlsberg Res. Commun., 42 (in press).
18. Westergaard, M. and von Wettstein, D. (1970) Compt. Rend. Trav. Lab. Carlsberg, 37, 195-237.
19. Moens, P.B. (1969) J. Cell Biol., 40, 542-551.
20. Solari, A.J. and Tres, L.L. (1970) in The Human Testis, Rosenberg, E. and Paulson, C. eds., Plenum, New York, pp. 127-138.

Chromosomes Today Volume 6, A. de la Chapelle and M. Sorsa eds.

FUSION MODEL OF TELOMERE REPLICATION AND ITS IMPLICATIONS FOR CHROMOSOMAL REARRANGEMENTS

BARRY M. DANCIS and GERALD P. HOLMQUIST
Department of Biology, Temple University, Philadelphia, Pennsylvania 19122 USA
and Department of Medicine, Baylor College of Medicine, Houston, Texas 07730 USA.

ABSTRACT

A model for telomere replication is presented based on the presence of terminal DNA hairpins and a mechanism of molecular recombination. A novel feature of the model is the requirement for transient telomere-telomere (t-t) fusions as replication intermediates. The model explains both the double nature of metacentric centromeres and the similarity of mouse satellite strand polarities, and suggests that Robertsonian translocations are reversible aberrants of the transient fusions required during replication. In addition, only our replication model correctly predicts the types of t-t fusions observed in senescing human cells. Finally, it is suggested that transient chromosome rings formed during replication are evolutionary descendents of bacterial ring chromosomes.

INTRODUCTION

The eukaryote's chromosome appears to contain a single linear molecule of DNA[1,2] stretching from one end of the chromosome to the other without any massive longitudinal foldbacks[3]. Complete replication of such a molecule (Fig. 1a) requires a special mechanism for replacing the RNA primer at the 5' end of the nascent strand with DNA[4]. A number of replacement mechanisms have already been proposed for both prokaryotes[4] and eukaryotes (Fig. 1b[5], c[6]), but the latter are not supported by recent cytological data (see below). This report presents a new model for DNA synthesis at the 5' termini of the double helix which is in accord with the cytological data and provides a simple explanation for the origin of certain chromosomal aberrations.

FUSION MODEL OF TELOMERE REPLICATION

The replication of chromosome ends, or telomeres, is envisioned to occur as shown in Fig. 2. First, a nick is introduced near the end of a DNA double helix where, as suggested by Bateman[6], the terminal nucleotides are joined into a "hairpin" by 5' to 3' covalent bonds (Fig. 2b). After unpairing (Fig. 2c), the single-strand tail invades a nearby, but not necessarily homologous, telomere containing the same DNA sequence and anneals with the complementary strand (Fig. 2d). Fusion of the two telomeres is completed by nicking the displaced strand followed by repair and ligation (Fig. 2e). The problem of replacing terminal RNA primers is eliminated by fusion, and replication occurs as in nonterminal parts of the chromosome (Fig. 2f). Fission of the fused telomeres begins with the introduction of a pair of symmetrical nicks (Fig. 2g),

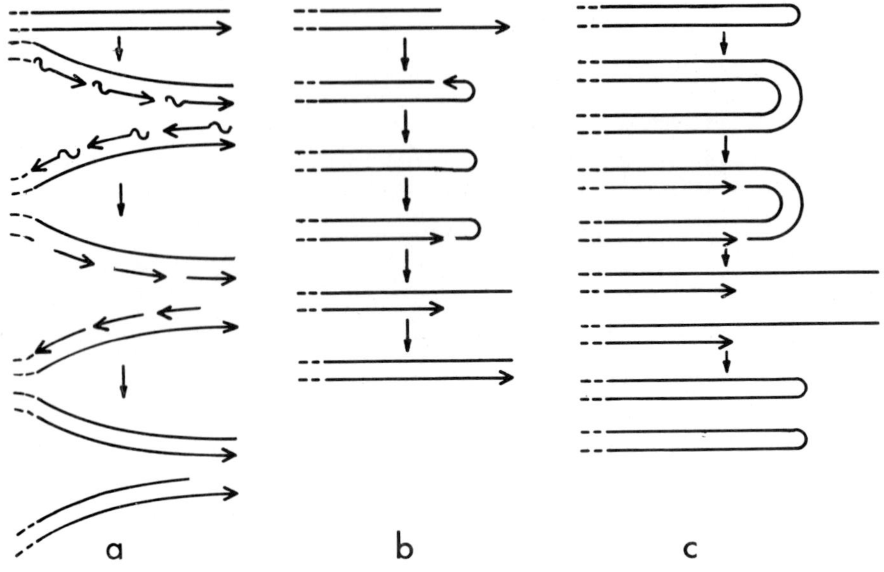

Fig. 1. Replication of linear DNA. DNA synthesis requires RNA primers (∿) and can only add nucleotides to 3' ends (arrowheads) of nascant strands (DNA strand at one end of double helix —, remainder of DNA strand – –). a. After removal of RNA primer, replication leaves gap at 5' end of nascent strand without a 3' end to prime gap-filling synthesis. b. Cavalier-Smith[5] proposed a terminal palindrome so that the strand opposite the gap can fold back as a hairpin. With proper nicks and ligations, the gap switches strands and leaves a 3' end to prime gap-filling synthesis. Only one of two sister chromatids is shown. Other chromatid has gap at opposite end of double helix. c. Bateman[6] proposed the terminal hairpin as a replicatable telomere. Staggered nicks are required to separate the fused sister telomeres.

followed by unpairing (Fig. 2h), selfannealing (Fig. 2i) and ligation (Fig. 2j). The mechanism of fusion (Fig. 2b-f) is similar to the initial reactions in the model of molecular recombination proposed by Meselson and Radding[7], while the mechanism of fission (Fig. 2h-k) is the same as that in Bateman's model of telomere replication except that here fission separates fused homologous and nonhomologous chromatids rather than fused sister chromatids.

ENZYMOLOGICAL EVIDENCE

Many reactions similar to those required in Fig. 2 have already been demonstrated *in vitro*. The introduction of nicks into one telomere (Fig. 2) is comparable to the reaction catalized by pancreatic, mung bean, and other DNases[8] while, the formation and stabilization of single-strand DNA during unpairing (Fig. 2c) may result from unwinding enzymes[9] and proteins[10,11] respectively, similar to those found in prokaryotes. Intact telomeres should preferably be supercoiled while nicked ones definitely are not. Supercoiled DNA is partially denatured[12], which facilitates the initiation of annealing with the invading single strand[13]. Annealing also removes supercoils from the unnicked

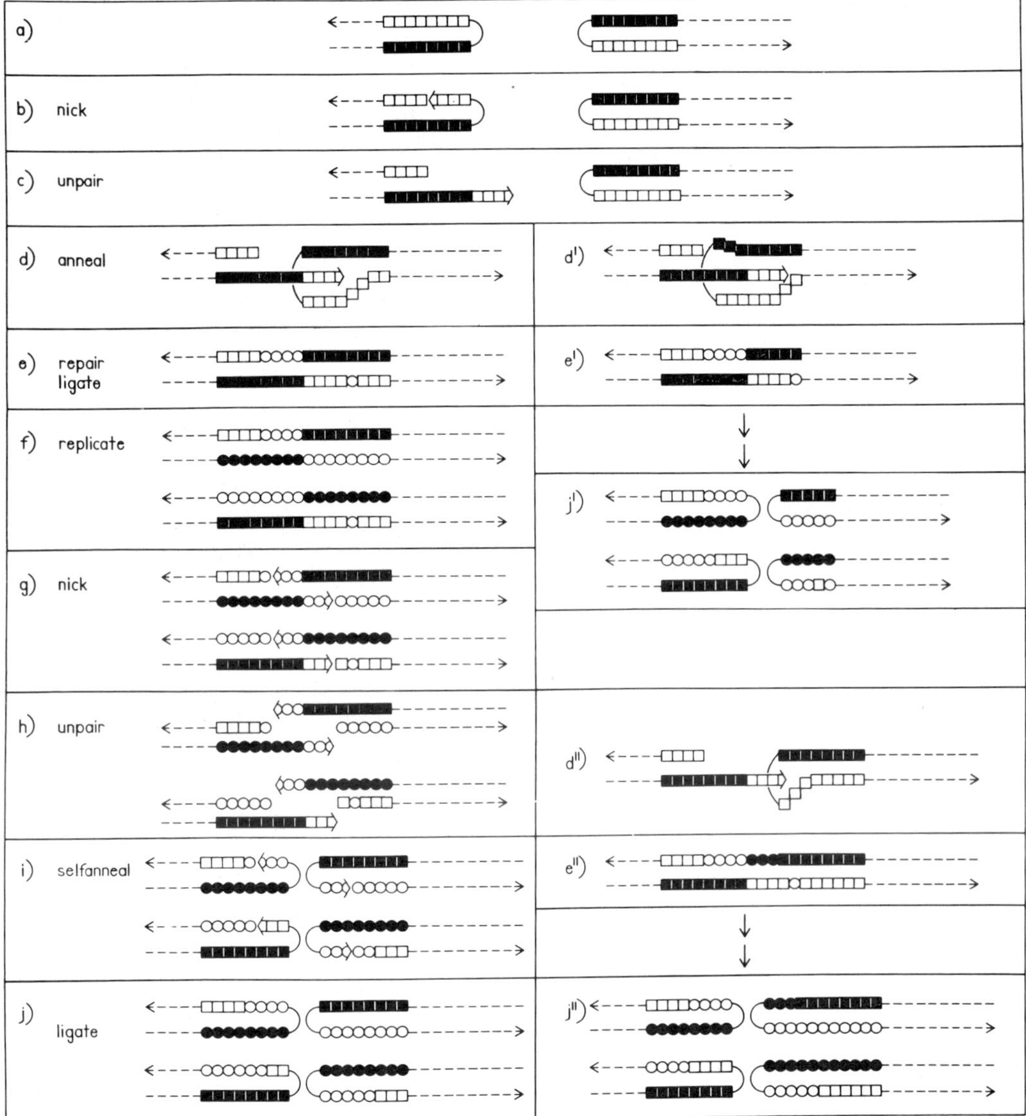

Fig. 2. Fusion model of telomere replication. Open squares represent one original strand of double helix at telomere, closed squares - complementary strand, circles - repair synthesis and replicated strands, dashed lines - remainder of indefinitely long chromosomes, curved lines - 5' to 3' phosphodiester bonds joining terminal nucleotides of the two strands of the double helix, arrowheads - 3' end of DNA strand. Potential nonhomology of fusing telomeres indicated by difference in length of dashed lines. If telomere sequence is repeated, unequal pairing during annealing results in a decrease (Fig. 2') or an increase (Fig. 2") in length of telomere sequence. a. Apposed telomeres; b-e. Fusion joins apposed telomeres; g-j. Fission restores telomeres.

telomere and thereby drives the reaction to completion. In this modified recombination mechanism, the invasion of an intact telomere by a nicked one would be energetically favored. Conversion of the D-loop annealed structure (Fig. 2d) to fused telomeres (Fig. 2e) is similar to excision repair[14] and may use similar enzymes. The process of fusion may be further energetically favored by straightening the bond between the terminal nucleotides in the intact telomeres. Following replication (Fig. 2f), fission (2g-j) may proceed as postulated by Bateman[6]. Symmetrical cleavage about the reflection point of a palindrome (Fig. 2g) is similar to the action of restriction endonucleases[15], and results in DNA duplexes with single-strand tails (Fig. 2h). As suggested by Cavalier-Smith, if all the telomeres are the same, fewer telomere-specific enzymes[5] will be needed during fission. In addition, self-annealing (Fig 2i) and ligation of the tails (Fig. 2j), which complete the fission process, have already been demonstrated *in vitro* using T4 DNA and polynucleotide ligase[16].

CYTOLOGICAL EVIDENCE

For fusion to occur, telomeres must first come together (appose) and then form chromatin connections (attach). When the DNA in the two telomeres becomes covalently linked (fused), the process is complete. Since movement along the nuclear membrane results in homologous telomere pairing just prior to meiosis[17] similar movement may facilitate apposition during somatic interphase, sometime prior to replication. Using a variety of techniques, different workers have concluded that telomeres are both bound to the nuclear membrane[18,19,20] and apposed in groups of two or more[21,22,23] during somatic interphase. If all telomeres within a species contain the same sequence, then telomere attachment and fusion does not have to be limited to homologous chromosomes. During meiosis[24,25,26,27], in Dipteran salivary glands[28,29,30,31] and during somatic interphase and early prophase[21,22], chromatin attachments between nonhomologous as well as homologous chromosomes have been observed. Since eukaroyte DNA is chromosome-size and not larger[1,2], then, contrary to DuPraw[34], our proposed telomere fusions would be transient, though telomere attachments may last throughout interphase and possibly longer. Thus, the fusion model of telomere replication is consistent with both enzymological and cytological observations.

PREDICTIONS

The fusion model predicts that all the telomeres are the same within a species. In *Drosophila melanogaster*, the *in situ* hybridization of a cloned fragment shows this sequence to be present in all telomeres and the chromatin strands attaching them[35], while in other *Drosophila* species, similarity of telomere sequences has been inferred from the observation of attachments between all pairs of nonhomologous telomeres[28,29]. Telomere specific DNA appears to be located at many, if not all, the telomeres in several species of plants[36,37] and animals[38,39,40,41] and its absence in other species may merely reflect a limitation of the detection methods used[42].

The telomere specific DNA in *Drosophila* contains about 12,000 base pairs[35], enough to form the supercoiled loops[43] which energetically favor annealing[13]. In addition, this DNA is repetitious [35] as are the satellite DNAs in mouse [44] and rye[36], and 5s ribosomal DNA in *Xenopus*[45] which are also located near, if not at, the telomeres. Such repeated sequences increase the possibilities for annealing (Fig. 2d, d', d'') and allow for variation in the size of the replicated telomere (Fig. 2j, j', j''). Possibly, the sequence annealed during fusion (Fig. 2d) differs from the one cleaved during fission (Fig. 2g-j). Then satellite and 5s ribosomal DNA sequences would be subterminal to the hairpin sequence at the very end of the double helix, but could still participate in telomere fusion. Such an arrangement permits the invading sequence and the terminal sequence to evolve separately in response to the different requirements of fusion and fission.

The process of fusion also suggests a mechanism for maintaining similar sequences in all telomeres. Any sequence differences which arose through mutation could result in mismatched bases during annealing (Fig. 2d) and self-annealing (Fig. 2i). Repair of the mismatches would reduce the sequence differences between nonhomologous[20] telomeres, while unequal crossing over during meiosis followed by selective loss of chromosomes whose telomeres are too small or too large would reduce the sequence difference[46] between homologous telomeres. In addition, multiple invasions during annealing (Fig. 3b) could produce rings of telomere sequences which can then recombine with, or integrate into, other telomeres and thereby help to maintain similarity of the sequences.

A set of cytological observations by Benn[47] serves as a test for the different telomere replication mechanisms. In senescing human primary cell lines, where the replication mechanism may be dysfunctional, 2.5% of the metaphase chromosomes are dicentrics formed by telomere-telomere fusions[47]. The proposed models of telomere replication can all predict dicentric formation when replication is dysfunctional, but different models predict different types. In Bateman's model, dicentrics form as a result of replication (Fig. 1b), and, therefore, only fused sister chromatids are possible, but none were observed[47]. Moreover, dicentrics of two fused nonhomologous chromosomes should not occur, but, experimentally, they are 2/3 of the ones detected. In Watson's[4] and in Cavalier-Smith's primary[5] models, dicentrics might result from post replication fusion of chromatids with single-strand tails. Since one, and only one, telomere of a sister telomere pair would have such tails (Fig. 1b), only half-dicentrics are expected; they were not observed[47]. In Cavalier-Smith's alternate model[5], the fusions should be rare and therefore produce more dicentrics involving one sister chromatid than those involving both; fused sister chromatids should also form. If, however, fusion precedes replication, as in Fig. 2, then the reverse should be true. In fact, only dicentrics involving both chromatids were observed[47]. Thus, Benn's data is most consistent with our model where fusion precedes replication. Telomere fusion might, however, be a rare event unrelated to replication. Even then, it must occur before replication to account for the observed dicentrics and the mechanism of fusion

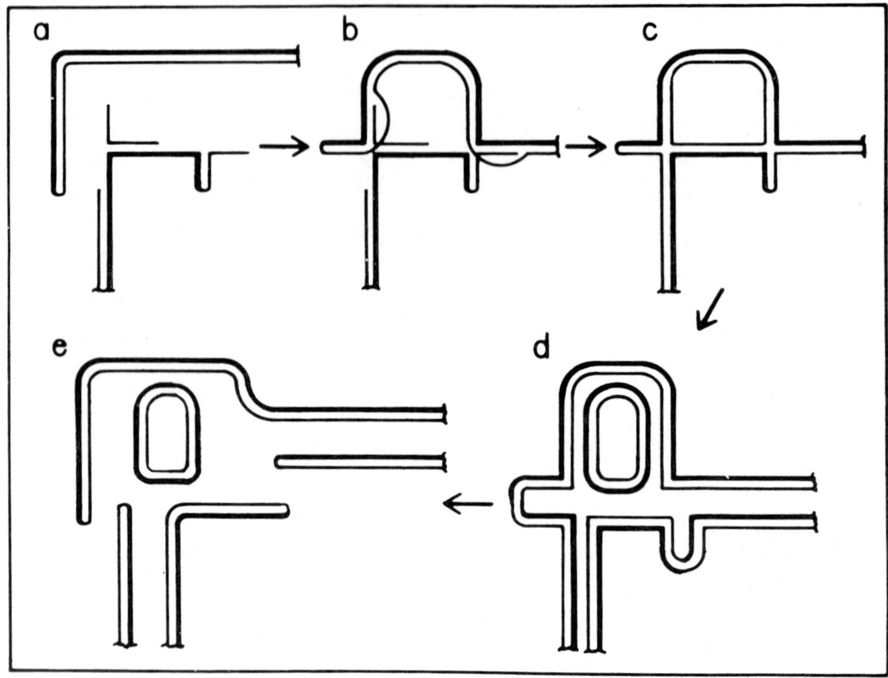

Fig. 3. Telomere replication: Fusion at two sites. Two telomeres after a. one is nicked twice and partially unpaired, b. invasion and annealing, c. repair to complete fusion, d. replication, e. fission. One strand of telomere DNA —; complementary strand ▬ ; remainder of chromosome ≷ . Replication (d) produces a free ring of telomere DNA which is insensitive to fission enzymes (e).

may be as shown in Fig. 2(b-c) or it may occur by the reverse of fission (Fig. 2 g-j) or by some other mechanism.

Normally, fission separates the fused telomeres. If fission does not occur, a metacentric (Fig. 4a) or a dicentric (Fig. 4b) chromosome should form, and the metacentric should show contralateral symmetry of the fused sequences (Fig. 2f, 4a, b, d). All mouse metacentric chromosomes, which probably arose by joining two rod chromosomes at their centric telomeres, show the expected contralateral symmetry of sequences in their pericentric heterochromatin[48,49]. Because no loss of chromosome material should occur during fusion (Fig.s 2, 4), the model also suggests that metacentrics have twice the centric structure of rod chromosomes. White, however, postulates that translocation between two acrocentrics gives rise to a small centric fragment in addition to a metacentric chromosome[50] (Fig. 4f). By White's model, fission should be impossible once the centric fragment is lost and all chromosomes should have the same centromere structure. The double nature, however, of the metacentric centromere compared to that of a telocentric has been inferred from experiments using x-rays[51,52] and has actually been observed in both the light[52,53,54,55,56] and electron

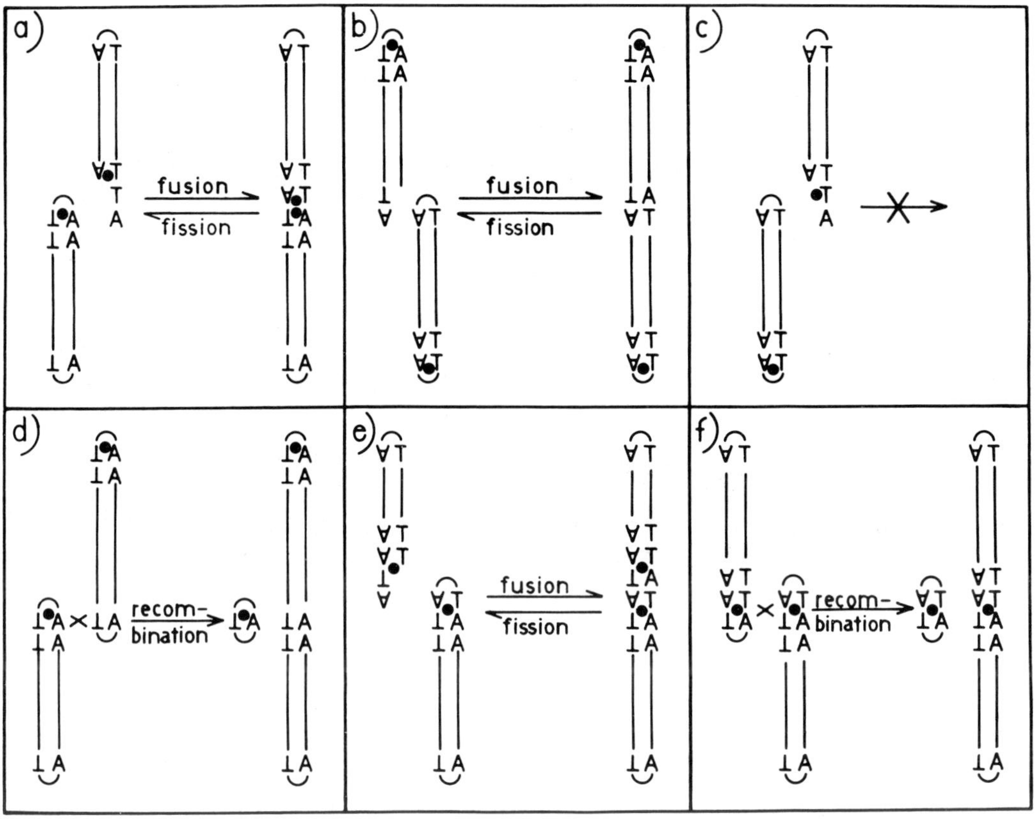

Fig. 4 Chromosomal rearrangements. Letters - telomere DNA sequence, top of letter - 3' end of sequence, bottom of letter - 5' end, paired straight lines - remainder of double helix in chromosome, curved line - 5' to 3' phosphodester bond, closed circle - centromere. Strand polarity determined by end of sequence closest to referent. With regard to tip of telomere, telomeres containing centromeres have the opposite strand polarity from telomeres lacking centromeres. In a.-c., e. one of two fusing telomeres has already been nicked and unpaired (see Fig. 2b, c). Fusion of telomeres with same strand polarities produces a metacentric (a) or a dicentric (b) using mechanism in Fig. 2. Telomeres with opposite strand polarity cannot fuse by that mechanism (c), but can fuse by molecular recombination (d). Acrocentrics can fuse by both methods (e, f). Loss of small centric fragment makes fusion by recombination irreversible (d, f). Telocentrics are defined not by the exact location of the centromere in the telomere but by the telomere sequence. With respect to tip of the telomere, telocentrics do not contain a reversal of strand polarities at the centromere, acrocentrics do.

microscopes[56]. In addition, the centric fragment hypothesized by White has not been observed even where fusions have occurred somatically[57,58].

The conversion of two rod chromosomes into a metacentric one or vice versa without altering the fundamental arm number (Robertsonian translocation) is a common feature of karyotype evolution. A typical example is the Bovoidae where, in the approximately 50 species examined, the diploid number of chromosomes varies between 30 and 60 and over

90% of the variations are due to such translocations[59]. Similar changes have even been observed to have occurred within a single organism. In *Salmo irideus*[57] and *salmo trutta*[58] the number of chromosomes in different cells of the same fish may vary by as much as seven while the fundamental arm number remains the same. We propose that Robertsonian translocations are inherently reversable chromosome mutations (Fig. 4) resulting from alterations in the fusion and fission mechanisms of the telomere replication model (Fig 2).

The DNA in fusing telomeres must have the same strand polarity as well as the same sequence in order for fusion to occur. In mouse[48,49], all the satellite sequences near, if not at, the centric telomeres have the same strand polarity. These properties are incorporated into Fig. 4, where the top of a letter represents the 3' end of a sequence and the bottom, the 5' end. Here, all the A-rich sequences have their 3' ends closest to the centromere. Thus, with regard to the tip of the telomere the strand polarity of the A-rich sequences in the centric telomeres is opposite that of the A-rich sequences in the acentric telomeres. Consequently, centric and acentric telomeres as depicted in Fig. 4 (a and b respectively) can fuse with themselves but not with each other (Fig. 4c).

In bacteria, the bearers of genes (genophores) are continuous circles of DNA which have no ends to complicate replication. Separation of replicated genophores may occur by partitioning their attachment points on the cell membrane into two daughter cells[60]. In ascomycetes and basidiomycetes, over 70% have mitotic figures consistent with the two track model of chromatid separation[61]. In this model, the chromatids of each haploid set are connected, and possibly fused, end-to-end into a ring. During mitosis, the paired replicate rings are each broken once and then appear to be pulled into separate cells by their membrane attachments at opposite poles of the dividing nucleus. Towards the end of mitosis, the rings reform. As an extension of this evolutionary trend, we have proposed that chromosomes of higher eukaryotes are joined end-to-end into rings only transiently during replication and attach to the nuclear membrane only during interphase.

A common view of chromosomes is that they evolved to provide linkage groups by which syntenic genes can segregate together[62]. However, since distal genes on the same chromosome often segregate independently, this linkage group hypothesis is questionable[62]. We propose that the evolution of segmented genophores from intact circular ones was a response to increasing genophore size and the resulting greater difficulty of disentagling and separating the larger replicates. Consequently, the function of binding the genophore to the membrane is now restricted to preventing the tangling of DNA[21], and to telomere fusion, while the disjunction of the genophore's multiple segment pairs is accomplished by, and required the evolution of, a spindle apparatus. Thus, transient fusions of the eukaryotic genophore, here deemed necessary for replication, remind us of when the genophore was always in the shape of a ring.

ACKNOWLEDGEMENTS

This work was supported by NIH grants GM 21671 (to BMD) and GM 23906 (to GPH) and Temple University Faculty Grant-in-Aid (to BMD).

REFERENCES

1. R. Kavenoff, L. C. Klotz and B. H. Zimm. (1973) Cold Spring Harbor Symp Quant. Biol. 38, 1-8.
2. T. D. Petes, C. S. Newlon, B. Byers, and W. L. Fangman. (1973) *ibid*. 38, 9-16.
3. B. A. Kihlman. (1975) Chromosoma 51, 11-18.
4. J. D. Watson. (1972) Nature (London), New Biol. 239, 197-201.
5. T. Cavalier-Smith. (1974) Nature (London), 250, 467-470.
6. A. J. Bateman. (1975) *ibid* . 253, 379.
7. M. S. Meselson and C. M. Radding. (1975) Proc. Natl. Acad. Sci. USA. 72, 358-361.
8. M. Laskowski, Sr. (1967) Advan. in Enzymology 29, 165-220.
9. M. Abdel-Monem, H. Durwald, H. Hoffmann-Berling. (1976) Eur. J. Biochem. 65, 441-449.
10. B. Alberts and L. Frey. (1970) Nature (London) 227, 1313-1318.
11. N. Sigel, H. Delius, T. Kornberg, M. Gefter, and B. Alberts. (1972) Proc. Natl. Acad. Sci. USA 69, 3537-3541.
12. T. A. Beerman and J. T. Liebowitz. (1973) J. Mol. Biol. 79, 451-470.
13. W. K. Holloman, R. Wiegand, C. Hoessli and C. M. Radding. (1975) Proc. Natl. Acad. Sci USA. 73, 2394-2398.
14. J. E. Cleaver. (1974) Advan. Rad. Biol. 4, 1-75.
15. D. Nathans and H. O. Smith. (1975) Ann. Rev. Biochem. 44, 273-293.
16. B. Weiss. (1976) J. Mol. Biol. 103, 669-673.
17. C. B. Gillies. (1975) Ann. Rev. Genetics 9, 91-109.
18. L. Vanderlyn. (1948) Bot. Rev. 41, 270-318.
19. J. A. Sved. (1966) Genetics 53, 747-755.
20. D. E. Comings and T. A. Okada. (1970) Esp. Cell Res. 63, 62-68
21. E. B. Wagenaar. (1969) Chromosoma 26, 410-426.
22. C. P. Fussell. (1975) *ibid* . 50, 201-210.
23. Y. Kitani. (1964) Japanese Journal of Genetics 38, 244-256.
24. E. B. Wagenaar and R. S. Sadasivaiah. (1969) Canad. J. Genet. Cytol. 11, 403-308.
25. T. Ashley and E. B. Wagenaar. (1974) *ibid* . 16, 61-76.
26. M. E. Drets and M. Stoll. (1974) Chromosoma 48, 367-390.
27. C. R. Ribbands. (1941) J. Genet. 41, 411-413.
28. H. Bauer. (1936) Proc. Natl. Acad. Sci. USA. 22, 216-222.
29. H. D. Berendes and G. F. Meyer. (1968) Chromosoma 25, 184-197.
30. B. P. Kaufmann and H. Gay. (1969) *ibid* . 26, 395-409.
31. M. Warters and A. B. Griffen. (1950) J. Hered. 41, 182-190.
32. G. Diaz and K. R. Lewis. (1975) Chromosoma. 52, 27-35.

33. A. Weimark. (1975) Hereditas 79, 293-300.

34. E. J. DuPraw. (1970) DNA AND CHROMOSOMES. (Holt, Rinehart & Winston, New York).

35. G. Rubins. (1977) Cold Spring Harbor Symp. Quant. Biol. 42, (in press)

36. A. E. Limin and J. Dvorak. (1976) Can. J. Genet. Cytol. 18, 491-496.

37. S. M. Stack, C. R. Clarke, W. E. Cary and J. T. Muffly. (1974) J. Cell Science 14, 499-504.

38. J. R. A. Eckhardt and J. G. Gall. (1971) Chromosoma 32, 407-427.

39. M. L. Pardue, D. D. Brown, and M. L. Birnstiel. (1973) *ibid*. 42, 191-203.

40. K. W. Jones. (1970) Nature (London) 225, 912-915.

41. M. L. Pardue and J. G. Gall. (1970) Science 168, 1356-1358.

42. W. Hennig, I. Hennig and H. Stein. (1970) Chromosoma 32, 31-63.

43. C. Benyajati and A. Worcel. (1976) Cell 9, 393-407.

44. W. G. Flamm, M. McCallum and P. M. B. Walker. (1967) Proc. Natl. Acad. Sci. USA. 57, 1729-1734.

45. D. D. Brown and C. S. Weber. (1968) J. Mol. Biol. 34, 661-680.

46. G. P. Smith. (1976) Science 191, 528-535.

47. P. A. Benn. (1976) Am. J. of Hum. Genet. 28, 465-473.

48. G. P. Holmquist and D. E. Comings. (1975) Chromosoma 52, 245-259.

49. M. S. Lin and R. L. Davidson. (1975) Science 185, 1179-1181.

50. M. J. D. White. (1961) THE CHROMOSOMES (Methuen, London) p. 94.

51. A. Lima-de-Faria. (1956) Hereditas 42, 85-160.

52. B. McClintock. (1938) Genetics 23, 315-376.

53. J. H. Tjio and A. Levan. (1950) Nature (London) 165, 368.

54. B. John and G. M. Hewitt. (1966) Chromosoma 20, 155-172.

55. D. I. Southern. (1969) *ibid*. 26, 140-147.

56. D. E. Comings and T. A. Okada. (1970) Cytogenetics 9, 436-449.

57. S. Ohno, C. Stenius, E. Faisst and M. T. Zenzes. (1965) *ibid*. 4, 117-129.

58. M. T. Zenzes and I. Voiculescu. (1975) Genetica 45, 531-536.

59. D. H. Warster and K. Benirschke. (1968) Chromosoma 25, 152-171.

60. A. Ryter, Y. Hirota and F. Jacobs. (1968) Cold Spring Harbor Symp. Quant. Biol. 33, 669-676.

61. A. W. Day. (1972) Canad. J. Bot. 50, 1337-1347.

62. J. R. G. Turner. (1967) Evolution 21, 654-657.

Chromosomes Today Volume 6, A. de la Chapelle and M. Sorsa eds.

VARIATION IN GENOME ORGANISATION IN RELATED SPECIES: AN ANNOTATION

GABRIEL DOVER

Department of Genetics, University of Cambridge, Cambridge, England.

ABSTRACT AND SUMMARY

A short review is presented of selected examples of variation in the genomes of related organisms with respect to the distribution and composition of (1) repetitive and single copy DNA; (2) satellite DNA and C-bands; and (3) G-bands. Comments in the body of the essay contrast the data on the relative fluctuations in these components that are derived from different species groups. Mechanisms of derivation and maintenance, and correlations with the phenotype are alluded to.

The conclusions are (with qualifications given in the text) that (1) the magnitude of sequence divergence in any one component closely reflects the genetic distance between organisms based on other criteria. (2) Most differences in the composition and organisation of the genome could well be the consequence of separation of species over time, and the quirks of behaviour of replicating DNA and chromosomes, rather than the cause of the separation. (3) The complexity and frequency of repetition of components can either be dependent variables (often involving intercalation of ancestral with new sequences) or independent variables despite a fine interspersion of sequences. (4) Small conserved portions of the genome (other than the structural genes), that are stable over time subsequent to species separation hint at a biological role. (5) The data is as yet too contradictory and of low resolution to point to a special role of a component of the genome in any adaptive feature of a species, including aspects of chromosome behaviour at meiosis.

ABBREVIATIONS: R-DNA = repetitive DNA; SC-DNA = single-copy DNA; IR-DNA = moderately repetitive DNA; HR-DNA = highly repetitive DNA; sat.DNA = satellite DNA; chrm(s). = chromosome(s); const.het = constitutive heterochromatin. The style is of necessity terse and in the form of annotated notes in places.

INTRODUCTION

It is a sobering thought that despite the ever increasing theoretical refinements of the dynamics of genetic changes in populations, we have no theory of speciation, for (to quote Lewontin)[1] '<u>we know virtually nothing about the genetic changes that occur in species formation</u>' (his emphasis) and further: 'that the problem of making quantitative statements about the multiplication of species has been that we have been unable to connect the phenotypic differences between populations and species with any particular genetic changes'

In the same way we are unable, despite the numerous examples of comparative studies cited and described below, to quantitatively relate differences in genome organisation between species with their phenotype; although attempts to define causal relationships are often made in this direction of which 2,3,4 are recent examples in specific genera and White's comprehensive book[5] is the final synthesis. Interesting attempts that speculate on the significance of gross differences in DNA amount of organisms across the kingdoms are also abundant[6,7,8,9,10].

For Lewontin the paradox is that the 'stuff of evolution' are the subtle changes in phenotype whilst dynamic theories are built around the genotype. This raises the question, so far as our problem is concerned, is whether the between species differences in numbers and types of chromosomes and differences in the heterogeneity of sequences of DNA that they contain,make some special contribution to the phenotype that is of adaptive significance in that these differences are physically linked to the genotype proper, i.e. the gene sequences. For example their presence might involve the control of expression of genes and the permissible permutations of genes in a population.

VARIATION IN REPETITIVE AND SINGLE COPY DNA IN CLOSELY RELATED SPECIES (NOT INCLUDING SATELLITE DNA)

It quickly became established that the magnitude of the differences between species in the composition of DNA reinforced phylogenetic distances based on other criteria. Early work centered on techniques that measured the extent (% homology) and stability (ΔTm) of heteroduplexes of total DNA between two organisms, in defined criteria of reassociation.

Primates show clear differences in the extent and fidelity of hybridisation of total DNAs between closely and distantly related species and phylogenetic trees can be constructed without too many surprises[11]. Measurements on ΔTm's between several genera of birds related to the house sparrow (Passer domesticus) tell a similar story[12]. Such studies can only be of limited interest in that they are of general low resolution and do not determine the relative contributions of the types of sequences that make up the differences in DNA amount between closely related species.

Informative comparisons between related and unrelated species of the fluctuations in amount of several compounds of DNA have been made without resort to hybridisation of the genomes. Comparisons of the mean and modal buoyant densities of main-band DNA of 12 unrelated mammals from 5 orders and two bird species has revealed a striking similarity in the composition and amounts of three main-band compounds[13]. Invertebrates (D.melanogaster) and lower eukaryotes (Saccharomyces cerevisiae) have essentially symmetrical bands in density gradients[13]. However, three main-band components (other than satellite DNAs) have been resolved in all six species of the melanogaster species sub-group, in actinomycin D-CsCl gradients[14]. The kinetics of DNA renaturation in 4 amphibian species[15] belonging to two different sub-classes, Anura and Urodela,indicate that within each sub-class the difference between species in DNA amount is due predominantly to changes in the amounts of R-DNA, accounted for by variation in the number of repeated sequences (maintaining the same average length of 380 b.p.) with variations in length only of the interspersed SC sequences (from

760 to 1600 b.p.). However the difference in DNA amount between the sub-classes is related to a proportional difference in both classes of DNA, and the genomes maintain the same general organisation. Duplications of long stretches of interspersed R- and SC-DNA may underlie the between sub-class difference; however another type of amplification mechanism is needed to explain the simultaneous increase in the frequency of R-DNA without altering its complexity, and in the complexity of SC-DNA without altering its frequency. Similarly two modes of change in the amount of IR-DNA have been found in other genera[16,17]. In 11 spp.of Cichorieae (Compositae) there is a variation in total DNA amount involving both a proportional increase in the amount of IR- and SC-DNA, and also an incremental jump in IR-DNA only[16]. Kinetic analysis of a three-fold difference in DNA amount of 9 spp.of sweet pea (Lathyrus)[17], shows that variation in DNA amount is correlated with variation in IR-DNA. The complexity of IR-DNA increases three-fold in 6 of the species which can be separated into 3 groups with respect to equal incremental increases in IR-DNA amount. There is only a slight increase in SC-DNA that is composed of a regular ratio of IR- and SC-DNA (Rees; personal communication).

The data from Anura and Urodela is an important step forward in the analysis of the more subtle changes than can occur in interspersed sequences. Similarly, recent studies in the Gramineae[18], sea urchins[19] and frog[20] have emphasised the extent of differential conservation of interspersed sequences that have accompanied species divergence. In the sea urchin, Strongylocentrotus purpuratus, interspersed R-DNA is in two size classes: long repeats (>2000 b.p.) that are thermally stable (conserved) and short repeats (~ 300 b.p.) that are thermally unstable (diverged). The absolute quantity of short repeats is correlated with the complexity of SC-DNA. As before this would necessitate a method of generation permitting a sequence to increase in frequency whilst its nearest neighbours increase in complexity only. The long repeat fraction is not correlated with SC-complexity and is probably of ancient origin and well conserved, in that it contains all degrees of repetition found in total DNA. In Xenopus laevis and X.borealis[20] the IR-DNA (both with long and short interspersed repeats) contains a representative of all sequences present in the other but to a lesser extent. In primates the proportion of high to low stability fractions in the R-DNA of each species is the same and there is an increase in the amount of conserved sequences not held in common as species distance increases[21], as is the case of rat/mouse comparisons[22], indicating that the thermally-stable fractions are continually being added, and are of recent origin in these two groups.

However we need to turn to the studies in the Gramineae and Plethodon for extensive and informative analysis of interspecific differences in genome organisation, of which only a part is described here.

A full two-way analysis of all possible reciprocal hybridisations of R-DNA between 4 spp.of Gramineae, wheat, oats, rye and barley has permitted the isolation and subdivision of R-DNA into 7 groups[18]: the identity of the group is defined by the species that share a common fraction of R-DNA, e.g. Gp.I sequences (the most ancestral) are those sequences held in common between each of the species; Gp.II are held in common between wheat/rye/barley but are not found in oats. Phylogenetic trees based on group identities agree well with the known relationships between the species. At or after each branch point new sequences of a group are added to account for the increase in sequence homology between subsequent species. Interestingly, each of the groups is composed of conserved and non-conserved sequences and furthermore there is a similar degree of heterogeneity within each group. However there is a lesser extent of divergence within a group within a species than there is between the same group between species. This would require at least two rounds of amplification each following mutational changes in closely related sequences both before and after species divergence. At the same time there could be a requirement for selection of sorts that maintains the same overall level of divergence in all the groups. Renaturation of R-DNA of wheat under high and low stringencies followed by S-1 enzyme digestion shows that short conserved sequences (350-650 b.p.) are interspersed with longer regions of diverged sequences (up to 6000-10,000 b.p.)[23]. (Unlike Xenopus, the short repeats show least divergence). Similarly Gp.I sequences are not clustered but are distributed over 65% of all the genomes, (600 b.p.interspersed with 800 b.p.) (Flavell, pers.comm.). Growth of the genome appears to involve an ordered finely precise intercalation of newly amplified sequences between the ancestral Gp.I sequences; strongly suggesting that amplification and interspersion are part and parcel of the same process. Alternatively, interspersion of new and old R-DNA sequences could be generated with amplification of one R-DNA/SC-DNA unit giving rise to a tandem array of two repeating sequences. A continual process of amplification of selected units would maintain an overall equality of divergence in the groups: processes not too dissimilar from those envisaged in the generation of sat.DNAs.

A similar process of intercalation and amplification of R-DNA is considered to give rise to the ordered increase in the size of the genomes of salamanders of the genus Plethodon[24].

The taxonomy and evolution of the Plethodon genus is well understood and the relationships between species derived from the extent and stability of heteroduplexes of R-DNA fits well with previous patterns. Cytophotometric analysis of the 14 chrms. shows that the relative lengths of arms for any one 'homologous' chrm.of the species remain constant. Selective amplification of sequences must occur such that the ensuing increases are distributed over the genome in the correct proportions. The same process could explain a high conservation of G-band patterns in related species (see later), whilst allowing genomes to differ in size

It might be of great significance that the small amount of residual homology in R-DNA (10%) in the most distantly related salamanders has been shown to be thermally stable and is of equal frequency (6000) and length in both species[25]. Its elution profile reveals considerable heterogeneity and is distributed both in the chrms.and in the genus. It also occurs in approximately half the amount in other genera of the same tribe, Plethodontini and in one tenth the amount in other tribes. In many respects it is similar to Gp.I of the Gramineae. Significant amounts of homology of conserved R-DNA under stringent conditions is known in species of crustaceae belonging to different orders[26].

There are recent reports of high thermal stability of fractions of residual homology of SC-DNA in the most distantly related species of a group. Two species of Lathyrus that share as little as 15% of SC-DNA (at a fixed stringency of incubation) nevertheless display only a 5% divergence in this shared fraction. Elution profiles indicate that three quarters of this have stabilities within range of the homoduplexes and is equally complex. Similarly, a small 12% homology of SC-DNA that resides in the genomes of distantly related S.purpuratus and Lytechinus pictus[28], is stable and three quarters of this fraction elute over a range of temperatures that overlap the elution temperature of the homoduplexes. Hybridisation of complementary DNA to total mRNA of sea-urchin gastrula to L.pictus shows that the conserved SC-DNA fractions is not enriched in structural genes. In the 4 spp.of sea urchin and the genus Dermestes (Fox, pers.comm.) the % homology and stability of SC-DNA between species reflect groupings based on traditional taxonomy. However, in all the groups Dermestes, Lathyrus, Plethodon and sea-urchin the divergence of SC-DNA is often at a slower rate than that of R-DNA.

VARIATION IN SATELLITE DNA AND C-BANDS IN CLOSELY RELATED SPECIES

Sat.DNA has received considerable attention in interspecific comparisons due to the ease of its isolation and its singular properties. Early work in rodents quickly established that differences between species involved changes in amount and composition of sat.DNA[29] whilst differences between sub-species involved changes in amount only[30], (depicted by C-band variation) of an identical satellite identified by renaturation kinetics and ability to bind anti-5-methyl cytosine antibodies in situ. Polymorphic variation in C-bands is known in several sub-species of Asian-type black rats (R.rattus)[31] (with a diminution of C-band occurring in evolution) and in subspecies of Peromysus[32].

Interspecific satellite relationships in the virilis group of Drosophila conform to the relatedness of species based on morphology, inversions and protein polymorphisms[33]. In this group, satellites are related in length and sequence and can be derived from each other by one or two base changes of the repeating heptanucleotide. One satellite, common to all species, is considered ancestral and has not been lost in subsequent speciation events. The seven sibling species of the melanogaster subgroup of Drosophila cannot be so simply related by a scheme that requires the retention of ancestral satellites in all subsequent speciation events and the addition of a new satellite at each branch point[14]. Nevertheless the relative spatial relationships of five of the species (see Fig.) is identical to a phylogeny based on chrm.

inversions[34]. The distribution of long pyrimidine tracts of sat.IV of D.melanogaster (sat.VII in the Fig.) in the other species of the group supports the relationships as shown (Sederoff, pers.comm.)

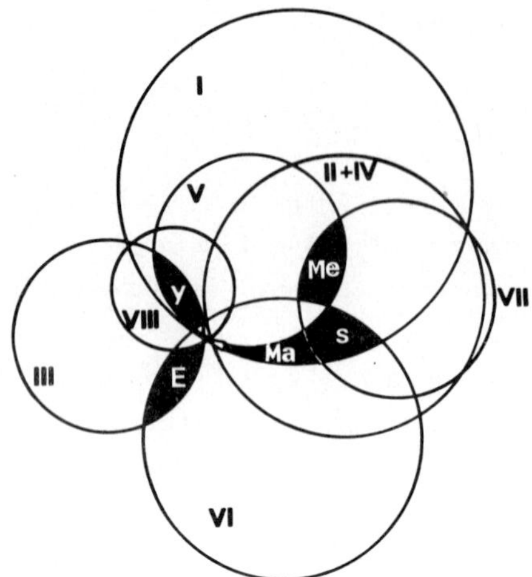

Relationships of five species of the melanogaster species subgroup of Drosophila, based on the presence of satellite DNAs; (resolvable in actinomycin-D/CsCl density gradients).

Me = D.melanogaster; S = D.simulans; Ma = D.mauritiana; Y = D.yakuba; E = D.erecta.

I-VII are satellite DNAs. The size of each circle is determined by the number of species sharing a satellite. Distances between species are not absolute although the spatial relationships are fixed in this arrangement.

Fig. (from S.R.B. Barnes amd G.A. Dover)

It might well be that the effective functional differences in satellites that are shared between species, or shared between chrms.of a species, reside in the differential patterns of distribution and in variation in sequence. Chrms.of D.melanogaster vary in a fine pattern of arrangement of contiguous blocks of satellite at the centromeres and there are indications of differences in the distribution of blocks in equivalent chrms.of D.simulans[35,36]. Higher order periodicities, similar to mouse sat. DNA[37], have been found in sat.III[38] and sat.IV[39], and both are heterogeneous with respect to sequence variants[40,41]. Similarly, recent data on the organisation of mouse, guinea-pig and African green monkey satellites has revealed the existence of segments within the blocks that are variants of a common ancestral sequence with respect to the organisation of higher periodicity[42,43,44]. No between species sequence-variants of sat.DNAs have been described so far, although a male-specific sequence variant of human sat.III has been located on the Y chrm., (and possibly also on the 9; Bostock, pers.comm.), but not at other sat.III localities in the genome. Restriction enzyme digests of total DNA of species of Apodemus indicate the same higher order periodicities in species that share a satellite, with some minor variation in frequencies[45].

Satellite differences in amount and composition that reflect known phylogenetic groupings have been described in species of Plethodon[46], primates[47], Dermestes[48], Dipodomys[3], Peromyscus (this volume); and preliminary investigations in the complex of species and subspecies in Glossina (Tsetse) suggest a similar story[49]. In Dipodomys the ratio of intermediate density DNA (1.702) to heavy density satellite (1.713) increases as the fundamental number of chrms.arms increases. The amount of sat.1.707 is not involved directly with this correlation; nevertheless the karyotypic differences between the species indicate loss and gain of sequences during the conversion of uni-armed to biarmed chrms. There are good indications that similar karyotypic processes involving fluctuating amounts of sat.DNA are occurring in other genera. A presumed loss of sat.DNA has occurred following conversion of chrm.17 of chimpanzee and orang-utang and 16 of gorilla (all having C-bands that hybridise to human sat.III) to the derived human submetacentric, 18, that is without sat.III sequences[50]. Similarly, chrms.of goat, sheep and ox have a much reduced amount of C-band material in meta-centrics when compared to their homologous acrocentric counterparts[51]. However, the karyotype of tobacco mouse, Mus poschiavinus, has recently evolved by fusion of Mus musculus telocentrics giving rise to metacentrics that show no reductions in the amount of const.het.[52].

It is difficult, as always, to assess cause and effect in these instances; it is not inconceivable that chrm.rearrangement would predominate in regions adjacent to or within const.het., and that this would involve a gain or loss as the case may be. A high breakage rate could be the result of out-of-phase replication of centromeric heterochromatin[53], or due to the high frequency of sister-chromatid-exchange at euchromatin/heterochromatin junctions[54,55,56]. This latter effect could account for C-band polymorphisms, and the rapid spread of a new sat.DNA throughout a comple-ment, e.g. it could explain the occurrence of autosomal and sex chrm.polymorphisms in the hamster Cricetus cricetus, which has large amounts of C-banded heterochromatin, and their absence in C.griseus whose chrms.have inconspicuous or no C-bands[57]. Perhaps these phenomena, and the phenomena of high-order periodicity, could be considered the aberrations of a quirk of DNA replication or exchange at differentially condensed areas of the chrm.that contains sat.DNA; the condensation itself presumably being a necessary prerequisite for other functions such as chrm.movement. Interestingly, it has been shown that cloned fragments of melanogaster satellites in recombination defective hosts are unstable in length, presumably the result of unequal intramole-cular recombination events[58].

In primates the acquisition of sat.DNAs during speciation has not resulted in the appearance of new heterochromatic arms of the sort described in spp.of Peromyscus[32] and Notomys[68] except possibly in the case at the African green monkey (Cercopithecus) (2n = 60; 110 arms) that can be rearranged to resemble that of the rhesus monkey (Macaca) (2n = 42; 84 arms) by simple deletion of heterochromatic short arms of some of the biarmed chrms; only one pair of small metacentrics remaining unmatched[59]. In the higher primates the distribution of four satellites indicates an addition of a new satellite sequence at or subsequent to points of divergence of the genera[47]. Ancestral satellites are not eliminated from the genomes although modulation in amount and pattern of distribution over the chrms.has occurred with time. For example human sat.III hybridises often in identical positions in equivalent chrms.of chimpanzee, gorilla and orang-utan but in varying amounts[50,60]. At other times no correspondence is found between sat.DNA locations and chrm.homology,(including the Y that is different in morphology, quinacrine and G-banding in all 4 spp.) The use of anti-5-MeC antibodies against heterochromatin regions of chimpanzee, gorilla and human chrms.reveals a diversity in size and location of concentrations of GC rich satellites (e.g.human satellites I and IV) and depicts a closer relationship between gorilla and human chrms.than either is to chimpanzee[61]; indicating the rapid evolution of sat.DNA.

The ability to identify a species by the amounts and distributions of selected types of satellites from a spectrum of types within a genus has led to much current speculation on their role in aspects of meiosis, and through that on the speciation process itself. It has been strongly proposed that the presence of telomeric sat.DNAs in species of *Atractomorpha*[2] regulates (by reduction) the observed differences in frequency of recombination; and that chiasma positions are altered accordingly. In this genus, the three sibling spp.*A.australis*, *A.species-1* and *A.similis* differ in karyotype only by the presence of heterochromatic blocks at telomeric positions and a large cryptic satellite has been located at the telomeres of nearly all chrms.in *A.similis*. In contrast, it should not be overlooked (to cite but one example) that species of the *paniculatum* group of the genus *Allium* (all 2n = 8 and with similar karyotypes) that vary in the amounts of distal and intercalary blocks of C-banded heterochromatin (10% - 30%) and that are also often polymorphic in this respect, show comparable patterns of chiasma frequency and distribution in all species[62]. It is not inconceivable that the presence of heterochromatin *per se* (with or without sat. DNA) is responsible for a redistribution of chiasmata (where this occurs) and hence affect their frequency. Heterochromatic 'B' chrms.increase or decrease chiasma frequency without the help of sat.DNA[63,64,65,66], and a perusal of the substantial table [67] of correlated variation of both constitutive and facultative heterochromatin content and recombination should suffice as an indication that there are no simple answers as yet to this intriguing problem.

The normal functioning of the meiotic process with respect to pairing and disjunction in bivalents that are heteromorphic for the presence of sat.DNA does not support, on *prima facie* evidence, the often proposed involvement of sat.DNA in these processes. F-1 hybrids of subspecies of *Mus musculus*[30], *Rattus rattus*[31] and *Peromyscus maniculatus*[32] exhibit no irregularities in meiotic chrm.pairing or segregation despite differences in amount of sat.DNA in the autosomes, and sometimes, in the sex chrms.: differences in two subspecies of *Mus musculus* extend to the X-Y chrms. [It has been proposed that the presence of sat.DNA in the W sex chrm.of more advanced orders of snakes and its absence in families of snakes that have primitive states of sex differentiation, is involved in the evolution of sex chrms.[4]. However sex chrm.polymorphisms in const.het.of the X and Y exist both within and between species of *Notomys*[68]; and furthermore the range of X and Y sex chrm.types that involve deletions of const. het.in *Bandicota bengal bengal* have no apparent effect on fertility[69].]

It is possible that the regularity of pairing in heteromorphic bivalents and subspecific F-1 hybrids is the result of a shared similar sequence albeit in different amounts. However there are observations of regular pairing and disjunction of X's in females of Drosophila having large deletions of basal heterochromatin containing the sat.DNAs[70]. The dramatic effects of 'B' chrms.in reducing pairing of homoeologous genomes in interspecific hybrids in plants[71,72,73] might be instructive to our problem. There is convincing evidence that 'B's are derived from 'A's in several genera[67,74,75] and the presence of the same spectrum of sequences in both[63-66] suggests that the 'B's are duplications of 'A's. The failure of 'B's to pair with their homoeologous 'A' counterparts could be the result of subsequent heterochromatisation *per se* (of a pairing site?) rather than the presence of a B-specific sat.DNA. However 'B's like sex chrms.are a veritable Pandora's box of tricks and it is not as yet wise to spin out too many generalities. With similar deference it should not be overlooked that many loci have been located in plants and animals that behave as single genes and that affect the pairing, exchange and disjunctional behaviour of chrms.[76,77]

It is clear that a meaningful interpretation of the patterns of distribution of satellites in related species and their possible involvement with an adaptive feature that might be involved in the process of speciation needs to await further study on the complex variants of satellites; on the fine-structure of heterochromatin organisation and on the effects and behaviour of aberrant chrms. In Drosophila at least 14 types of heterochromatin have been resolved as a result of differential response to Hoechst 33258,Q,C and N banding agents[78,79]; of which 4-7 are found in individual species. It needs to be seen what relationships exist between satellites and their variants and the permutations of heterochromatin composition.

VARIATION IN G-BANDS BETWEEN CLOSELY RELATED SPECIES

Although the general theme of comment presented above has centered on the karyotypic differences between related species with respect to sat.DNA and its equivalent chromosomal features, it might be of more biological relevance that the majority of karyotypes, in so far as G-band patterns depict this, are homologous in related species. [There are examples of a constancy of C-band patterns themselves. For example three species in the genus Allium that have common morphological characters and the ability to cross hybridise show evolutionary conservation of C-bands located at telomeres[80]. Furthermore, a within-species variability in C-bands can sometimes be non-randomly distributed. In Allium pulchellum basic patterns are confined to particular populations[62] and in Anemone blanda there is a concurrence of certain chrm.variants such that two karyotypes exist in the species; (the patterns within individuals remaining constant). The two types are fully cross compatible and the two karyotypes cannot be explained by inbreeding or selection against heteromorphic pairs of chrms[81].]

In mammals the striking constancy of the X chrm.in form and banding has, in itself, led to much speculation on the selective pressures involved in its preservation[82]. Chrms.of 4 spp.of Hominidae can be ascertained equivalent identities by their reaction to a range of banding techniques[60] and chrms.1 and X are identical. The use of cell hybrid techniques has assigned structural loci of 12 enzymes in man to homologous locations in ape and the 5S and rRNA genes are in the same positions in each species[50]. Regions of known genetic identity in mouse and rat show no band homologies although evolutionary conservation can be seen in the identical trypsin banding patterns in 40% of each genome of rat and mouse[83]. This is much less extensive than primates which have diverged over a similar span of time of 10^6 years; suggesting that the difference in generation times of the two groups is significant. In rats there is a uniformity of G-bands in all species and subspecies[84] despite high variability in C-bands[31]; G-band patterns reveal characteristic chrm.rearrangements in different geographical types leading to the suggestion that the highly polymorphic karyotypes found in the Asian-type black rat were established prior to the differentiation of karyotypic differences between the species belonging to other geographical types.

In other genera extensive G-band homologies have been found, sometimes extending to distantly related groups of organisms. Phylogenetic studies of bird karyotypes extending over 12 orders have revealed conservation of some morphologically distinct chrms.and chrm.arms in shape and G-band, including G-band conservation in the Z[85] Similarly 10 species of eurydid turtles show little intragenic variation in both G- and C-bands[86]. The chrms.of sheep, goat and ox exhibit remarkable evolutionary conservation of G-band patterns with few minor exceptions permitting the precise matching of chrms.across the genera[51]. Nevertheless the ox contains 14% more DNA than sheep and goats, and the proportion of centromeric sat.DNAs in each species is similar suggesting that the evolution of the karyotypes must have involved numerous minute interstitial deletions or additions of fractions of DNA; of the sort presumably taking place in the changing compositions of genomes in species of Gramineae, Plethodon and Lathyrus.

COMMENT

Interpretations of the variation in genome organisation are given in the summary. Whether any light has been shed on the process of speciation is a matter of personal persuasion. It could quite well be, to return to Lewontin, 'that the acquisition of reproductive isolation does not require a major overhaul of the genotype'; and that the genomic changes involved in the inception of differentiation of populations are cryptic and difficult to locate. Alternatively, we may view the prevalence of polymorphic variation of C-bands and sequence-variants of sat.DNA as the stuff on which an incipient process of divergence may feed. These are early days.

REFERENCES

1. Lewontin,R.C.(1974) The Genetic Basis of Evolutionary Change, Columbia,pp.1-346
2. Miklos,G.L.G. and Nankivell,R.N.(1976) Chromosoma 56,143-167
3. Hatch,F.T.,Bodner,A.,Mazrimas,J.A.and Moore,D.H.(1976) Chromosoma 58, 155-168
4. Singh,L,Purdom,I.F.and Jones,K.W.(1976) Chromosoma 59, 43-62
5. White,M.J.D.(1973) Animal Cytology & Evolution, Cambridge
6. Hinegardner,R.(1976) in: Molecular Evolution,ed.F.Ayala Sinauer,pp.1-267
7. Wilson,A.C.(1976) in: Molecular Evolution, ed.F.Ayala Sinauer,pp.1-267
8. Rees,H.and Jones, R.N.(1972) Int.Rev.Cytol.32, 53-92
9. Sparrow,A.H.and Nauman,A.F.(1976) Science 192, 524-529
10.Nagl,W.(1976) Nature 261, 614-615
11.Kohne,D.E.,Chiscon,J.A.and Hoyer,B.H.(1972) J.Human Evol.1, 627-644
12.Shields,G.F.and Straus,N.A.(1975) Evolution 29, 159-166
13.Thiery,J.P.,Macaya,G.and Bernard,G.(1976) J.Mol.Biol.108, 219-235
14.Barnes,S.R.B.and Dover,G.A.(1977) (in preparation)
15.Baldari,C.T.and Amaldi,F.(1977) Chromosoma 61, 359-368
16.Bachmann,K.and Price,H.J.(1977) Chromosoma 61, 267-275
17.Narayan,R.K.J.and Rees,H.(1976) Chromosoma 54, 141-154
18.Flavell,R.B.,Rimpau,J.and Smith,D.B.(1977) Chromosoma (in press)
19.Britten,R.J.,Graham,D.E.,Eden,F.C.,Painchaud,D.M.and Davidson,E.H.(1976)J.Mol.Evol.9.
20.Galau,G.A.et al (1976) in: Molecular Evolution,ed.F.Ayala Sinauer,pp.1-267
21.Gillespie,D.(1977) Science 196, 889-891
22.Rice,N.R.(1972) in: Evolution of Genetic Systems,ed.H.H.Smith,Brookhaven Symp.23
23.Flavell,R.B. and Smith,D.B.(1976) Heredity 37, 231-252
24.Mizuno,S.and Macgregor,H.C.(1974) Chromosoma 48, 239-296

25.Mizuno,S.,Andrews,C.and Macgregor,H.C.(1976) Chromosoma 58,1-31
26.Graham,D.E.and Skinner,D.M.(1973) Chromosoma 40,135-152
27.Narayan,R.K.J.and Rees,H.(1977) Chromosoma (in press)
28.Angerer,R.C.,Davidson,E.H.and Britten,R.J.(1976) Chromosoma 56, 213-226
29.Hennig,U.and Walker,P.M.B.(1970) Nature 225,915-919
30.Dev,V.G.et al (1975) Chromosoma 53, 335-344
31.Yosida,T.H.and Sagai,T.(1975) Chromosoma 50, 283-300
32.Pathak,S.,Hsu,T.C.and Arrighi,F.E.(1973) Cytogenet.Cell Genet.12, 315-326
33.Gall, J.G..and Atherton,D.D.(1974) J.Mol.Biol.85, 633-664
34.Lemeunier,F.and Ashburner,M.(1976) Proc.Roy.Soc.B.193, 275-294
35.Peacock,W.J.,Appels,R.and Steffensen, (1977) (in preparation)
36.Brutlag,D.,Appels,R.,Dennis,E.S.and Peacock,W.J.(1977) J.Mol.Biol.112,31-47
37.Southern,E.(1975) J.Mol.Biol.94, 51-69
38.Manteuil,S.,Hamer,D.H.and Thomas,C.A.Jr.(1975) Cell 5, 413-422
39.Shen,C.K.and Hearst,J.E.(1977) J.Mol.Biol.112, 495-506
40.Carlson,M.and Brutlag,D.(1977) Cell 11, 371-381
41.Brutlag,D.,Carlson,M.,Fry,K.and Hsieh,T.S.(1977) (in preparation)
42.Horz,W.and Zachau,H.G.(1977) Sur.J.Biochem.73, 383-392
43.Altenburger,W.,Horz,Z.and Zachau,G.(1977) Eur.J.Biochem.73, 393-400
44.Fittler,F.(1977) Eur.J.Biochem.74, 343-351
45.Cooke,H.J.(1975) J.Mol.Biol.94, 87-99
46.Macgregor,H.C.,Horner,H.,Owen,C.A.and Parker,I.(1973) Chromosoma 43,329-348
47.Jones,K.W.(1976) in: Chromosomes Today 5. John Wiley, 305-313
48.Rees,R.W.,Fox,D.P.and Maher,E.P.(1976) in: Curr.Chromosome Res.,ed.K.Jones and P.E.Brandham. Elsevier, 33-41
49.Amos,A.,Southern,D.and Dover,G.A.(1977) (in preparation)
50.Mitchell,A.R.et al (1977) Chromosoma 61, 345-358
51. Evans,H.J.,Buckland,R.A.and Sumner,A.T.(1973) Chromosoma 42, 383-402
52.Zech,L.,Evans,E.P.,Ford,G.E.and Gropp,A.(1971) Expt.Cell Res.70, 263-268
53.Hsu,T.C.and Markvong,A.(1975) Chromosoma 51, 311-322
54.Carrano,A.V.and Wolff,S.(1975) Chromosoma 53, 361-369
55.Bostock,C.J.and Christie,S.(1976) Chromosoma 56, 275-287
56.Hsu,T.C.and Pathak,S.(1976) Chromosoma 58, 269-273
57.Gamperl,R.,Vistorin,G.and Rosenkranz,W.(1976) Chromosoma 56, 259-265
58.Brutlag,D.,Fry,K.,Nelson,T.and Hung,P.(1977) Cell 10, 509-519
59.Stock,A.D.and Hsu,T.C.(1973) Chromosoma 43,211-224
60.Bobrow,M.and Madan,K.(1973) Cytogenet.Cell Genet.12, 107-116
61.Schnedl,W. et al (1975) Chromosoma 52, 59-66
62.Vosa,C.G.(1976) Chromosoma 57, 119-133
63.Dover,G.A.(1975) Chromosoma 53, 153-173
64.Dover,G.A.and Henderson,S.A.(1976) Nature 259, 57-58
65.Chilton,M.D.and McCarthy,B.J.(1973) Genetics 74, 605-614
66.Rimpau,J.and Flavell,R.B.(1975) Chromosoma 52, 207-217
67.John,B.(1973) Chromosoma 44, 123-146
68.Baverstock,P.R.,Watts,C.H.S.and Hogarth,J.T.(1977) Chromosoma 61, 243-256
69.Sharma,T.and Raman,R.(1973) Chromosoma 41, 75-84
70.Yamamoto,M.and Miklos,G.L.G.(1977) Chromosoma 60, 283-296
71.Dover,G.A.and Riley,R.(1972) Nature 240, 159-161
72.Vardi,A.and Dover,G.A.(1972) Chromosoma 38, 367-385
73.Evans,G.M.and Macefield,A.J.(1972) Nature 236, 110-111
74.Hewitt,G.M.(1973) Cold Spring Harb.Symp.38, 183-194
75.Dover,G.A.(1976) in: Curr.Chrom.Res.ed.K.Jones and P.E.Bradham. Elsevier, 67-76
76.Dover,G.A.and Riley,R.(1977) in: The Meiotic Process. Proc.Roy.Soc.B.277, 313-326
77.Baker,S.B.and Hall,J.C.(1976) in: Genetics & Biology of Drosophila,1a,ed.Ashburner & Novitsky, 1-483
78.Gatti,M.,Pimpinelli,S.and Santini (1976) Chromosoma 57, 351-375
79.Holmquist,G.(1975) Nature 257, 503-505
80.El-Gadi,A.and Elkington,T.T.(1975) Chromosoma 51, 19-23
81.Marks,G.E.(1976) in: Chromosomes Today V,ed.Pearson & Lewis,179-184
82.Ohno,S.(1973) in: Cold Spring Harb.Symp.38, 155-164
83.Nesbitt,M.N.(1974) Chromosoma 46, 217-224
84.Yosida,T.H.and Sagai,T.(1973) Chromosoma 41, 93-101
85.Takagi,N.and Sasaki,M.(1974) Chromosoma 46, 91-120
86.Bickham,J.W.and Baker,R.J.(1976) Chromosoma 54, 201-219

EVOLUTION OF KARYOTYPES

Chromosomes Today Volume 6, A. de la Chapelle and M. Sorsa eds.
© 1977 Elsevier/North-Holland Biomedical Press, Amsterdam, The Netherlands

THE EVOLUTION OF KARYOTYPES - AN INTRODUCTION

RALPH RILEY
Plant Breeding Institute, Cambridge, England

The organisers of this Sixth International Chromosome Conference at Helsinki were particularly wise in deciding to devote a part of the proceedings to the subject of the evolution of karyotypes for we are at a point of new departure in this subject which has for long been a component of classical chromosome cytology. In the past decade we have seen the introduction of various banding procedures that have enabled longitudinal differentiation within the chromosome to be visualised and to this has been added *in situ* nucleic acid hybridisation. The fractionation and cloning of DNA will add to our armoury the use of new nuclear probes.

So we are now becoming possessed of new techniques that will enable us to study the ways by which karyotypes evolve. But before we turn to a consideration of this subject in depth in the rest of the session it is perhaps desirable that we start with a definition of our field study. I am not clear when the word "karyotype" originated but I note that it was not used in the 1936 edition of "Recent Advances in Cytology" - that distinguished work of our Honorary President, C.D. Darlington - but that it is defined in the glossary of his 1949 work with Kenneth Mather "The Elements of Genetics". As generally used in classical cytology the term "karyotype" was employed to define the character of a nucleus in terms of the number, sizes and shapes of its chromosomes as seen at mitotic metophase. An alternative usage of "karyotype" is to describe, again in terms of numbers, sizes and shapes, the mitotic chromosome complement of an individual, a group or a species of organisms. The shapes of chromosomes were originally described in terms of the positions of the primary and secondary constrictions, so longitudinal differentiation of the chromosomes was always used as a component character of the karyotype. Thus the inclusion of descriptions of other kinds of longitudinal differentiation - heterochromatin, banding or *in situ* molecular probes - was easily accommodated.

In evolutionary studies the karyotype has been considered in two ways. The first is similar to much of the study of general evolution carried out not by geneticists or cytologists but by comparative morphologists and histologists; considered in this way the karyotype is just another morphological character that can be used comparatively, like the structure of an anatomical organ. However, since the karyotype is constructed, in part, of molecules that determine other morphological and physiological characters it may be considered to provide more significant evidence on the course of evolution than can be gleaned from the study of peripheral characters. Secondly, the evolution of the karyotype has for long been directly studied, particularly in relation to the structural modification of chromosomes. Evolution of the karyotype may involve interchanges, inversions, deletions, duplications and centric divisions and fusions, as well as changes arising due to aneuploidy and polyploidy.

All of this is well trodden ground. For the future, with a rapid extension of our understanding of the structure of chromatin and DNA, we may expect a new rigour to be introduced into studies of karyotype evolution. For example in the wheat group which includes species in *Triticum* and rye, *Secale*, between 75 and 80 per cent of the DNA is highly reiterated and rye is distinguished from wheat by having a species-specific group of reiterated DNA. My colleagues R.B. Flavell and D.B. Smith have shown that this rye-specific reiterated DNA group occurs on every chromosome of the rye complement. Questions are immediately raised about the origin of this group of DNA and how, in karyotype evolution, it came to be distributed throughout the rye chromosome complement.

Such questions will only be answered by molecular studies of the karyotype and it is in the use of this methodology that future advances in the understanding of the evolution of the karyotype must principally depend.

Chromosomes Today Volume 6, A. de la Chapelle and M. Sorsa eds.
© 1977 Elsevier/North-Holland Biomedical Press, Amsterdam, The Netherlands

THE ROLE OF ROBERTSONIAN CHANGE IN KARYOTYPE EVOLUTION IN HIGHER PLANTS

KEITH JONES
Jodrell Laboratory, Royal Botanic Gardens, Kew, England

ABSTRACT

A combination of Robertsonian fusion and polyploidy has been the principal method of karyotype evolution in Cymbispatha. Chromosome size and symmetry has increased and stable isochromosomes are produced. It is suggested that a similar mechanism may be involved on a wider scale in the evolution of karyotypes in higher plants.

INTRODUCTION

Although the chromosome has the same organisation in plants and animals and is subject to the same sorts of structural and numerical mutation in both, there are important differences in the utilization of these changes in the two kingdoms. Polyploidy, a very common factor in the evolution of plants, is of much more restricted occurrence in animals - a well-known distinction which may be understood in terms of such factors as differences in complexities of embryo development, life span, sexual differentiation, choice of mates, etc. It is less easy to understand the reasons for the success of Robertsonian changes in animals and their apparent rarity in plants. There are ample records of the role of chromosome fusions and fissions as the accompaniment to or the cause of population and species differentiation in animals and, in some groups like the placental mammals, they seem to represent the only significant major structural changes in the evolutionary lineage. It cannot be doubted then that they have a highly significant impact, whether this be in the modification of patterns of recombination, the development of new adaptive combinations or indeed in position effect. Why then should this type of change not have been equally influential in plant evolution? Are Robertsonian events so rare and sporadic in plants or is it the case that they are more difficult to detect? These are questions to which I will apply myself in this paper but first some comment needs to be made on the differing views held by plant and animal cytologists on what chromosome types are considered to be primitive and derived.

It seems that in general the small acrocentric is viewed by the animal cytologist as the type of chromosome from which larger metacentrics are derived. In contrast it is the metacentric, the symmetrical chromosome, which is more the archetype for those who work with plants. Both views seem to have become accepted dogma. But in these days when the whole matter of chromosome evolution has been re-opened by the rapid advances in knowledge brought about by banding studies, DNA measurement and above all by the detailed investigation of unique and repetitive nucleotide sequences, dogma should have little place in the formation of our ideas. Hypotheses are acceptable providing that they are always put to the test, but when by repetition alone they harden into dogma their only effect is to stultify constructive thought and critical analysis. Specifically taking Robertsonian change as the subject for discussion it can be seen as a process which at one time might be fusion and at another fission. The necessary preliminary for centric fusion is what may be called the preadaptation of acrocentricity. The result is to join the bulk of the genetic material of two chromosomes by union with a common centromere and a move towards greater symmetry. Successive changes involving different pairs of chromosomes can result in a major transformation of karyotype pattern. If a symmetrical karyotype is the consequence any further visible structural changes must be a move towards asymmetry. These may include asymmetrical pericentric inversions which can produce acrocentrics capable once again of centric fusions now bringing together the genetic material of four ancestral chromosomes. And so the process may continue over the span of evolutionary time with cycles of symmetry and asymmetry probably associated with periods of chromosome doubling. If this happens there seems little point in declaring either the acrocentric or metacentric as primitive unless it is wished to comment on the probable position of the centromere when it first became discrete. On that point evidence is hard to come by and the wish to accept the archetypal nature of the metacentric in the plant field seems to be influenced more by the aesthetic appeal of symmetry than by observation. There are perhaps more logical grounds for assuming the first localisation of centromeric activity at or close to chromosome ends in the way envisaged by Varaama[1]. Neither point of view can yet be proven and opinions should remain open.

In concluding this introduction it is necessary only to reiterate that the reversibility of so many types of chromosome change makes the detection of their direction no easy matter. Corroborative evidence is often required either from parallel changes in the chromosomes themselves or from biochemical and phenotypic characters. I will now consider chromosome evolution in the genus *Cymbispatha* - a member of the Tradescantia family (Commelinaceae) and finally pass on to wider considerations.

Chromosome variation in Cymbispatha

Species of *Cymbispatha* were first described under *Tradescantia* and have been so dealt with by earlier chromosome investigators. These species are however quite unlike the Tradescantias which are so familiar to many as material for chromosome study and demonstration, and they are also distinct in their chromosomes. I therefore prefer to keep them in a separate genus. They are low growing, rather straggling tuberous perennials with small pink, blue or white flowers. Only one species is known to be annual and self-fertile - the rest are outcrossing. The species occur in upland regions of Mexico, Guatemala and adjacent countries in Central and South America. The plants studied at Kew were collected as vegetative material from natural populations in Mexico during several expeditions. C. commelinoides has been collected more frequently than C. geniculata and C. plusiantha each of which has been found in a single restricted locality.

The chromosome numbers known for the genus are given with karyotype constitution in Table 1. Complements consist entirely of metacentrics or acute acrocentrics or mixtures of both, the metacentrics being about twice the length of the acros. In some cases the acrocentrics have no visible second arm and could better be called telocentric but for convenience the distinction will not be made in this brief paper.

Species	2n.	Karyotype		n.f.
		M	A	
C. geniculata	14	-	14	14
C. plusiantha	12	2	10	14
C. sp.	36	6	30	42
C. standleyi	16	12	4	28
C. commelinoides	14	14	-	28
"	22	20	2	42
"	28	28	-	56
"	30	26	4	56

TABLE 1. The chromosome constitution of Cymbispatha

Somatic chromosome numbers form an irregular series, 2n = 12, 14, 16, 22, 28, 30, 36, with occasional and sporadic intermediates at the higher levels. In contrast, the number of major chromosome arms (n.f.) is, in all cases, a multiple of 7 producing the series 14, 28, 42 and 56. It is evident from this that Robertsonian changes and periods of chromosome doubling have been major events in karyotype evolution in the genus. On the basis of n.f. the species must be considered diploid, tetraploid, hexaploid and octoploid in their genetic constitution if not in chromosome number. For the sake of convenience in this condensed account the disguised polyploids will be referred to with their level of genetic polyploidy stated within inverted commas.

Meiotic studies both confirm and amplify the situation indicated by the n.f. The two diploids are bivalent-forming and differ by a single Robertsonian change in the homozygous state (Figs. 1, 2). C. geniculata (2n = 14A), the annual and self-fertile species was found as a single population of seedlings near the Guatemalan border which was polymorphic for one pair of acrocentrics which had developed a larger second arm by pericentric inversion. C. plusiantha (2n = 2M10A) was found about 100 miles to the north. Its chromosome constitution has been given earlier by Handlos under the name Tradescantia tonalamonticola[2].

C. commelinoides was reported as 2n = 14M by Celarier (as T. commelinoides) who found a single quadrivalent at meiosis[3]. The plants which I examined came from a restricted area just to the south of Mexico City. All were 2n = 14M and each genotype examined was quadrivalent-forming with up to 3 quadrivalents per p.m.c. (Fig. 4) The frequency of quadrivalents showed that they were the result of pairing between full structural homologues rather than relatively interchanged chromosomes.

In one genotype only, either one or rarely two, univalents formed univalent rings typical of isochromosomes in 32 per cent of the metaphase p.m.c. (Fig. 3) sometimes in the presence of three quadrivalents. It is assumed that this plant reveals in an unstable way a constitution hidden by stability in the rest of this cytotype and it is concluded that the 2n = 14M karyotype consists of three groups of four homologues each and one pair of what should strictly be called pseudo-isochromosomes.

The 2n = 20M2A plants seem to have a wide distribution extending north-west from Mexico City certainly as far as Durango. Many multivalents occur at meiosis with quadrivalents and hexavalents particularly common. The two acrocentrics pair together or form a multivalent, usually a quadrivalent, with two metacentrics.

The 2n = 26M4A plants occur to the south of Mexico City in Oaxaca and Puebla. These too show extensive multivalent pairing with associations of 4, 6 and 8 common. The four acrocentrics can pair together as a quadrivalent (Fig. 5) or may associate with two metacentrics to form a hexavalent.

The 2n = 28M individuals have been found only in the region of San Cristobal de las Casas to the south of the 2n = 26M4A forms. Apart from the absence of the acrocentrics multivalent pairing is similar in type and frequency.

In each of the 'octoploids' there is evidence of interchange heterozygosity in the form of occasional associations of up to 12 chromosomes.

Under experimental conditions at Kew spontaneous hybrids were formed between the 2n = 14 and 2n = 22 plants and between the 2n = 22 and the 2n = 30 with evidence of occasional non-reduction in the 2n = 14 parent. Pairing in these hybrids demonstrated a high degree of structural homology between the chromosome sets of the parents and evidence of probable inversion differences between those of the 2n = 22 and the 2n - 30 cytotypes.

The single plant with 2n = 6M30A formed an occasional multivalent between its metacentric chromosomes but was otherwise bivalent-forming.

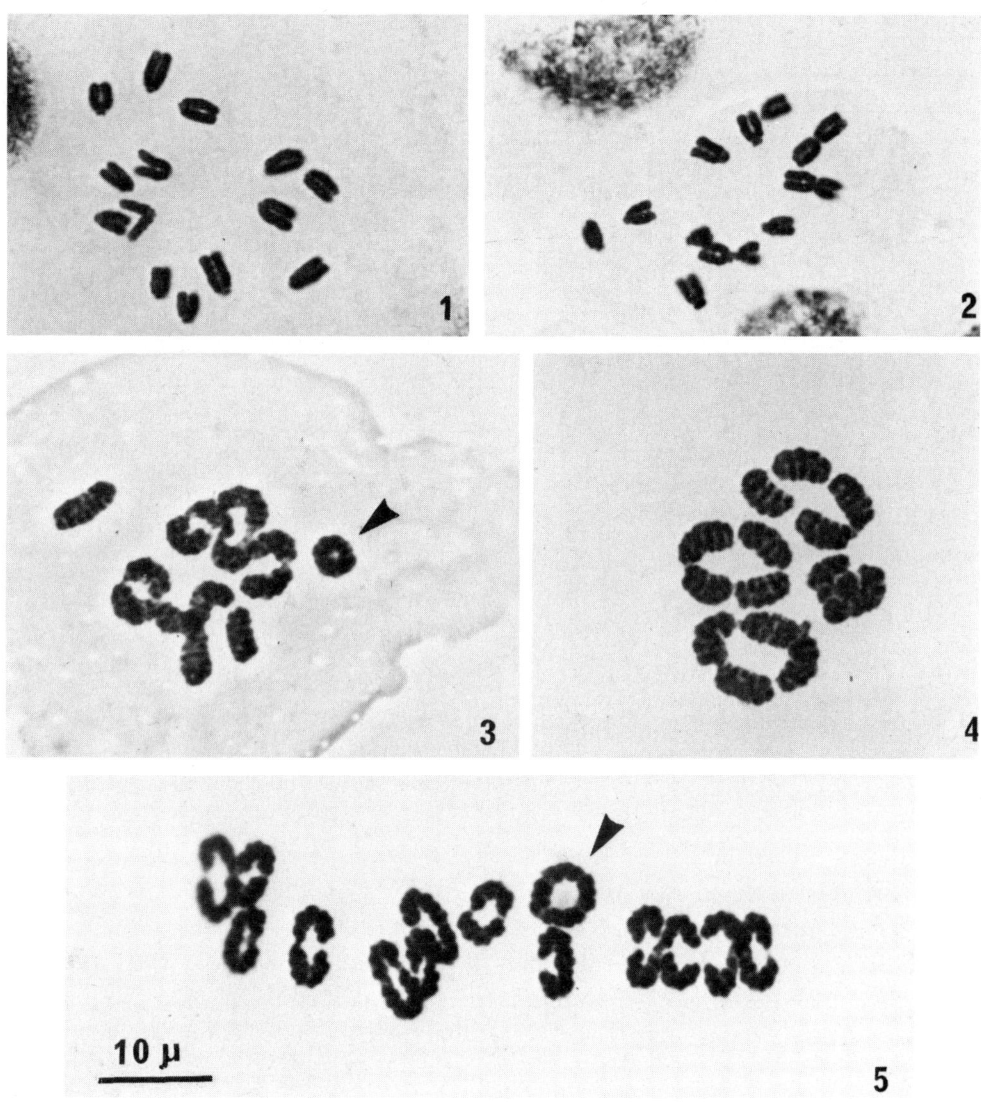

FIG. 1. Cymbispatha geniculata 2n = 14A. FIG. 2. C. plusiantha 2n = 2M10A.
FIG. 3. C. commelinoides 2n = 14M. 1iv 1iii 2ii 3i. Ring univalent arrowed.
FIG. 4. C. commelinoides 2n = 14M. 3iv 1ii. FIG. 5. C. commelinoides 2n = 26M4A.
1viii 2iv 7ii. Ring of 4 acrocentrics arrowed.

C. standleyi (2n = 12M4A) was reported on by Mattsson under the erroneous name of T. commelinoides[4]. It is quite distinct from what we know at Kew as C. commelinoides and the plant, which we also have in the Kew collection, comes from Guatemala. As was shown by Mattsson it forms only bivalents with highly proximal chiasmata - a feature which would in any event inhibit any potential multivalent pairing.

The evolution of Cymbispatha

The type of pairing found in the cytotypes of C. commelinoides indicates an ancestry of autopolyploidy with variable degrees of fusion occurring within one group of homologues subsequent to the initiation of the polyploid state.

The origin of such plants must lie in diploids which may have been of either of the two karyotype constitutions so far known or in other undiscovered ones in which the 14 arms showed greater degrees of fusion. This we cannot say but if on current evidence fusion rather than fission is the prime basis for variation in the genus fewer mutational events would be required at the diploid level. It might be further suggested that the impact of Robertsonian change on recombination, or other factors of adaptive significance, would be more emphatic and more subject to selection in diploids. In the hypothetical scheme of evolution (Fig. 6) a fusion series culminating with 2n = 6M2A is postulated, chromosome doubling then directly producing a 2n = 12M4A tetraploid. This condition would allow the fusion process to continue but now only between homologues producing chromosomes similar to isos. This is seen as the manner of origin of the 2n = 14M types whose chromosome number would now necessitate its being regarded as a diploid with x = 7. This condition must be preceded by heterozygosity for the fusion (2n = 13M2A) and this could produce the 2n = 26M4A directly by doubling, and the 2n = 20M2A by crossing to the 2n = 14M with an unreduced gamete. The evolutionary origins of the 'hexaploid' and 'octoploid' may not be precisely as shown for there are other alternatives but these suggestions are adequate to explain their constitutions.

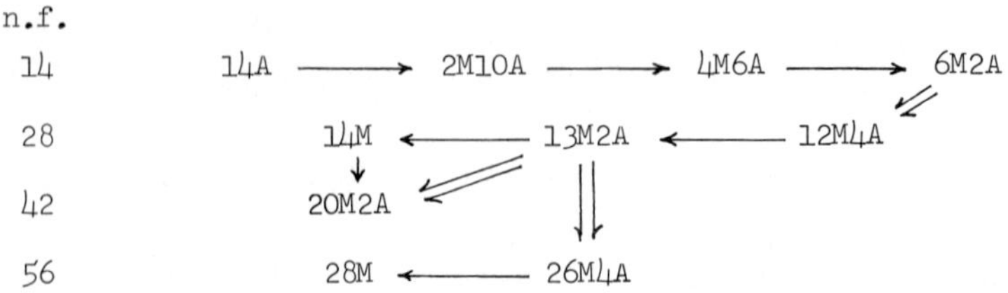

FIG. 6. The evolution of C. commelinoides

What we see in Cymbispatha is chromosome evolution probably following the pathway of fusion, certainly including the formation of stable isochromosomes, which increases chromosome size and symmetry. Polyploidy initially restores chromosome number to that of the probable ancestor and then carries it further. It is likely that fusions were most important for the differentiation of diploids with increase of chromosome number being more significant for the differential spread of the various forms consequent to initial tetraploidy. In Cymbispatha the survival of some of the important stages in the evolution of this situation and the maintenance of multivalent pairing make it possible to know the real constitution and probable origins of the plants. In their absence the interpretation of chromosome number and karyotype could be quite incorrect and indeed neither Celarier nor Mattsson had any reason for believing that their plants were anything other than diploid.

Isochromosomes

Isochromosomes have very rarely been reported as stable members of the normal complement of plant species. They occur in Nicandra physaloides[5] and in Philadelphus species[6]. In the strictest sense of their definition they arise from sister reunion of unstable telocentrics but they can also arise from fusion of homologous acros and the different origins can be detected by comparison of chromosome number in related species. To be stable they presumably require a pairing homologue and to be restrained from 'inside' pairing - conditions which are fulfilled in Cymbispatha. It is implicit here that stable isos are expected to be indicative not only of fusion but also of polyploidy, either present or ancestral. In the Jodrell Laboratory Mrs. Bhattarai has found isos in a species of the Gibasis linearis alliance with 2n = 22 but which is known to be a genetic tetraploid with fusion in its ancestry. Mr. A. Martinez also in my laboratory found typical iso rings in diploid hybrids of Tradescantia species with complements of 2n = 12M. I have found further evidence in other hybrids in the same genus. These discoveries raise the possibility that fusion and polyploidy have played a role in the evolution of some species which are presently regarded as diploid and primitive in their karyotype symmetry.

Not only hybridisation but also haploidy can reveal isochromosomes by depriving them of their partners and promoting 'inside' pairing. If we examine haploid meiosis using, as we should, the same principles of analysis that we apply to non-haploids then we may recognise them. Typical iso rings seem to be present in haploid barley (cf. Figs. 10 & 11, in Ref. 7). I think that they also occur in rye-wheat hybrids. As indicators of important evolutionary events they deserve our attention.

The wider occurrence of Robertsonian change in plants

Although there are many observations of sporadic fusions or fissions as rare mutations in natural populations there are as yet few known examples where it can be said that these changes have played a significant part in either population or species differentiation. At the population level my colleague Miss C. Brighton has recently discovered differences within *Crocus minimus* due to several probable fissions. Dr. P. Goldblatt of St. Louis tells me that he has found populations of *Homeria tenuis* which seem to be distinguished by fusions. On the level of speciation the Cycads[8] and more particularly the Podocarpaceae[9] have made use of extensive Robertsonian changes whilst in the genus *Lycoris*[10] (Amaryllidaceae) arm transpositions seem to be as extensive as in *Cymbispatha*. In the Commelinaceae fusion has been involved in speciation in *Gibasis scheideana*[11] and probably in other members of the genus and there can be little doubt that it has played a part in the evolution of *Zebrina*. *Alisma*, *Vicia*, *Fritillaria*, *Lilium*, *Nothoscordum*, *Hyacinthus* are other candidates for this type of structural change but I believe there will be others when the possibilities of Robertsonian change are taken into account in the analysis of karyotype differences.

FINALE

One of the major problems in the understanding of karyotype evolution is the very wide variation in chromosome dimensions and DNA content. If on current evidence we exclude polynemy as playing any substantial part in this we must find the explanation in some method of addition of chromatin to the chromosome axis. Perhaps the easiest mechanism for increase of size is centric fusion which can under the right circumstances eventually affect all the chromosomes of a complement and within any mechanical limits imposed by the cell can continue for several or, in the long term, many cycles. Chromosome doubling can maintain chromosome number and provide the redundancy of genetic material which may be necessary for the emergence of new gene loci[12] but also for the further repatterning of the chromosomes. Sparrow and Naumann have recently shown that the DNA per genome in a very wide range of organisms shows not a continuous range of variation but fits a line of progressive doubling[13]. It is not known to what extent fusion has played a part in this but it is a possibility which needs to be taken into account. Centric fusion and polyploidy can interact in a highly significant manner which may be advantageous for evolutionary progress. But for the observer it is this interaction which may conceal the occurrence of both types of event. If we can disentangle some of the complexities of chromosome evolution we may find surprising situations and answers to paradoxes. Could it be for example that the excessive levels of polyploidy characteristic of the Pteridophytes have been tolerated equally by Angiosperms but modified and made harder to detect by fusion? It is this sort of speculation that we can engage in when we feel free of the constraining influences of dogma.

REFERENCES

1. Varaama, A. (1954) Cytological observations on Pleurozium schreberi with special reference to centromere evolution. Ann. Bot. Soc. 'Vanamo' 28: 1-59.

2. Handlos, W. (1970) Cytological investigations of some Commelinaceae from Mexico. Baileya 17: 6-33.

3. Celarier, R.P. (1955) Cytology of the Tradescantiae. Bull. Torrey Club. 82: 30-38.

4. Mattsson, O. (1963) Telocentric chromosomes in Tradescantia commelinoides. Bot. Tidsskr. 59: 195-208.

5. Darlington, C.D. and Janaki-Ammal, E.K. (1945) Adaptive isochromosomes in Nicandra. Ann. Bot. N.S. 9: 267-282.

6. Janaki-Ammal, E.K. (1958) Iso-chromosomes and the origin of triploidy in hybrids between old and new world species of Philadelphus. Proc. Ind. Acad. Sci. 48: 251-258.

7. Sadasivaiah, R.S. (1974) Haploids in genetic and cytological research in K.J. Kasha ed. Haploids in Higher Plants: 355-386. University of Guelph.

8. Marchant, C. (1968) Chromosome patterns and nuclear phenomena in the Cycad families Stangeriaceae and Zamiaceae. Chromosoma (Berl.) 24: 100-134.

9. Hair, J.B. and Beuzenberg, E.J. (1958) Chromosomal evolution in the Podocarpaceae. Nature 181: 1584-1586.

10. Bose, S. and Flory, W.S. (1963) A study of phylogeny and of karyotype evolution in Lycoris. Nucleus 6: 141-156.

11. Jones, K. (1974) Chromosome evolution by Robertsonian change in Gibasis (Commelinaceae). Chromosoma (Berl.) 45: 353-368.

12. Ohno, S. (1970) Evolution by Gene Duplication, pp. 160. Springer-Verlag.

13. Sparrow, A.H. and Nauman, A.F. (1976) Evolution of chromosome size by DNA doublings. Science 192: 524-529.

Chromosomes Today Volume 6, A. de la Chapelle and M. Sorsa eds.

EVOLUTIONARY DNA VARIATION IN LATHYRUS

H. REES and R. K. J. NARAYAN
Department of Agricultural Botany, University College of Wales, Aberystwyth.

ABSTRACT

Divergence and evolution of Lathyrus species are associated with substantial variation in nuclear DNA amount. The "supplementary" DNA contributing to the variation is of uniform composition and organisation. Each additional picogram of DNA within euchromatin is accompanied by 1.2 picograms in heterochromatin. Each additional picogram of non-repetitive DNA is accompanied by an increment of 4 picograms of moderately repetitive DNA. There is evidence for a comparable uniformity in the composition of supplementary DNA in other groups of both plants and animals. The quantitative variation in nuclear DNA was also accompanied by qualitative change within base sequences. The rate of qualitative change was the same for both non-repetitive and moderately repetitive fractions.

1. INTRODUCTION

Among species within many genera of Higher Plants there is a massive variation in nuclear DNA amount. The variation is often independent of numerical chromosome change and is the result of amplification, or deletion, of DNA base sequences within the chromosomes. Investigations in the genus Lathyrus provide information about changes in the composition and organisation of chromosomes which accompany the quantitative variation in nuclear DNA content.

2. CHROMOSOME SIZE

Lathyrus species are all diploids with 2n = 14 (fig. 1). There is, however, a two to three fold variation in chromosome size among species which, in turn, is closely and directly correlated with nuclear DNA content. Comparisons of chromosome sizes within and among species indicate that the quantitative DNA changes affected all chromosomes within the Lathyrus complement[1].

3. HETEROCHROMATIN AND EUCHROMATIN

Table 1 and Fig. 2a shows that with increasing DNA amount there is a disproportionate increase in heterochromatic as distinct from euchromatic chromosome segments. Spot microdensitometry[2] shows that the DNA density of heterochromatin at interphase is 1.6

times greater than that of euchromatin. Knowing the areas of interphase occupied by heterochromatin and euchromatin we can estimate how much of the total DNA in each

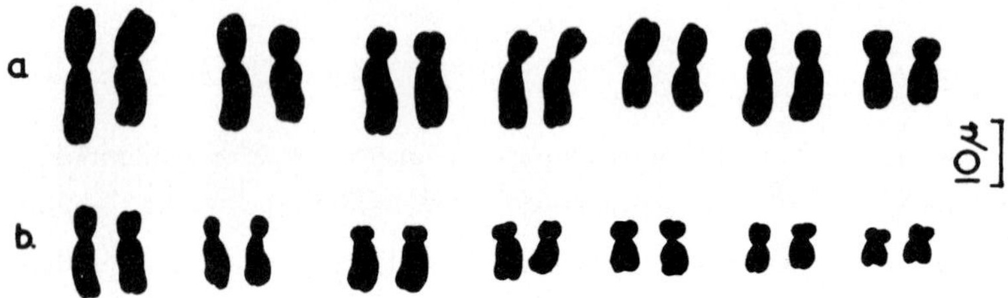

Fig. 1. The chromosomes of L. hirsutus and L. angulatus at metaphase in root meristems.

species is located in each fraction (table 1). We have plotted these estimates against the total DNA amounts in fig. 2b. The linear regressions are significant at the one

TABLE 1. Nuclear components of Lathyrus species. DNA amounts, other than percentages, in g x 10^{-12}. The fast fraction is that which reassociates before Cot x 10^{-2}; the moderate fraction, between Cot x 10^{-2} and Cot 10^{3}; the non-repetitive fraction, beyond Cot 10^{3}. The total DNA amounts for L. angulatus and L. tingitanus are revised values (cf. Narayan and Rees, 1976).

	Total DNA	% heterochr. at interphase	DNA in heterochr.	DNA in euchr.	Non-repetitive DNA	Mod-erate fraction	Fast fraction	% repetitive DNA
L. angulatus	10.90	14.68	1.60	9.30	4.36	4.80	1.74	60
L. articulatus	12.45	15.30	1.90	10.55	5.48	6.35	0.62	56
L. nissolia	13.20	21.04	2.77	10.43	5.41	6.73	1.06	59
L. clymenum	13.75	21.23	2.91	10.84	5.23	7.01	1.51	62
L. ochrus	13.95	25.0	3.48	10.47	5.58	7.39	0.98	60
L. aphaca	13.97	-	-	-	5.17	6.74	1.81	63
L. cicera	14.18	19.45	3.36	10.82	5.96	7.51	0.71	58
L. sativus	17.15	26.0	4.50	12.65	5.83	9.95	1.37	66
L. odoratus	17.16	-	-	-	5.67	9.78	1.72	67
L. hirsutus	20.27	37.0	7.49	12.78	6.17	12.56	1.62	70
L. tingitanus	22.18	33.3	7.39	14.81	7.90	12.54	1.78	64
L. sylvestris	24.26	35.86	8.7	15.56	7.27	15.52	1.45	70

percent level. There is, also, a significant heterogeneity in slopes ($P = < 0.01$). While increase in the total DNA is achieved by increase in both heterochromatin and euchromatin

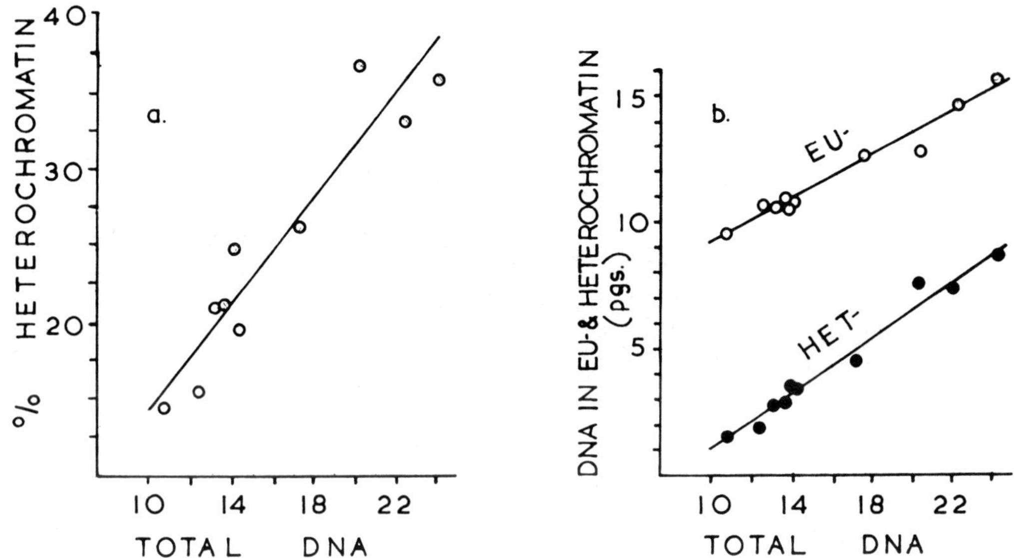

Fig. 2a. The amount of heterochromatin, as percentage of the area of root meristem nuclei at interphase, plotted against total nuclear DNA amount ($\times 10^{-12}$ g).
b. DNA in heterochromatin and in euchromatin plotted against total nuclear DNA.

the rate of increase in heterochromatin is significantly greater. For each additional picogram of "euchromatic DNA" there is an increment of 1.2 picograms of DNA located in heterochromatin.

From the fact that the proportion of heterochromatin varies among species with different DNA amounts we conclude that the composition of the supplementary DNA is not representative of the total. From the linear nature of DNA increase in both euchromatin and heterochromatin we conclude that the supplementary DNA in itself, however, is of a markedly constant composition with regard to the euchromatic and heterochromatic constituents. It is clear that a particular fraction of the chromosomal DNA is implicated by amplification (or deletion). We may account for the particularity on a structural basis, in the sense that heterochromatic segments are more vulnerable to amplification or deletion than euchromatic segments. Alternatively, and more feasibly, the explanation may be related to function, in the sense that the only DNA variation tolerable, from the standpoint of fitness, is that involving particular segments of specific composition.

That the DNA variation is to some degree restricted to particular segments is supported by the results of C banding by Giamsa staining. Fig. 3 shows that increase in heterochromatin is mainly located in the vicinity of the centromeres.

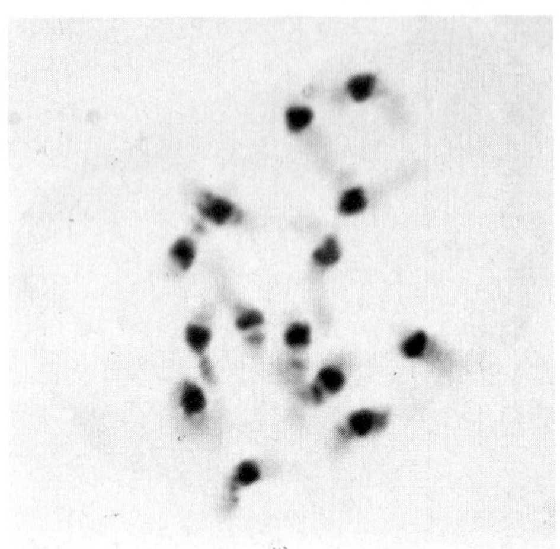

Fig. 3. The concentration of heterochromatic C-bands around the centromeres of L. tingitanus. Photograph kindly supplied by Dr. S. C. Verma.

The situation in Dermestes beetles[3] is much the same as in Lathyrus in that the supplementary DNA is made up of consistent fractions of "heterochromatic DNA" and "euchromatic DNA". We estimate the ratio to be 3.2 to 1. A proportional increase of DNA sequences in heterochromatin and euchromatin has been reported also in the chromosomes of Salamanders[4].

3. REPETITIVE VERSUS NON-REPETITIVE DNA

From Cot re-association (table 1) we find,

(i) a disproportionate increase in repetitive as compared with non-repetitive DNA with increasing nuclear DNA amount (fig. 4a). The changing composition accompanying the change in quantity re-inforces the evidence, based on the distribution of euchromatin and heterochromatin, that amplification (or deletion) involves a particular nuclear DNA fraction.

(ii) that increase in total DNA is accounted for mainly by increase in the moderately repetitive fraction (fig. 4b). There is, nevertheless, a significant, but lower, rate of increase in the non-repetitive component ($P = < 0.01$). Previous results, based on a smaller sample of species, did not reveal this change in non-repetitive DNA[2]. There is no significant variation in the "fast" fraction.

We draw particular attention to the striking linearity of the regressions for both the moderate and non-repetitive components. It means that the supplementary DNA with respect to these components as for heterochromatin and euchromatin, is of constant composition. For every increment of 1 picogram of non-repetitive DNA we have

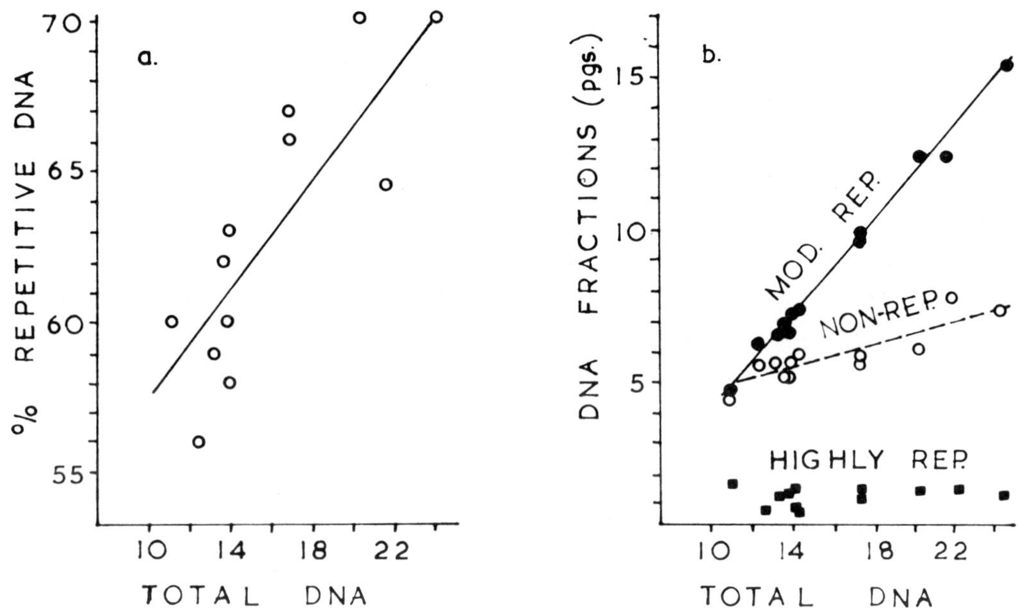

Fig. 4a. The percentage repetitive DNA plotted against the total nuclear DNA amount.
b. The amount of non-repetitive, of moderately and highly repetitive DNA (the "fast" fraction) plotted against total DNA. (Data from Narayan and Rees, 1976).

approximately 4 picograms of moderately repetitive DNA.

4. QUALITATIVE CHANGE

We have seen that the divergence and evolution of Lathyrus species were accompanied by substantial quantitative change in moderately repetitive DNA, relatively little change in the amount of non-repetitive DNA. This, it could be argued, might be expected on the grounds that the non-repetitive fraction incorporates the structural genes, comprised therefore of base sequences which are relatively intolerant of alteration. Relative constancy in amount, however, is no guarantee of constancy in composition. In order to compare the magnitude of qualitative change between, on the one hand, the non-repetitive and, on the other, the moderately repetitive fractions among species we have adopted the method of DNA/DNA hybridisation first developed by Britten, Graham and Neufeld[5]. 0.15 μgs of non-repetitive DNA and of moderately repetitive DNA from L. hirsutus (which has a high total nuclear DNA content of 20 picograms) were labelled, in turn, with ^{125}I[6], mixed with a large quantity, 1.8 μgs, of sheared, unlabelled, total DNA from L. hirsutus and from six other Lathyrus species. The mixtures, in 0.12 M phosphate buffer, were rendered single stranded by heating to 100°C and allowed to re-associate at 62°C, to Cot 2×10^4 for the non-repetitive, to Cot^{500} for the repetitive DNA. The re-associated DNA was trapped in a hydroxyapatite column and the amount of labelled L. hirsutus DNA annealed with the unlabelled DNA determined by measuring the radioactivity in the duplexes trapped in the column. The reassociation percentages for hirsutus/hirsutus (homologous) mixtures were taken as indices of complete homology (1.0). The reassociation percentages for the hirsutus/other species (heterologous) mixtures were transformed to indices of homology using the homologous reassociation percentages as standards (see Narayan and Rees[7] for details). In Fig. 5b the

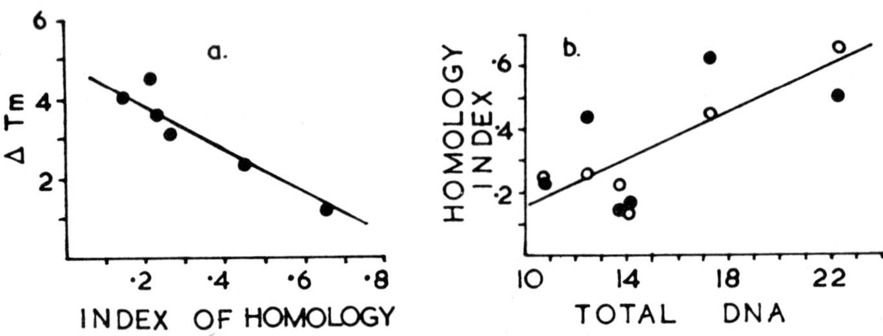

Fig. 5a. The Δ Tm of heterologous duplexes (from moderately repetitive DNA) plotted against the index of homology based on the degree of reassociation of DNA/DNA "hybrids". b. The index of homology for both non-repetitive (open circles) and repetitive DNA (closed circles) plotted against the total nuclear DNA amount. Data from Narayan and Rees (in press; Chromosoma, 1977).

indices of homology, for both repetitive and non-repetitive heterologous mixtures are plotted against the total nuclear DNA amounts of species whose DNA was hybridised with that of L. hirsutus. It will be seen that the lower the DNA amount in species represented in heterologous mixtures the lower is the index of homology with L. hirsutus DNA.

Divergence in DNA composition goes hand in hand with divergence in quantity. It will also be seen that the rate of change in homology is the same for both repetitive and non-repetitive components. This is confirmed by a regression analysis of variance which shows a significant joint linear regression (P = < 0.01) but no significant heterogeneity either in slopes or means. We conclude from this that base sequences which make up the moderately repetitive DNA are no less conserved than those incorporated in the non-repetitive DNA (cf. Mizuno and Macgregor[8]).

Another way of measuring the divergence in DNA composition among species is by comparing the thermal stability of homologous (hirsutus/hirsutus) duplexes with that of heterologous (hirsutus/other species) duplexes. This is achieved by establishing the Δ Tm, the drop in the temperature required to render half the DNA of heterologous duplexes single stranded, relative to the Tm, at which half the DNA of homologous duplexes is "melted". The Δ Tm is directly related to the degree of divergence in base sequencies between the single strand components of the duplexes. Fig. 5a shows a close and significant correlation (P = < 0.01) between the Δ Tm's and the indices of homology derived from reassociation. The correlation confirms and reinforces the conclusions based on the reassociation experiments[7].

In retrospect the results are, perhaps, not altogether surprising. If, as is generally assumed, much of the "extra" repetitive and, for that matter non-repetitive, DNA serves a regulatory function we must suppose that the regulation, in part at least, is achieved through the activity of base sequences of specific composition. It would follow that conservation of these sequences is no less essential than of sequences which make up the structural genes.

5. DISCUSSION

It is generally assumed that much of the quantitative DNA variation among closely related species, as in Lathyrus concerns base sequences which serve a regulatory role in the organisation, transmission and expression of chromosome material. There are good grounds for the assumption. For example, supplementary DNA, in the form of supernumerary heterochromatic segments in grasshoppers, affects chiasma frequencies in spermatocytes[9]. Supplementary DNA, comprised of both euchromatin and heterochromatin, in the form of B chromosomes influences a whole range of characters in many species[10]. In Lathyrus itself the quantitative DNA variation is known to affect the duration of mitotic cycles[11].

In Lathyrus we have seen that each "instalment" of supplementary DNA is a complex of uniform composition with respect to the relative amounts of euchromatin and heterochromatin, of non-repetitive and repetitive DNA sequences. A comparably extensive survey by Bachmann and Price[12] indicates that the same uniformity in composition applies to supplementary DNA in species of Cichorieae (Compositae). Fig. 6 shows a linear increase in both non-repetitive and repetitive DNA with increasing DNA amount. The pattern

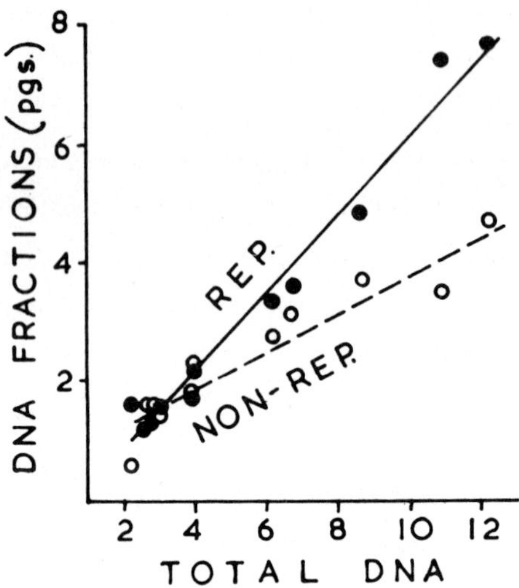

Fig. 6. The amounts of non-repetitive and repetitive DNA plotted against the total nuclear DNA amount in species of Cichorieae (Compositae). Data from Bachmann and Price, 1977.

differs from that in Lathyrus only in the relative proportions of non-repetitive and repetitive fractions in each DNA increment. For each extra picogram of non-repetitive we have 2 picograms of repetitive DNA. It would appear therefore that the nature of the DNA increments varies from group to group but that within each taxonomic group the ratio of the one fraction to the other is characteristic. If we are justified in attributing a regulatory function to this supplementary DNA, its composition in different groups of plants and animals may reflect upon the composition of the regulatory elements, in respect both of the degree of sequence repetition and of heterochromatic and euchromatic ingredients.

REFERENCES

1. Rees, H. and Hazarika, M. H. (1969) Chromosome evolution in Lathyrus. "Chromosomes Today, Volume Two". Oliver and Boyd, Edinburgh.

2. Narayan, R. K. J. and Rees, H. (1976) Nuclear DNA variation in Lathyrus. Chromosoma, 54, 141-154.

3. Rees, R. W., Fox, D. P., and Maher, E. P. (1977) DNA content, reiteration and satellites in Dermestes. Current Chromosome Research. Elsevier/North-Holland, Amsterdam.

4. Macgregor, H. C., Horner, H., Owen, C. A. and Parker, I. (1973) Observations on centromeric heterochromatin and satellite DNA in Salamanders of the genus Plethodon. Chromosoma, 43, 329-348.

5. Britten, R. J., Graham, D. E. and Neufeld, R. (1974) Analysis of repeating DNA sequences by reassociation. Methods in enzymology, Vol. 29E, 363-418. Academic Press, New York.

6. Commerford, S. L. (1971) Iodination of nucleic acids in vitro. Biochemistry, 10, 1993-1999.

7. Narayan, R. K. J. and Rees, H. (1977) Nuclear DNA Divergence among Lathyrus species. Chromosoma, in press.

8. Mizuno, S. and Macgregor, H. C. (1974) Chromosomes, DNA sequences and evolution in Salamanders of the genus Plethodon, Chromosoma, 48, 239-296.

9. John, B. (1973) The cytogenetic systems of grasshoppers and locusts: II. The origin and evolution of supernumerary segments. Chromosoma, 44, 133-146.

10. Rees, H. (1974) B chromosomes. Sci. Prog. Oxf. 61, 535-554.

11. Evans, G. M., Rees, H., Snell, C. L. and Sun, S. (1972) The relationship between nuclear DNA amount and the duration of the mitotic cycle. Chromosomes Today, 3, 24-31. Longman, London.

12. Bachmann, K. and Price, H. J. (1977) Repetitive DNA in Cichorieae (Compositae). Chromosoma, 61, 267-275.

Chromosomes Today Volume 6, A. de la Chapelle and M. Sorsa eds.

REPETITIVE DNA AND HETEROCHROMATIN AS FACTORS OF KARYOTYPE EVOLUTION IN PHYLOGENY AND ONTOGENY OF ORCHIDS

WALTER NAGL and INGRID CAPESIUS
Department of Biology, The University, D-6750 Kaiserslautern,
and institute of botany, The University, D-6900 Heidelberg
(Federal Republic of Germany)

ABSTRACT

Speciation and the formation of higher taxonomic orders are characterized by changes in total nuclear DNA content, DNA base composition, percentage highly and intermediately reiterated DNA, and percentage heterochromatin. It is concluded that the amplification and diversication of regulatory DNA sequences are the molecular basis of speciation and cladogenesis. The same or similar mechanisms are suggested to be the molecular basis also of somatic differentiation, as the data from the genus Cymbidium indicate. A "DNA optimization model" is put forward which illustrates speciation and evolution as a decicion process in order to obtain an optimum of regulatory DNA sequences, and which interpretes DNA variation during somatic development as a predefined process depending on the evolutionary state of nuclear DNA in a given species.

INTRODUCTION

There are two events in living matter which are only poorly understood: speciation in phylogeny, and cell differentiation in ontogeny. Both phenomena are evidently closely related and have been seen as the two sides of the same coin[1]. Recently evidence has been obtained for the view that cladogenesis (e.g. speciation) is related to some change in the gene regulation system rather than to mutations of protein-coding genes[2,3]. One of us has suggested that the basis of somatic cell differentiation may be similar to that of speciation, and that polyploidy and differential DNA replication may be the primary cause of differentiation and morphogenesis[4,5].

In this paper we report some data on phylogenetic and ontogenetic DNA variation in orchids. Although the data presented are rather a small condensation of a much larger body of results, they confirm the "DNA optimization model" of Nagl[5]. Because of limited space the data are mainly represented in the form of figures and tables. A more detailed manuscript is in the press[6].

MATERIAL AND METHODS

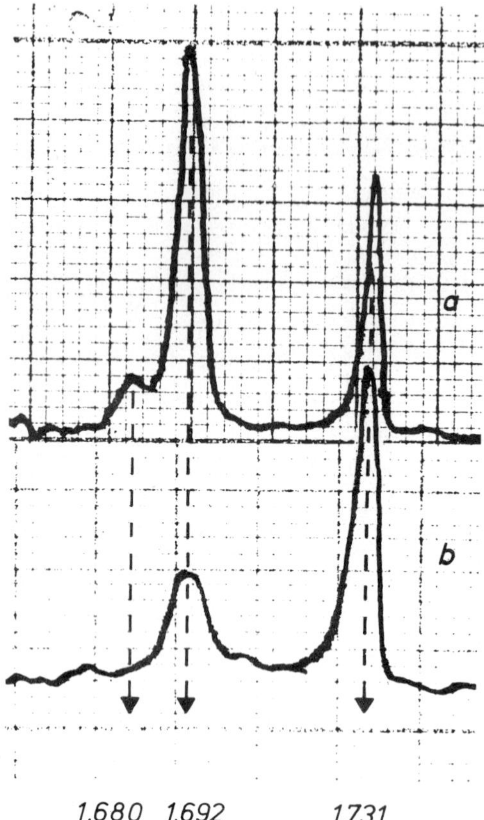

Fig. 1. Photoelectric scans of DNA extracted from isolated nuclei (a) and chloroplasts (b) and centrifuged to equilibrium in CsCl. Species: Cymbidium pumilum; Marker: M. lysodeikticus DNA; Centrifugation: 20 h at 44 000 rpm at 20°C

Most of the methods employed have been published in previous papers[8-12]. In situ hybridization of tritiated main band DNA, satellite DNA, complementary RNA and poly-uridylic acid to cytological preparations have been made by the techniques given by Jones[13].

Here we like to emphasize that the satellite DNA of Cymbidium, which is unique among monocots, is really a nuclear satellite. Fig. 1 shows the scans of DNA centrifuged to equilibrium in CsCl after extraction from isolated nuclei and chloroplasts. We also wish to indicate that the "thermal satellite" of Cymbidium DNA is comparable with the satellite revealed by CsCl gradient ultracentrifugation. Fig. 3 shows the derivative melting profile of total nuclear DNA, Fig. 2 the profiles of separated main band and satellite DNAs. Similar results have been recently obtained in different mammals.

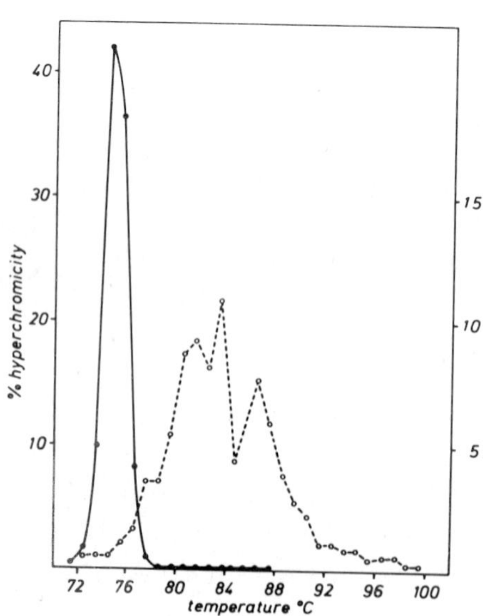

Fig. 2. Derivative melting profiles (% hyperchromicity per °C) of Cymbidium main band (broken line) and satellite DNA (through line). Fig. 3 shows a melting profile of total nuclear DNA.

RESULTS

Phylogenetic DNA variation

DNA content, base composition, and melting profiles. Table 1 summarizes the characteristics of the nuclear DNA of the four orchid species studied. All the species belong to the subfamily Orchidoideae; Cattleya is classified in the tribe Epidendreae (subtribe Epidendrinae), the other three species are Vandeae (subtribes Oncidiinae: Brassia, Cymbidiinae: Cymbidium, and Sarcanthinae: Phalenopsis), according to Engler's Syllabus.

TABLE 1
CHARACTERISTICS OF ORCHID DNA
The data given were obtained with DNA extracted from well-growing liquid cultures of protocorms. Because of chromosome instability and ontogenetic DNA variation different values may be obtained under different culture conditions.

Parameters	Brassia[a]	Cattleya[b]	Cymbidium[c]	Phalenopsis[d]
2C DNA content (pg)	7.05	2.51	8.70	2.34
Total repet.DNA (%)	±45.00	±45.00	±45.00	±45.00
Highly repet.DNA (%)	20.00	18.00	12.00	13.00
Buoyant density (g/cm^3)	1.692	1.694	1.692	1.694
Satellite DNA (g/cm^3)	1.705	-	1.680	1.705
Tm of total DNA (^oC)	83.0	86.1	84.0	85.2
Derivative Tm peaks (^oC)	80.1	84.0	75.0	-
	82.4	88.5	81.5	-
	86.0	-	83.3	-
	-	-	87.0	-
Average GC content (%)	33.8	40.7	26.8	35.9
Nuclear structure[e]	d/c	d/c	c	d/c
Heterochromatin (%)	±3.0	±4.0	±7.0	±6.0

[a] Brassia maculata.
[b] Cattleya schombocattleya.
[c] Cymbidium pumilum cv. "Gareth Latangor"; the DNA of C. ceres is very similar[10]; the DNA of the hybrid "In memoriam Cyrill Strauss" differs in its 2C value (10.7 pg) and high percentage repetitive DNA.
[d] Phalenopsis amabilis.
[e] d = diffuse euchromatin, c = chromocenter heterochromatin.

The data indicate that splitting of the genera studied must be accompanied, or even caused, by changes in the total nuclear DNA contents, and by changes in the average base composition, e.g. by gain or loss of DNA fractions which build individual peaks in derivative thermal melting profiles (Figs. 3 and 4).

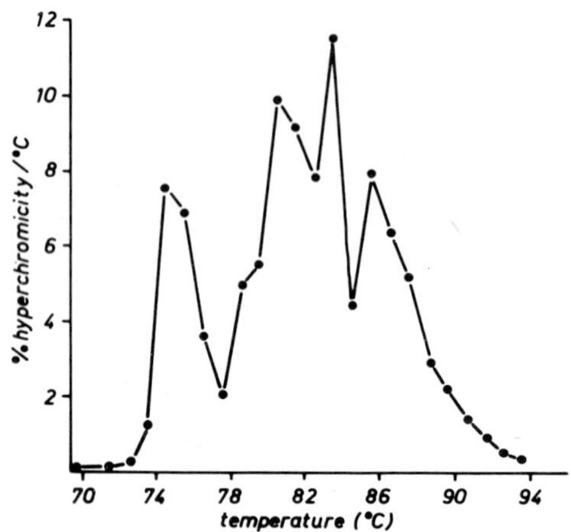

Fig. 3. Derivative melting profile of Cymbidium DNA. Notice the prominent AT-rich fraction at left.

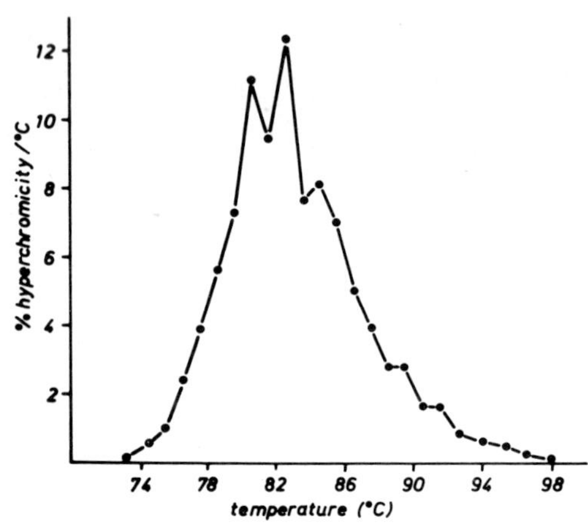

Fig. 4. Derivative melting profile of Brassia DNA. Notice the similarity to Cymbidium DNA except the AT-rich flank.

Repetitive DNA and satellite DNA. As Fig. 5 shows the four species differ somewhat with respect to the amount of highly repetitive DNA, but apparently display nearly the same percentage total repetitive DNA. This indicates that the variation in nuclear DNA amount and in average GC content is brought about by gain or loss of all types of sequences, which are probably interspersed with each other to a high content. This assumption is confirmed by the fact that only Cymbidium DNA shows a small satellite upon ultracentrifugation in neutral CsCl (the satellite DNA being composed of least two fractions[10] and AT-rich[9]), while Brassia DNA (Fig. 6) and Phalenopsis DNA show a GC-rich heavy shoulder fraction which can be demonstrated by repeated centrifugation of the GC-rich fractions. Cattleya, which is classified within a different tribe, displays a homogeneous peak in DNA/CsCl gradients. In situ hybridization of Cymbidium satellite DNA and poly(uridylic) acid to nuclear preparations isolated from protocorms indicates location of the DNA satellite preferably in the region of the heterochromatic chromocenters[6] (Figs. 7 and 8).

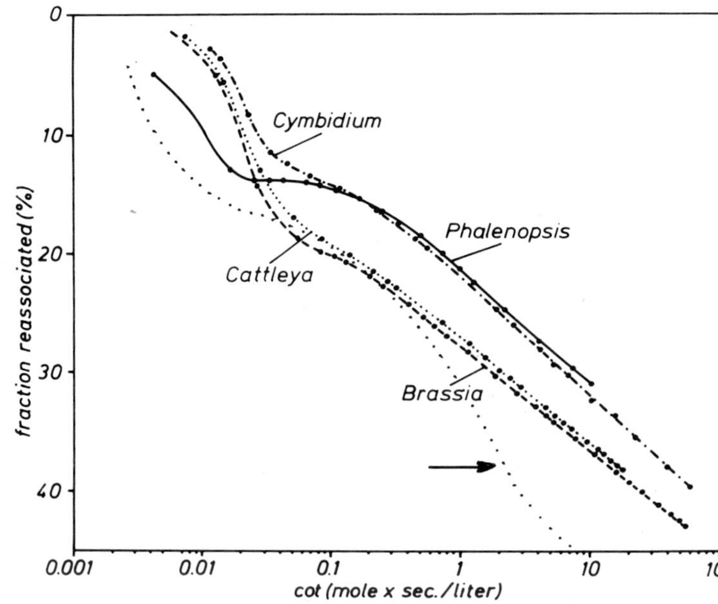

Fig. 5. Reassociation kinetics of orchid DNAs up to a cot of 50 and 100, respectively, including most of the repetitive DNA. The dotted curve (arrow) is a preliminary result and shows the cot-curve of DNA extracted from differentiating protocorm cells of Cymbidium "In memoriam Cyrill Strauss". The other curves are the result of at least three repeats with DNA from different extractions.

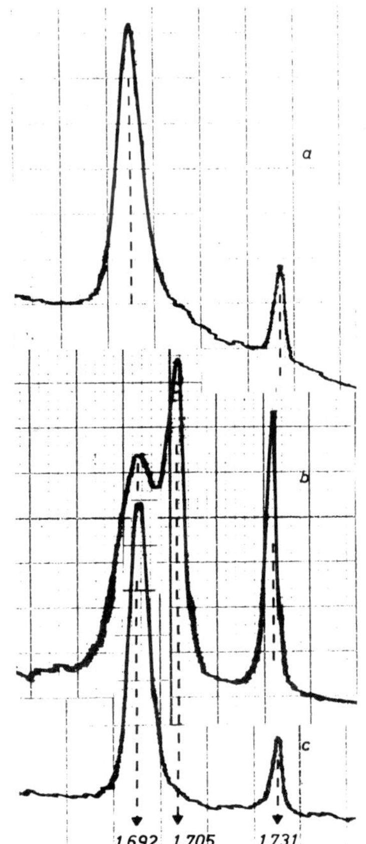

Fig. 6. Photoelectric scans of Brassia DNA after ultracentrifugation with a Beckman Model E analytical ultracentrifuge. (a) Total nuclear DNA, (b) purified GC-rich fractions, (c) pure main-band DNA; the marker DNA is from Micrococcus lysodeicticus.

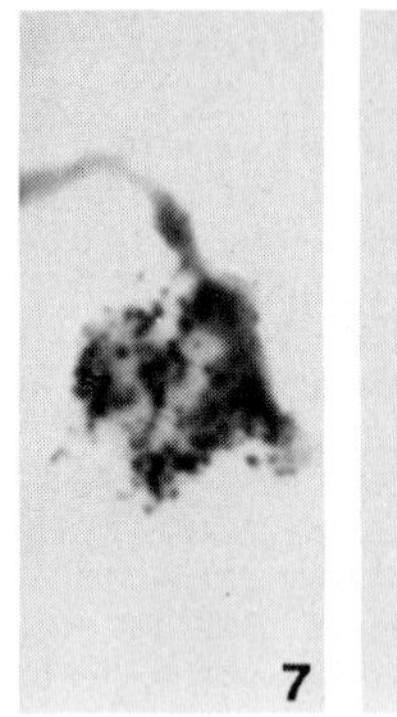

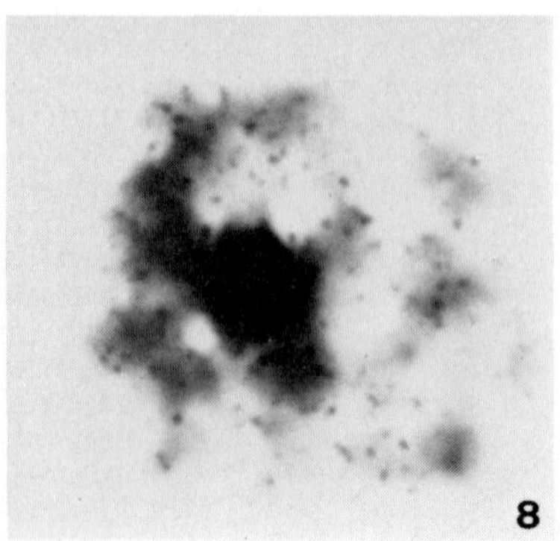

Figs. 7.and 8.
Isolated Cymbidium nuclei; autoradiograms after in situ hybridization with tritiated satellite DNA (Fig. 7) and tritiated poly(uridylic) acid (Fig. 8). Notice that the satellite DNA is located in the chromocenters of the diploid nucleus, while most of the AT-rich DNA, which hybridized with poly(U) of low specific activity (50 Ci/µM), is located more at the surface, or in the vicinity, of the chromocenters.
For interpretation see below and Capesius and Nagl[6].

Nuclear structure and heterochromatin content. Tanaka[14] made the first attempt to classify the nuclei of orchids according to their structure. He found that progressive taxonomic groups which show large variation in morphology and rapid speciation (e.g. Orchis, Dendrobium, Oncidium, Vanda and relatives) show the chromocenter type, while most of the taxonomic groups which show small variation and which seem to be stable and slowly speciating, exhibit the diffuse (chromonematic) type or prochromosome nuclei (for detailed discussion show Capesius and Nagl[6]). According to Tanaka's[14] terminology, three of our species display the simple chromocenter type, while Cymbidium shows the complex chromocenter type. Our electron microscopic studies confirm this (Figs. 9 and 10): Cymbidium nuclei possess a variable number of differently sized and structured chromocenters[12], while the nuclei of Brassia, Phalenopsis, and Cattleya show only a few small chromocenters. The euchromatin is, on the contrary, somewhat denser in the latter genera than in Cymbidium.

The relationships between 2C value, proportion repetitive DNA and nuclear structure does not follow the common pattern, which is the following: the higher the C value, the more repetitive DNA is found within the genome, and the more likely is a chromonematic (or reticulate), chromocenter-less structure[15,16]. It may be speculated that Cattleya and Phalenopsis progress towards low C values with little heterochromatin (no satellite DNA, at best heavy shoulder DNA), while Brassia progress towards high DNA values with little heterochromatin. The DNA of all three underwent high sequence divergence, so that no satellite DNA can be found. Cymbidium, on the other hand, must have recently endured DNA amplification, so that the uncommon AT-rich DNA satellite and much heterochromatin is visible.

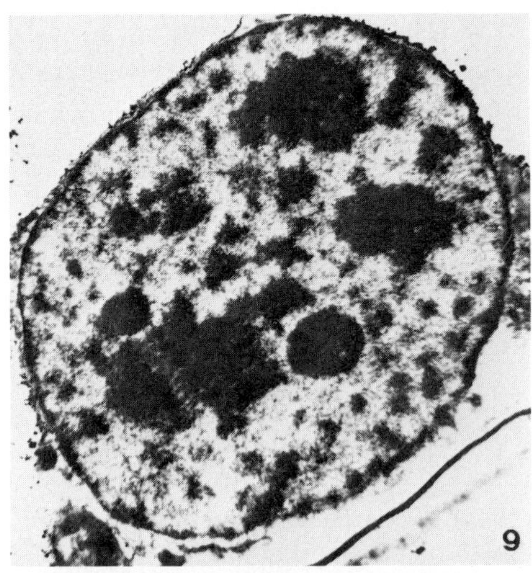

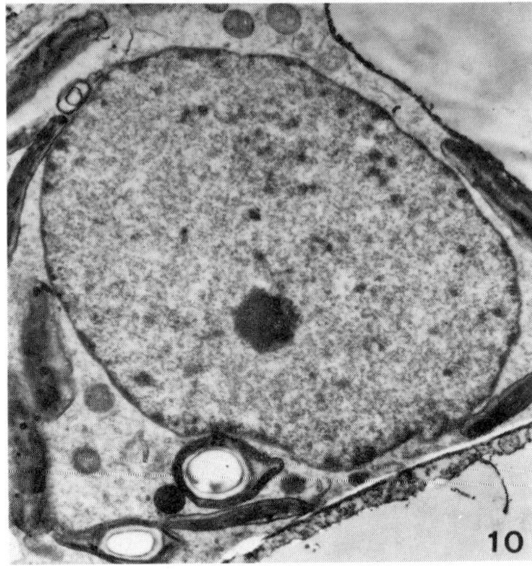

Figs. 9 and 10. Ultrastructure of orchid nuclei.
Fig. 9. *Cymbidium* nucleus belonging to the complex chromocenter type.
Fig. 10. *Brassia* nucleus belonging to the simple chromocenter type with diffuse euchromatin.
Uranyl acetate - lead citrate staining, 2 100 x.

Although we do not have analyzed the organization of the orchid genomes, the biophysical, biochemical and cytological data are consistent with the suggestion that splitting in orchids is characterized by an amplification-diversification cycle of the DNA sequences, as it has also been assumed to be the case in other groups of plants and animals[1,17,18]. *Cymbidium* may well be an exciting genus which we can observe during the DNA optimization process as supposed by Nagl[5].

Ontogenetic DNA variation

DNA content, base composition and melting profiles. Various species and cultivars of *Cymbidium* are best studied in this direction, but see also the data on *Vanda*[19]. It was shown in a series of investigations that the nuclear DNA content increases during cell differentiation by endopolyploidization in many plants and animals[5]. In addition, differential DNA replication (extra and under-replication) are apparently more frequent than thought so far. In *Cymbidium* protocorms and roots, good evidence hass been obtained for a DNA amplification - DNA release cycle in certain differentiating cells by scanning cytophotometry, structural analysis, autoradiography, melting profiles and renaturation kinetics[7,8,12,15]. Figs. 11 - 13 show nuclei passing through the amplification - release cycle. The biochemical demonstration of the latter is very difficult, as all types of cells with nuclei differing in the cell cycle undergoing, and in the stage of the cell cycle being, occur together in a protocorm, which permanently changes its developmental state. As we do not yet have found the way of separation of the different nuclei by centrifugation, approximate data were obtained by extracting the DNA from aseptic protocorm cultures, which were treated with different phytohormones. This has been found to lead to preferential stimulation and inhibition, respectively, of either the mitotic, endomitotic, or amplification cycle[7,15]. Fig. 14 shows the highly reproducible derivative melting profiles of DNAs from gibberellic acid- and 2,4-dichlorophenoxy acetic acid-treated cultures. The variable amount of differently AT-rich fractions is well reflected by the differential fluorescence of the chromocenters[12] (Fig. 13).

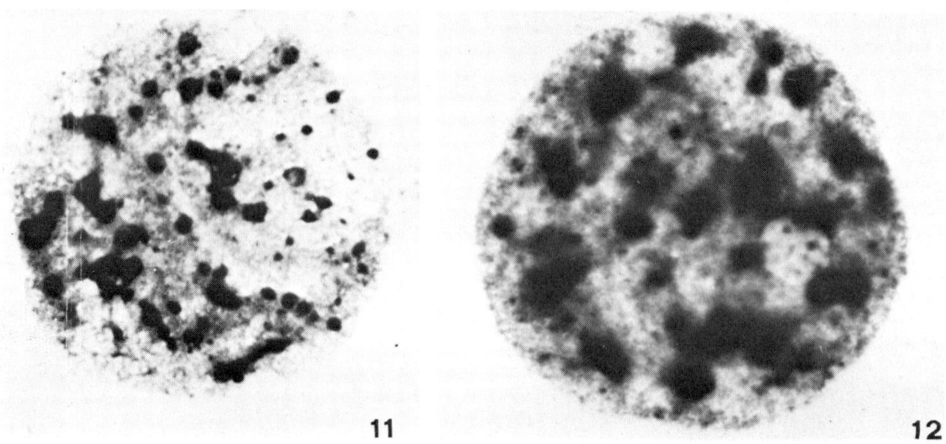

Figs. 11 and 12. Giemsa-C-stained endopolyploid nuclei from a Cymbidium protocorm. About 1 300 x. Courtesy of D. Schweizer, Vienna. Fig. 11. Nucleus with proportional amount of eu- and heterochromatin. Fig. 12. Nucleus with disproportionately much heterochromatin due to DNA amplification.

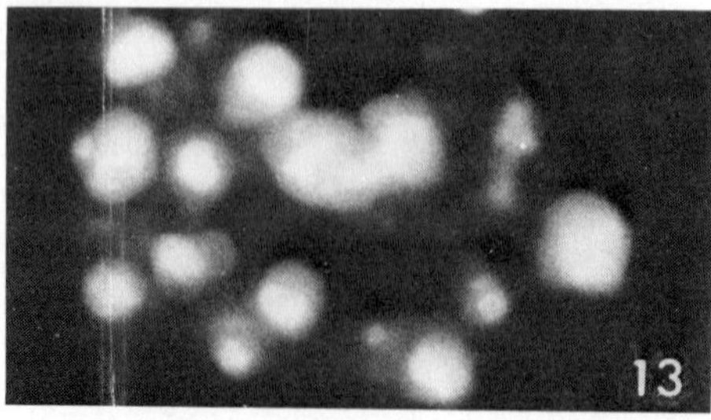

Fig. 13. Quinacrine-stained endopolyploid nucleus from a Cymbidium protocorm. Fluorescence microphoto, about 1 000 x. Courtesy of D. Schweizer, Vienna. Notice the differential fluorescence of the chromocenters indicating a composition of various DNA fractions differing in AT content.

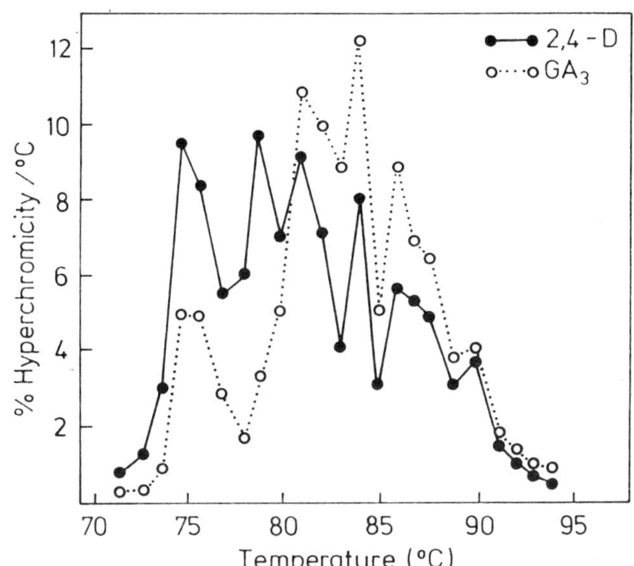

Fig. 14. Derivative melting profiles of Cymbidium DNA extracted from protocorm cultures treated with either gibberellic acid (GA3) or 2,4-dichlorophenoxyacetic acid (2,4-D). As the profiles are reproducible and the Tm values of the peaks significantly different from each other, this diagram gives evidence for differential DNA replication during somatic development. Reproduced, with permission, from Nucleic Acids Res. 3, 2033-2039 (1976)[11].

Repetitive DNA and satellite DNA. As indicated in Fig. 5, there seems to be ontogenetic variation in the percentage of various cot-families. The variable amount of satellite DNA obtained in a series of experiments[9] also indicates that the DNA is differentially replicated during ontogenesis.

Nuclear structure and heterochromatin content. Figs. 11 and 12 clearly show that the amount of Giemsa-positive heterochromatin in Cymbidium nuclei can vary considerably. Separate Feulgen-DNA measurements in eu- and heterochromatin have indicated that the heterochromatin might be replicated five cycles more often than the euchromatin[8]. Unpublished electron microscopic, as well as pulse-chase autoradiography[15] led to the assumption that the amplification cycle is only found during differentiation. Thereafter the extra-DNA is released from the nuclei and degraded.

CONCLUSIONS

Space does not allow to discuss the results here; the reader is referred to a forthcoming full paper[6]. Our conclusions in brief are the following. (1) The evolution of complex organisms was possible due to the evolution of gene controlling DNA[2]. (2) Endopolyploidization and differential DNA replication are understood as evolutionary strategies bringing somatic cells to a differentiated state. This somatic DNA variation may either substitute for inadequate phylogenetic one, or allow independent phylogenetic and ontogenetic changes, e.g. because of different nucleotypic requirements during p oliferation and differentiation[4]. (3) A "DNA optimization model" was therefore put forward, which interpretes phylogenetic and ontogenetic karyotype and genome variation like a computer process and additional predefined processes in order to achieve the optimal species-specific and tissue-specific amount of template and control DNA[5]. (4) It is concluded that the mechanisms (e.g. polyploidization, amplification, DNA loss, karyotype repatterning) and the necessary decisions leading to DNA variation in the one or other direction are very similar in speciation and cell specialization (differentiation).

ACKNOWLEDGEMENT

The studies on orchid DNA and differential DNA replication were supported by grants from the Deutsche Forschungsgemeinschaft.

REFERENCES

1. Galau, G.A., Chamberlin, M.E., Hough, B.R., Britten, R.J. and Davidson, E.H. (1976) in Molecular Evolution, Ayala, J.F. ed., Sinauer, Sunderland, pp. 1-20.
2. Britten, R.J. and Davidson, E.H. (1969) Science 165, 349-357.
3. Ayala, F.J. ed., Molecular Evolution, Sinauer, Sunderland, pp. 1-
4. Nagl, W. (1976) Nature 261, 614-615.
5. Nagl, W. (in press) Endopolyploidy and Polyteny in Differentiation and Evolution, North-Holland, Amsterdam, appr. 350 p.
6. Capesius, I. and Nagl, W. (in preparation) Plant Syst.Evol.
7. Nagl, W. and Rücker, W. (1972) Z. Pflanzenphysiol. 67, 120-134.
8. Nagl, W. (1972) Cytobios 5, 145-154.
9. Capesius, I., Bierweiler, B., Bachmann, K., Rücker, W. and Nagl, W. (1975) Biochim.Biophys.Acta 395, 67-73.
10. Capesius, I. (1976) FEBS Lett. 68, 255-258.
11. Nagl, W. and Rücker, W. (1976) Nucleic Acids Res. 3, 2033-2039.
12. Schweizer, D. and Nagl, W. (1976) Exptl.Cell Res. 98, 411-423.
13. Jones, K.W. (1973) in New Techniques in Biophysics and Cell Biology Vol.I, Pain, R.H. and Smith, B.J. eds., Wiley, London, pp. 29-66.
14. Tanaka, R. (1971) Bot.Mag. (Tokyo) 84, 118-122.
15. Nagl, W. (1976) Zellkern und Zellzyklen, Ulmer, Stuttgart, pp.1-486.
16. Nagl, W. (1977) Protoplasma 91, 389-407.
17. Bendich, A.J. (1977) in Molecular Biology of the Mammalian Genetic Apparatus, Ts´o, P.O.P. ed., North-Holland, Amsterdam, pp. 63-77.
18. Flavell, R.P., Rimpau, J. and Smith D.B. (1977) Chromosoma, in press.
19. Alvarez, M.R. (1968) Amer.J.Bot. 55, 1036-1041.
20. Zuckerkandl, E. (1976) J.Mol.Evol. 9, 73-104.

APPENDIX

THE DNA OPTIMIZATION MODEL FOR SPECIATION AND CYTODIFFERENTIATION

WALTER NAGL
Department of Biology, The University, D-6750 Kaiserslautern,
Federal Republic of Germany

The general line of the experiments described in the preceding paper is hardly understood, if the working hypothesis is not known to some extent. Therefore, the DNA optimization model will briefly be described on hand of a few examples. It must be emphasized that this model is primarily designed to explain the variation in nuclear DNA content and DNA composition during speciation and cell differentiation in progressive species and genera, which are subjected to the necessity to find an optimal DNA content and genome organization in order to be adapted to their new situation with respect to regulatory sequences and nucleotype.

Fig. 1 shows the process of karyotype change which may evolve during speciation and (in one case) during anagenesis. The consequently predefined process following during somatic cell differentiation is shown for three cases. As selection pressure affects ontogenesis and the soma, but evolution takes place at the level of phylogenesis and germ line cells,one must not wonder that the same inputs and decisions occur during phylogenesis and ontogenesis. The terminal results may, however, be very different, as a highly specialized cell such as a gland cell is subjected to a quite different criterion of selection than a proliferating stem cell. For instance, an annual plant obtaines high selective advantage, if unnecessary DNA (what ever this is) is lost, because this allows rapid growth and short generation time,and thus the adaptation to unfavorable habitats. On the contrary, the cells which support and nourish the embryo (i.e. suspensor or endosperm cells) need high synthetic capacity, which is best guaranteed by a high amount of template and regulatory DNA. These cells, therefore, show somatic DNA increase by endopolyploidization, polytenization or DNA amplification. This suggestion is confirmed by the negative relationship which is found between 2C nuclear DNA content and the tendency to, and the highest level of, somatic polyploidy[1,2]. A particularly clear example for this relationship has been found

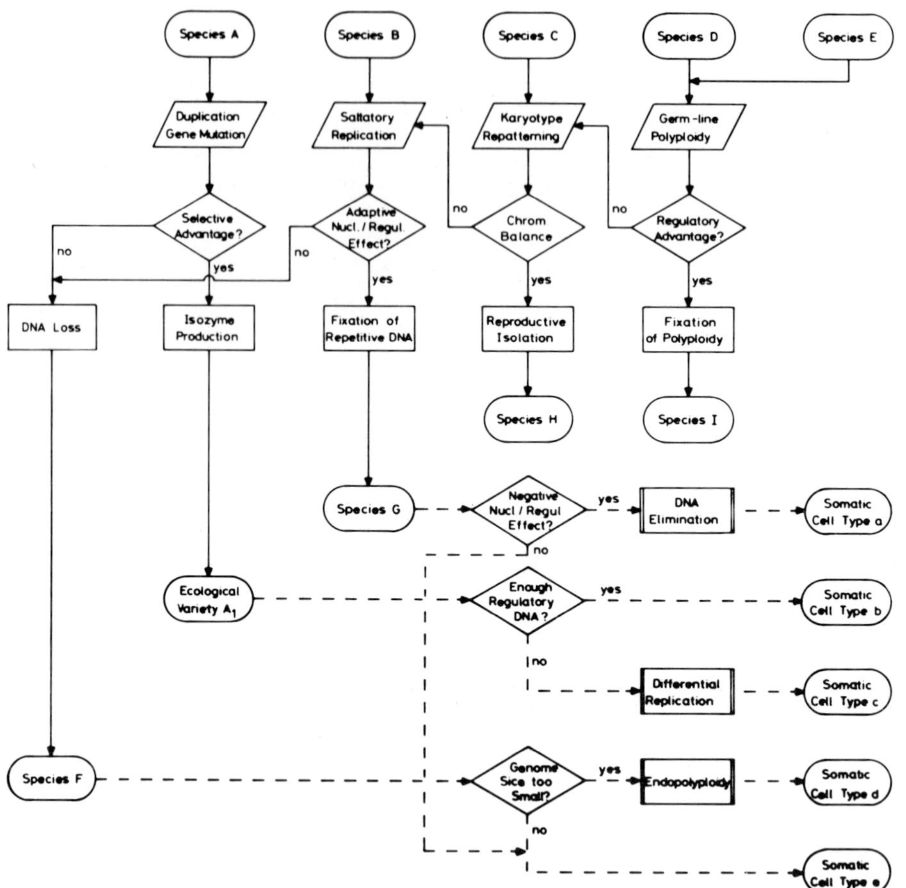

Fig. 1. Flow diagram to illustrate some possible changes in karyotype and genome organization during phylogenesis (particularly cladogenesis) and their possible consequences during certain steps of somatic cell differentiation. The symbols are borrowed from electronic data processing and have the following meanings: ⬭ terminal, interrupt; ▱ input, output; ◇ decision; ▭ process; ▭ predefined process; → flow line; ⤚ junction. - Reproduced from Nagl: Endopolyploidy and Polyteny in Differentiation and Evolution, North-Holland, Amsterdam[3].

by Bier[4] in ovarian development of insects. DNA amplification in oocytes and/or trophocytes, polyploidization of trophocytes, but also chromosome elimination during early somatogenesis can be easily explained by the DNA optimization model[2,3].

1. Nagl, W. (1976) Nature 261, 614-615.
2. Nagl, W. (in press) Nucleus (Delhi), Suppl.
3. Nagl, W. (in press) Endopolyploidy and Polyteny in Differentiation and Evolution, North-Holland, Amsterdam, appr.350 pages.
4. Bier, K. (1969) Zool.Anz., Suppl., 33, 7-29.

Chromosomes Today Volume 6, A. de la Chapelle and M. Sorsa eds.

KARYOTYPE VARIABILITY IN *Sorex araneus* L. (INSECTIVORA, MAMMALIA)*

KARL FREDGA and JANERIK NAWRIN
Institute of Genetics, University of Lund, S-223 62 Lund (Sweden)

ABSTRACT

G-band patterns were studied in the chromosomes of the common shrew. Four chromosome races were established, 3 in Sweden and 1 in Poland, all characterized by various combinations of 10 telocentric chromosomes into metacentrics (Nos 3,5,6,7,8). Pair No 4, which may also be involved in Robertsonian polymorphism, was identical in all races, as were autosomes Nos 1, 2 and 9, and the sex chromosomes.

Possible hybridization between the chromosome races is discussed as well as the immigration of the species into Sweden after the last glaciation. A survey of the chromosomes in the entire *S. araneus-arcticus* group is given.

INTRODUCTION

The genus *Sorex* with about 40 species inhabits the northern hemisphere, southward to Central America in the New World, and southward to Israel, Asia Minor, Kashmir and northern Burma in the Old World[1]. The type species of the genus is *Sorex araneus*, the common shrew. Among the *Sorex* species, the *araneus-arcticus* group is characterized by the uncommon XX/XY_1Y_2 sex chromosome mechanism, and this had led to the assumption that the species belonging to this group have a common phylogenetic origin[2]. The number of species, subspecies and geographical races within this group has been debated by zoologists for many years[3], and evidently the most promising way, and probably the only way, to solve these problems is by detailed chromosome studies.

So far, 7 species have been recognized within the XX/XY_1Y_2 group by their clearly different karyotypes (Table 1). *Sorex coronatus* is synonymus to *S.gemellus*, and is also known as *S.araneus*, chromosome race A. *S.granarius* has previously been regarded as a race of *S.araneus*. The karyotypes of the *S.*"*arcticus*" complex from Siberia are clearly different from the karyotype of *S.arcticus* from North America, and the Siberian shrews should probably be subdivided into different species[6] (*S.sibiriensis*, *S.irkutensis*, etc.). Until more detailed chromosome studies have been performed, we prefer to follow the opinion of Král and Radjabli[24], who conclude that the karyotypes of the Siberian populations so far examined can be arranged into a continuous chain and therefore do not represent separate species. For a more detailed discussion of the classification and nomenclature of the entire

---

*Dedicated to Prof. Charles E. Ford on the occasion of his 65th birthday

TABLE 1

Sorex SPECIES WITH SEX CHROMOSOMES XX/XY_1Y_2

The distribution, number of autosomes (2na), number of autosome arms (NFa), and designation of chromosome races[4] and types[5,6] are given for the 7 species.

Species	Distribution	2na	NFa	Race	Type	References
S.coronatus	W.Europe	20	38	A	A	5-11
S.araneus	Europe-Siberia	18-30	36	B,C	B_1-B_4	2,4,9,12-20
S.granarius	Spain, Portugal	32	34		H	21
S.arcticus	N.America	26	34		C	5,22
S."arcticus"	W.Siberia	30-34	52-56		D	2,23,24
S.daphaenodon	"	24	42		G	2
S.caucasicus	Caucasus	22	42		F	25

Sorex araneus-arcticus group, the reader is referred to Hausser[6].

The common shrew is the only species within the genus *Sorex* showing a Robertsonian system of chromosome polymorphism. The number of autosomes (2na) may vary between one animal and another, but the number of autosome arms (NFa) is constant. Since up to 6 pairs of biarmed autosomes (Nos 3-8) may be involved in this variation, the chromosome number of *S.araneus* may vary from 20 to 32 in the female and from 21 to 33 in the male. Within the same population, however, usually only 1 to 3 of the metacentric elements take part in the polymorphism, and those usually involved are Nos 6,7, and 8. Often a population is monomorphic with 0,2, or 5 telocentric chromosomes.

Even with the conventional acetic orcein squash technique it was discovered that karyotypes of *S.araneus* from different areas of its wide distribution range varied, not only in chromosome number, but also in size and arm ratio of certain chromosome pairs[2,4,5,15,17]. Evidently, an original set of telocentric chromosomes have combined into metacentrics by different series of Robertsonian fusions. Detailed chromosome measurements were performed by various authors in order to elucidate the exact arm combinations and the relations between the karyotypes of different populations, but because of the small and uniform size of many of the arms involved the problem could not be definitely solved. Two chromosome races were discovered in Sweden, one in the south and one in the north, tentatively called race B and C by Fredga[4]. Simultaneously Meylan and Hausser[5] proposed a classification of the main karyotypes of the entire *Sorex araneus-arcticus* group, in which *Sorex arcticus* from North America was designated as type C, and more recently Hausser[6] designated the various chromosome races within *Sorex araneus* as B_1-B_4 (cf Table 1). To avoid confusion, we have decided in the present paper not to designate any chromosome races by capital letters.

The present study was undertaken to explore the exact relationships between the chromosome races in Sweden, to locate the border area where the northern and southern chromosome races meet, and to find out whether in this area the two races interbreed or behave like biological species. Since it is very difficult, although not impossible, to breed shrews in captivity, we hoped to obtain an answer also to the last question by chromosome studies of wild-trapped animals. Detailed analyses of G-banded chromosomes should reveal any specimens having metacentrics from both races.

MATERIALS AND METHODS

In the years of 1973 to 1976 common shrews were trapped by the authors in different parts of Sweden. G-banded preparations were analyzed from 7 specimens from southern Sweden (6 from Skåne and 1 from Öland), 4 specimens from central Sweden (Jämtland), and 3 specimens from northern Sweden (Lapland). In addition, G-banded preparations were analyzed from 4 common shrews from Bialowieza, E Poland, which were kindly provided by Dr Stanislaw Fedyk. In the summer of 1977 a larger sample of common shrews was collected in the expected border area in central Sweden. So far, the G-band pattern has been analyzed in only one of these animals (from Uppland), but unbanded preparations have been analyzed preliminarily in 42 specimens, and the karyotypes of more than 125 specimens altogether from 50 localities in Sweden have been studied (including those reported earlier[4]).

The chromosomes were mainly studied in direct preparations from spleen, but also in preparations from tissue cultures initiated from biopsies of lung or heart. Preparations were made according to the conventional air-drying technique and stained with orcein or Giemsa. For G-bands the technique of Wang and Fedoroff[26] was applied with minor modifications[27]. (Technical details will be reported elsewhere.)

RESULTS

The exact arm combinations were determined by G-banding in animals from southern, central and northern Sweden and from eastern Poland. Each chromosome arm was designated by a small letter (a-u) as suggested by Halkka et al.[20] (cf Fig. 1 B). It was established that common shrews from southern and northern Sweden differ with respect to 4 metacentric chromosome pairs (and not 3 as previously assumed based on measurements only[4]). The arm combinations may be seen from Figure 1 A and C. Chromosome No 4 is the same, and No 5 in the southern race corresponds to No 3 in the northern. Even in unbanded preparations it is easy to distinguish these two races: The southern race is characterized by 2 relatively large m pairs with arms of similar length (Nos 3 and 4), No 5 is bigger than 6, and Nos 7 and 8 are very similar in size and shape. The northern race has only 1 relatively large m chromosome (No 4), No 5 is similar to 6, and 7 is bigger than 8 due to its longer long arm.

As a rule, southern shrews are monomorphic with 2na=18 and no t chromosomes. The only exception is one out of 3 specimens from the isle of Öland; this specimen had

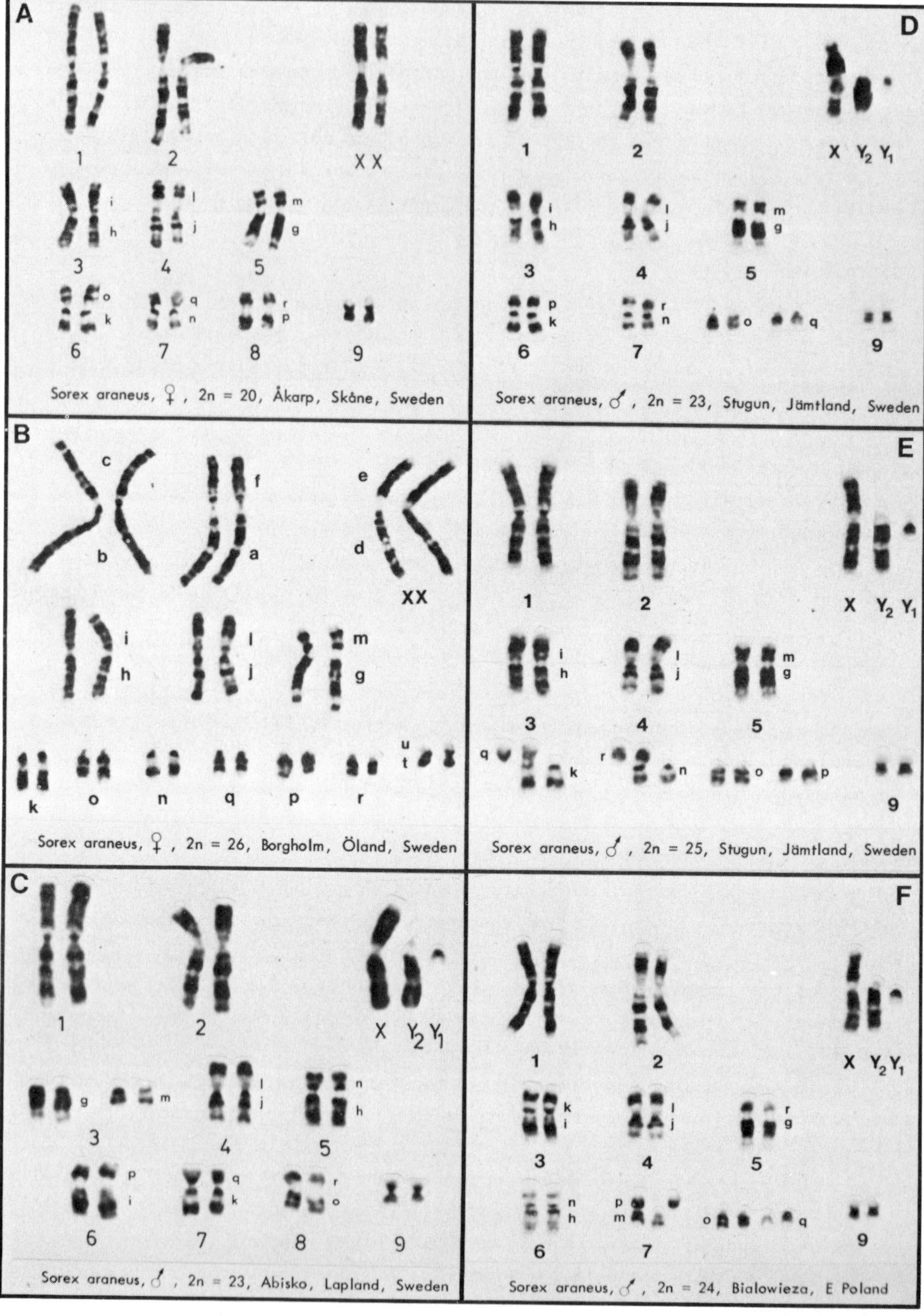
A
Sorex araneus, ♀, 2n = 20, Åkarp, Skåne, Sweden
D
Sorex araneus, ♂, 2n = 23, Stugun, Jämtland, Sweden
B
Sorex araneus, ♀, 2n = 26, Borgholm, Öland, Sweden
E
Sorex araneus, ♂, 2n = 25, Stugun, Jämtland, Sweden
C
Sorex araneus, ♂, 2n = 23, Abisko, Lapland, Sweden
F
Sorex araneus, ♂, 2n = 24, Bialowieza, E Poland

2na=24 with the small autosomes 6-8 represented by 6 pairs of telocentrics (Fig. 1 B).

The karyotypes of shrews from central Sweden differ from both southern and northern ones, in 3 and 4 chromosome pairs, respectively. New arm combinations are present in the 3 pairs Nos 6-8 (Fig. 1 D). The central chromosome race is characterized by having the same 2 pairs of relatively large metacentrics as the southern race (Nos 3 and 4), but combined with Robertsonian polymorphism involving 1-4 pairs of autosomes (Nos 5-8). Within the distribution range of the central race, the number of t chromosomes increases from 2 in the south up to 14 in northern populations. Two animals trapped outside Uppsala (Uppland) had 2na=19 and were heterozygous for pair No 8 (as confirmed by G-banding in one individual). Animals with 2na=18 and no t chromosomes are regarded as belonging to the southern race, but this has to be confirmed by G-band analyses of such animals from the border area between the southern and central races. Animals with no t autosomes could well belong to the central race, since in unbanded preparations even of the highest quality it is impossible to distinguish unambiguously between these two races, except by the criterium of presence or absence of t autosomes (as was done in Fig. 2).

Among 6 animals from one locality in Jämtland (Stugun), one exhibited arm combinations characteristic of both the central and northern races, thus indicating hybrid origin. Autosome No 6 had the arm combination kq and not kp. This specimen was heterozygous for Nos 6 and 7 and the arms o and p appeared as 4 telocentrics (Fig. 1 E). This individual is the only one yet discovered which might be of hybrid origin. Of the animals collected in 1977, so far analyzed in unbanded preparations only, none had 3 relatively large m chromosomes (Nos 3 and 4) as should be expected in an F1 hybrid between the central and the northern races, which have 4 and 2 chromosomes of this kind, respectively.

A fourth type of arm combinations was found in Polish shrews. They had only one of the six elements in common with the southern Swedish race (No 4), and two elements in common with the central (Nos 4 and 8) and northern (Nos 4 and 6) races. In our experience, the arms o and q usually appear as telocentrics both in central Sweden and in Poland (Fig. 1 D,F). In Figure 1 F, which shows the chromosomes of a Polish shrew heterozygous for No 7 and homozygous for the telocentric elements corresponding to No 8, the chromosomes are arranged and numbered in the same way as by Ford and Hamerton[16].

Our results are summarized in Table 2, from which it is clear that all the chromosome races studied have only one element in common, viz. No 4, the arms of which nearly always are fused. Does this mean that j and l were the first two telocentrics

←

Fig. 1. G-banded karyotypes of Sorex araneus from southern (A,B), northern (C) and central (D,E) Sweden, and from eastern Poland (F). Chromosome arms are designated by small letters. Note that among the pairs Nos 3-8 only one, No 4, has the same arm combination (lj) in all populations.

TABLE 2

ARM COMBINATIONS OF CHROMOSOME PAIRS NOS 3-8 IN Sorex araneus FROM DIFFERENT GEOGRAPHIC REGIONS

Figures: designations of chromosome pairs applied in the individual races; small letters: designations of chromosome arms. Identical arm combinations are framed

Geographic region	Chromosome No and arm combination						
Southern Sweden (Skåne)	3 hi	4 jl	5 gm	6 ko	7 nq	8 pr	
Central Sweden (Jämtland)	3 hi	4 jl	5 gm	6 kp	7 nr		8 oq
" " (" , hybrid?)	3 hi	4 jl	5 gm	6 kq	7 nr	8 op	
Northern Sweden (Lapland)	5 hn	4 jl	3 gm	7 kq	6 ip	8 or	
Eastern Poland (Bialowieza)	6 hn	4 jl	3 ik	5 gr	7 mp		8 oq

to become fused, or is this arm combination especially advantageous for the species? The combination gm is found in all Swedish shrews but not in the Polish. The combination hn, found in Lapland, Finland[20] and Poland, may be characteristic of eastern populations and is likely present also in Siberian populations of Sorex araneus. Král and Radjabli[19,24] have studied the chromosomes of common shrews from Novosibirsk with a G-band technique, and our interpretation of their results, "translated" to the present nomenclature, is: No 3 ik, 4 jl, 5 go, 6 hn, 7 mp, and 8 qr.

Autosomes Nos 1,2 and 9, and the sex chromosomes show the same band patterns in all populations studied.

DISCUSSION

The karyotype polymorphism of Sorex araneus is evidently much more complicated than could be expected before the era of chromosome banding, and there must be many contact zones in the large range of distribution of the species. During our attempts to find out what happens in the contact zone between the northern and southern races in Sweden, we discovered a new main type, the central race. It is too early to answer the question whether the different chromosome races interbreed, and if so whether their progeny is fertile or not. If two races differ with respect to four metacentric pairs, an F1 hybrid should form a ring of eight during meiosis. If the elements orient themselves at anaphase in a "proper" zig-zag configuration (like in Rhoeo) balanced gametes could be produced. In the contact zone between the northern and central Swedish races perhaps the problem has been solved by dissociation of some or all of the actual metacentric chromosomes into telocentrics. However, in the contact zone between the central and southern races the problem has not been solved in this way, telocentric elements being rare in this area.

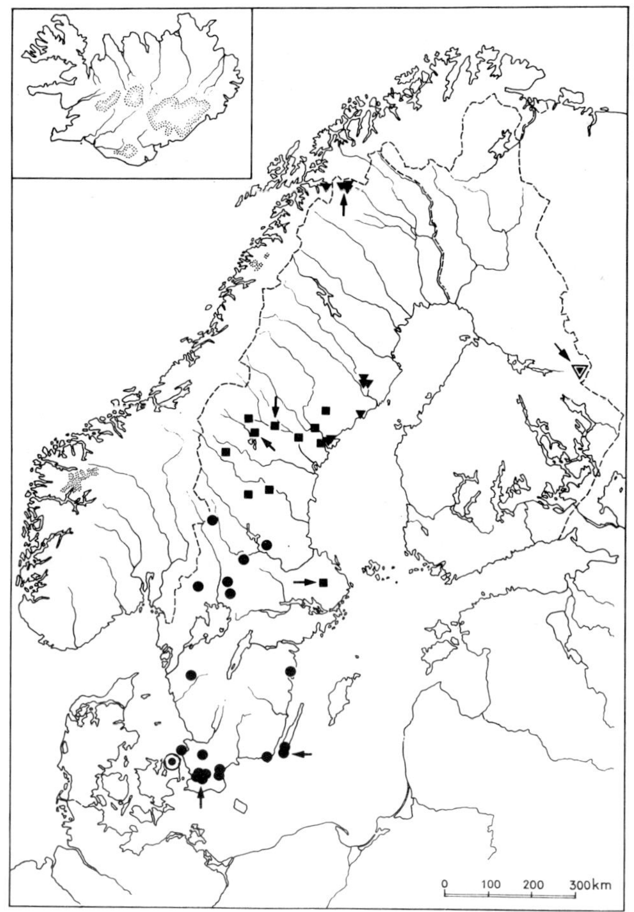

Fig. 2. Map of Fennoscandia showing the distributions of the chromosome races in southern (●), central (■) and northern (▼) Sweden. Arrows indicate that G-banded preparations were analyzed from these localities. Open symbols from Halkka et al.[20] (Finland) and from Meylan[14] (Denmark).

The distributions of the 3 chromosome races in Sweden are clear from Figure 2. Already in 1922 Sven Ekman[28] predicted that Sorex araneus had colonized Scandinavia from two directions after the last glaciation, from north-east via Finland and from south-west via the land connection with Denmark. Previous chromosome studies[4] supported this hypothesis, but evidence was lacking until Halkka et al.[20] showed that shrews in Finland were of the same chromosome race as those in northern Sweden. Meylan[14] had earlier reported 2na=18 in common shrews from Denmark (Zealand), which means that they likely had the same karyotype as those in southern Sweden. Thus, chromosome studies have contributed in solving an old immigration problem of fundamental theoretical interest.

What is the origin of the central chromosome race? We suggest that it originates from shrew populations which have "hibernated" during the last glaciation of Scandinavia in ice-free refuges at the west coast of Norway and then colonized Sweden from the west when the inland ice melted. Chromosome studies with modern banding techniques of shrews from these regions are urgently needed in order to test our suggestion. The central chromosome race has spread south-eastwards, at least as far as to Uppsala (Uppland), and most likely the type specimen of _Sorex araneus_, described in 1758 by Linnaeus, belonged to this chromosome race.

SUMMARY

By means of G-banding 4 chromosome races of _Sorex araneus_ were established, 3 in Sweden and 1 in Poland. All were characterized by different combinations of chromosome arms in 3,4 or 5 metacentrics. One specimen had arm combinations characteristic of two of the races, thus indicating hybrid origin. Robertsonian polymorphism is common in the northern and central races in Sweden but not in the southern.

ACKNOWLEDGEMENTS

We are grateful to Dr Stanislaw Fedyk who kindly supplied one of us (K.F.) with 8 living shrews during a visit in Bialowieza. - This work was supported by grants from the Swedish Natural Science Research Council and from the Nilsson-Ehle foundation.

REFERENCES

1. Walker, E.P. (1964) Mammals of the world. I, Johns Hopkins Press, Baltimore, pp. 1-646.
2. Fedyk, S. and Ivanitskaya, E.Y. (1972) Acta Theriol., 17, 475-492.
3. Hoffmann, R.S. (1971) Z. Säugetierkunde, 36, 193-200.
4. Fredga, K. (1973) Hereditas, 73, 153-157.
5. Meylan, A. and Hausser, J. (1973) Z. Säugetierkunde, 38, 143-158.
6. Hausser, J. (1976) Thése, Université de Geneve, Faculté des Sciences, No. 1732, pp. 1-89.
7. Bovey, R. (1948) Arch. J. Klaus-Stif., 23, 506-509.
8. Bovey, R. (1949) Rev. Suisse Zool., 56, 371-460.
9. Meylan, A. (1964) Rev. Suisse Zool., 71, 903-983.
10. Ott, J. (1968) Rev. Suisse Zool., 75, 53-75.
11. Olert, J. (1973) Thesis, Veröffentlichungen der Universitet Innsbruck 76, Innsbruck, pp. 1-73.
12. Sharman, G.B. (1956) Nature, 177, 941-942.
13. Ford, C.E., Hamerton, J.L. and Sharman, G.B. (1957) Nature, 180, 392-393.
14. Meylan, A. (1965) Rev. Suisse Zool., 72, 636-646.
15. Orlov, V.N. and Kozlovsky, A.I. (1969) Citologia, 11, 1129-1136.

16. Ford, C.E. and Hamerton, J.L. (1970) Symp. zool. Soc. Lond., No. 26, 223-236.

17. Kozlovsky, A.I. (1970) Citologia, 12, 1459-1464.

18. Kozlovsky, A.I. (1972) Citologia, 14, 761-768.

19. Král, B. and Radjabli, S.I. (1974) Zoologické Listy, 23, 217-227.

20. Halkka, L., Halkka, O., Skarén, U. and Söderlund, V. (1974) Hereditas, 76, 305-314.

21. Hausser, J., Graf, J.-D. and Meylan, A. (1975) Bull. soc. vaud. sc. Nat., 72, 241-252.

22. Meylan, A. (1968) Rev. Suisse Zool., 75, 691-696.

23. Kozlovsky, A.I. (1971) Zool. Zhurnal, 50, 756-762.

24. Král, B. and Radjabli, S.I. (1976) Zoologické Listy, 25, 327-334.

25. Kozlovsky, A.I. (1973) Zool. Zhurnal, 52, 571-576.

26. Wang, H.C. and Fedoroff, S. (1972) Nature New Biol., 235, 52-53.

27. Mandahl, N. and Fredga, K. (1975) Hereditas, 81, 211-220.

28. Ekman, S. (1922) Djurvärldens utbredningshistoria på skandinaviska halvön, Albert Bonniers Förlag, Stockholm, pp. 1-614.

Chromosomes Today Volume 6, A. de la Chapelle and M. Sorsa eds.

GENETIC CONTROL OF TESTICULAR DIFFERENTIATION IN MAMMALS

C. E. FORD
Medical Research Council External Staff, Sir William Dunn School of Pathology, University of Oxford, South Parks Road, Oxford, England.

ABSTRACT

Exceptional fertile female Wood Lemmings that bear daughters only and sterile human females with Pure Gonadal Dysgenesis have two properties in common: both have a normal XY karyotype and both lack the H-Y antigen. A third common property, determination by an X chromosome mutation, is probable in both and suggests another example of Ohno's rule that genetic information on the mammalian X chromosome is conserved. Loci on both X and Y may then be implicated in the determination of testicular development and the presence of H-Y antigen. Which, if either, of these loci bears structural information and which is regulator remains unknown.

The purpose of this short communication is to draw attention to a remarkable parallelism in the cytogenetics of sex determination in mammals and to consider its implications.

To recapitulate briefly a story that is now widely known: the Wood Lemming (Myopus schisticolor Lilljeborg) is a small palaearctic rodent subject to marked fluctuation in numbers from year to year with females that regularly outnumber males by 3 to 1; when bred in captivity some females give litters of normal size consisting solely of daughters; the exceptional females have the standard 32,XY somatic karyotype of normal males, but 32,XX oocytes that presumptively originate from primitive XY germ cells by double non-disjunction; these females lack the H-Y antigen[1-3].

Fredga and his colleagues[1] postulated the presence of a mutant gene (*) on the X chromosome of exceptional XY females (subsequently written X*Y) that has the effect of "repressing the male-determining effect of the Y", so that testicular differentiation of the indifferent gonads in early X*Y embryos is suppressed and ovarian differentiation ensues instead. According to this hypothesis all progeny will receive X* and will be females, whether karyotypically X*X or X*Y. The hypothesis also accounts for the 3:1 ratio in wild populations[4].

It is not necessary to require of an efficient hypothesis that it should account also for the regulation from XY to XX in the germ line of the exceptional females; from an evolutionary point of view it is plausible to suppose that the primary event was the postulated sex-linked mutation and that the original XY females had XY oocytes. Now that it has been shown in the mouse that XY oocytes can be functional (in fertile female XX/XY chimaeras[5,6]) we may assume with some confidence that such oocytes in the Wood Lemming would also be functional. But initially, even supposing normal numbers of oocytes were ovulated, a quarter of the fertilised eggs would be inviable YY zygotes so that litter size would be reduced. On the other hand, given an initial advantage to the species of the conversion of some expected males into sub-fertile females, there would inevitably be strong positive selection pressure in favour of any further genetic changes that promoted increased fertility. Regulation from XY to XX in the germ line therefore could have arisen secondarily and independently. Indeed, the recent identification of a high frequency (compared with other mammalian species) of sex chromosome aneuploids, and some normal XY sons, among the progeny of certain XY females[2,7] suggests that the regulatory process is still imperfectly adjusted and that some XO, XXY and XY oocytes may be produced. (The alternative explanation for the occasional normal sons is to assume that the postulated original X-linked mutation is incompletely penetrant.)

The condition known as XY gonadal dysgenesis, sometimes as Pure Gonadal Dysgenesis, though rare, is well known in human cytogenetics. The affected individuals are females and usually present with primary amenorrhoea; they are commonly moderately tall, eunuchoidal, and lack any of the dysmorphic stigmata associated with Turner's syndrome; they have an apparently normal 46,XY male karyotype and if exploratory laparotomy is performed it is found that the gonads are represented by sterile streaks of connective tissue. (The condition is not to be confused with the testicular feminisation syndrome, in which a female phenotype is also associated with a normal 46,XY karyotype, but which is distinguished by the presence of cryptic testes.) Though most cases are sporadic, and may be of heterogeneous origin[8], familial cases do occur. Most of the still limited evidence of familial transmission could be explained by sex-linked inheritance, but sex-limited autosomal dominant inheritance is not excluded and there is one aberrant pedigree[9]. The first cases examined have been found also to be H-Y antigen negative[10].

Disparate though they may seem at a first comparison, the two types of exceptional female, in the Wood Lemming and in man, have two striking points in common. Both have an apparently normal male karyotype; and both are H-Y antigen negative. Both may furthermore be determined by a mutant gene on the X chromosome, though in neither condition is this yet established. The outstanding difference between them, fertility of the XY female Wood Lemming, sterility of the human XY GD female, may prove to be trivial. There is no direct evidence on the status of the gonads of XY GD cases during foetal life or childhood. But the human XO females that also typically have streak gonads when adult are known to have ovaries during foetal life, sometimes to retain them to birth[12] and exceptionally even to bear children[13]. It is also relevant to recall that XO mice have ovaries and are fertile but have a shorter reproductive life than their XX sisters[11]. There is therefore some basis for the speculation that ovaries with primordial follicles may be present initially in the human XY GD foetus and may exceptionally be retained into post-natal, possibly even adult life.

Ohno[14] drew attention to the conservation of genetic loci on the X chromosome during the evolution of mammals, which is another way of saying that apparently homologous mutations if sex-linked in one species of mammal will be sex-linked in others. There is still no evident exception to this rule. Though the case is still incomplete, there are therefore gounds for postulating that the XY female Wood Lemmings and human females with XY gonadal dysgenesis are determined by homologous mutations, and, more generally, that there is a locus on the X chromosome in mammals that is concerned with the primary signal that the indifferent gonad should differentiate as a testis.

If this be so there will no longer be need to think of the mutant in the Wood Lemming as an active repressor of testicular development, as originally postulated by Fredga et al.[1]. Rather would the locus concerned be part of a regulatory system that must also include, minimally, a 'masculinising locus' on the Y chromosome. A breakdown at any point in this system would result in both absence of H-Y antigen and failure of the gonads to develop as testes, so that neither could be said to be causal of the other. That the gonads should then undergo at least initial ovarian development instead would be in accordance with the concept of ovarian differentiation as the natural destiny of the indifferent gonad in mammals in the absence of a specific signal to differentiate as a testis.

Whether the one locus is the structural locus for the H-Y antigen and the other a positive regulator (activator) locus, and indeed whether there is any virtue at all in the postulate of an X chromosome locus immediately concerned with male sex determination in mammals, perhaps the future will tell.

These ideas were developed largely in correspondence with Karl Fredga and Alfred Gropp in the closing months of 1976. They subsequently appeared in part in the communication by Wachtel et al.[3]

REFERENCES

1. Fredga, K., Gropp, A., Winking, H. and Frank, F. (1976) Nature, 261, 225-227.
2. Gropp, A., Winking, H., Frank, F., Noack, G. and Fredga, K. (1977) Cytogenet. Cell Genet., 17, 343-358.
3. Wachtel, S.S., Koo, G.C., Ohno, S., Gropp, A., Dev, V.G., Tantravahi, R., Miller, D.A. and Miller, O.J. (1976) Nature, 264, 638-639.
4. Bengtsson, B.(1976) Personal communication.
5. Ford, C.E., Evans, E.P., Burtenshaw, M.D., Clegg, H.M., Tuffrey, M. and Barnes, R.D. (1975) Proc. Roy. Soc. Lon. B. 190, pp. 187-197.
6. Evans, E.P., Ford, C.E. and Lyon, M.F. (1977) Nature, 267, 430-431.
7. Gropp, A. (1976) Personal communication.
8. Gray, J. (1976) Personal communication.
9. Simpson, J.L. (1977) in Genetic Aspects of Sex Differentiation, Academic Press, New York (in press).
10. Miller, O.J., Koo, G.C., Wachtel, S.S. and Breg, W.R. (1976) Paper presented at Vth International Congress of Human Genetics, Mexico City, October, 1976.
11. Lyon, M.F. and Hawker, S.G. (1973) Genet. Res. 21, 185-194.
12. Carr, D.H., Hagger, R.A. and Hart, A.G. (1968) Am. J. clin. Path. 49, 521-526.
13. Lajborek-Czyz, I. (1976) Clin. Genet. 9, 113-116.
14. Ohno, S. (1967) Sex chromosomes and sex-linked genes. Monographs in Endocrinology, vol. 1, Springer-Verlag, Berlin.

Chromosomes Today Volume 6, A. de la Chapelle and M. Sorsa eds.

CHROMOSOMES OF PEROMYSCUS (RODENTIA, CRICETIDAE) VIII. GENOME CHARACTERIZATION OF FOUR SPECIES

MARION WATT HAZEN[1,3], FRANCES E. ARRIGHI[1,3] AND DENNIS A. JOHNSTON[2,3]
Department of [1]Biology and [2]Biomathematics, The University of Texas System Cancer Center M. D. Anderson Hospital and Tumor Institute, and [3]The University of Texas Health Science Center at Houston Graduate School of Biomedical Sciences, Houston, Texas 77030

ABSTRACT

In situ hybridization experiments, using ^{3}H-cRNA complementary to a satellite DNA fraction from *P. eremicus* and fixed chromosome preparations of four *Peromyscus* species (*P. eremicus*, *P. crinitus*, *P. collatus* and *P. stephani*) showed that silver grains were localized over the short (heterochromatic) arms of both *P. eremicus* and *P. collatus* chromosomes, but were virtually absent from *P. crinitus* and *P. stephani* chromosomes. Also, these four species are compared with respect to the distribution of constitutive heterochromatin (C-band positive material) and nucleolus organizer regions (NORs), and to their DNA density profiles in neutral CsCl gradients.

INTRODUCTION

The American rodent genus, *Peromyscus*, is a New World taxon consisting of approximately 60 species with a wide spectrum of variation in both chromosome morphology[1,2,3] and DNA content[4,5]. This genus is distributed throughout North and Central America and inhabits a variety of ecological niches. The diploid number for all species studied is 48, although the total number of chromosome arms varies from 56 (e.g., *P. stephani* and *P. crinitus*) to 96 (e.g., *P. eremicus* and *P. collatus*).

Two *Peromyscus* species (*P. eremicus* and *P. crinitus*) were examined by Pathak *et al.*[4], using the newly-available banding techniques to determine the mechanisms of karyotype evolution which were responsible for the variability present in the genus. The G-banding patterns of the long arms of these two species were identical and the short arms of *P. eremicus* were found to be composed of additional heterochromatin (C-band positive material). It was hypothesized, therefore, that the DNA content of *P. eremicus* cells should be greater than that of *P. crinitus* cells. This has been confirmed using flow microfluorometry (FMF) analysis[5].

In the present investigation, we have characterized the C-band distribution patterns of *P. collatus* and *P. stephani*, and compared them with those of *P. eremicus* and *P. crinitus*. In addition, we have compared the distribution of ribosomal cistrons, the DNA buoyant density profiles in neutral CsCl and the localization of sites complementary to ^{3}H-cRNA transcribed from *P. eremicus* satellite DNA to the distribution of constitutive heterochromatic regions in these four species.

MATERIALS & METHODS

Cell culture and cytological preparations. Fibroblast cultures were initiated from ear biopsies from the following animals and maintained in McCoy's 5a medium supplemented with 20% fetal calf serum:

P. eremicus: ♀ (TCH-2352) - Mohave Co., Arizona - Sept. 1972

P. crinitus: ♀ (TCH-2350) - Washington Co., Utah - Sept. 1972

P. collatus: ♀ (TCH-2953) - Turner Island (Gulf of California) - Dec. 1975

P. stephani: ♂ (TCH-2690) - San Esteban Island (Gulf of California) - Dec. 1975

For cytological preparations, the cells were harvested after 1-3 hr colcemid treatment and fixed for 20 minutes in 40% acetic acid (for squash slides) or a modified Carnoy's fixative of 3:1 methanol-glacial acetic acid (for air-dried slides). Squash preparations for in situ hybridization studies were made on subbed slides and the coverslips were flipped off after freezing the slides on dry ice. The slides were rinsed twice in 95% ethanol and stored for in situ hybridization experiments. Conventional air-dried preparations were used for C-banding and silver staining.

Visualization of constitutive heterochromatin. For P. eremicus, P. crinitus and P. collatus, air-dried slides were treated according to the C-band procedure described by Hsu[6] with the following modification: A 0.01N NaOH solution was made by mixing 10 ml of 0.07N NaOH, made in water, with 60 ml of 2X SSC. For P. stephani cells, it was necessary to use a saturated $Ba(OH)_2 \cdot 8H_2O$ solution to denature the chromosomes instead of NaOH.

Staining for ribosomal cistrons. Air-dried slides were stained with 50% aqueous $AgNO_3$ as described by Bloom and Goodpasture[7] and modified by Lau and Arrighi[8].

DNA isolation and characterization. Native DNA was isolated and purified according to the method of Marmur[9] as modified by Stefos and Arrighi[10] from liver tissues (P. eremicus) or cultured cells (all four species). Buoyant densities were determined by centrifuging the DNA samples to equilibrium in neutral CsCl in a Beckman Model E analytical ultracentrifuge. Single sector cells were used in an AnF 410 rotor at 42,040 rpm at 25°C for 22-24 hours. The original refractive index of the DNA/CsCl solution was brought to 1.4020 with solid CsCl, and bacteriophage 2C (ρ = 1.742 g/ml) was used as a marker.

Neutral CsCl buoyant density profiles were obtained by scanning the UV negatives with a Dicomed model 50B image dissector[11]. The average density of each band was calculated from the digitized image. The densities of the DNA peaks were calculated according to the following equation[12]:

$$\rho_s = \rho_m + \frac{\omega^2}{2\beta_o}(r_s^2 - r_m^2)$$

where ρ_s = buoyant density of sample, ρ_m = buoyant density of marker, ω = angular velocity in radians/second, β_o = conditional constant of the Cs salt, r_s = distance (cm) from the center of rotation to the sample band and r_m = distance (cm) from the center of rotation to the marker band.

Satellite DNA isolation and in situ hybridization. Satellite DNA was isolated from native purified *P. eremicus* DNA in a Ag^+-Cs_2SO_4 gradient, using Ag^+ at a molar ratio of Ag^+ to DNA-phosphate of 0.2, as described previously[13]. The method for *in situ* hybridization was that described by Pardue *et al.*[14]. In each hybridization, the radioactive probe was ^{3}H-cRNA transcribed from *P. eremicus* satellite DNA. Four sets of slides (one set of slides from each species) were used as the immobile components for the hybridization experiments. Each set of slides was denatured using 0.07 N NaOH for 2 min. at 25°C or 95% formamide for 2.5 hrs. at 65°C. All slides were then exposed to the same ^{3}H-cRNA solution and treated in the conventional manner for *in situ* hybridization.

RESULTS

C-band distribution patterns. C-banded metaphase plates of four species of *Peromyscus* are presented in Fig. 1 for comparison. Figure 1a shows metaphase chromosomes of *P. eremicus*; 1b, *P. collatus*; 1c, *P. crinitus* and 1d, *P. stephani*. It is evident that the amount of constitutive heterochromatin varies among these four species. *P. crinitus* and *P. stephani* show very little heterochromatin, which is restricted primarily to the centromeric areas of the autosomes. One pair of *P. crinitus* autosomes also has telomeric C-bands. In *P. crinitus*, the short arms of the X chromosomes are heterochromatic, and this may be true for *P. stephani*. The long arms of the Y chromosomes are heterochromatic in both *P. crinitus*[15] and *P. stephani* (Fig. 1d). In *P. eremicus* and *P. collatus*, constitutive heterochromatin is found in the short arms of all but three pairs of chromosomes. The X chromosomes of both species shows large blocks of interstitial C-band positive material. In *P. eremicus* (Fig. 1a), this material is restricted to the short arm, while in *P. collatus* (Fig. 1b), it is located in the long arm of the X chromosome.

Location of ribosomal cistrons. Figure 2 shows the location of ribosomal cistrons as visualized after staining with 50% aqueous silver nitrate and incubation for 15-18 hours at 50°C. This silver-staining procedure is believed to preferentially stain certain protein(s) associated with the nucleolus organizer regions (NORs)[8]. All four species examined had 8-10 NORs per cell.

In the *P. eremicus* cell seen in Fig. 2a, 9 of the NORs are located in the short arms, while 1 (arrow) is found in a long arm. At present, we do not know if this 1 NOR in the long arm represents a translocation which has occurred in this cell line or a polymorphism present in the species. Eight of the NORs (solid lines) in *P. collatus* (Fig. 2b) are located interstitially in the short arms, while 2 are located in the distal regions of the long arms (arrows).

In *P. crinitus* (Fig. 2c), 6 of the NORs are found at the ends of the short arms of biarmed chromosomes. The other 4 NORs are localized near the centromeres of small acrocentric chromosomes. The P. stephani cell shown (Fig. 2d) contains 9 NORs, but the number of NORs may vary from 8 to 10 per cell. Here, only 2 of the silver-staining regions are located on biarmed chromosomes.

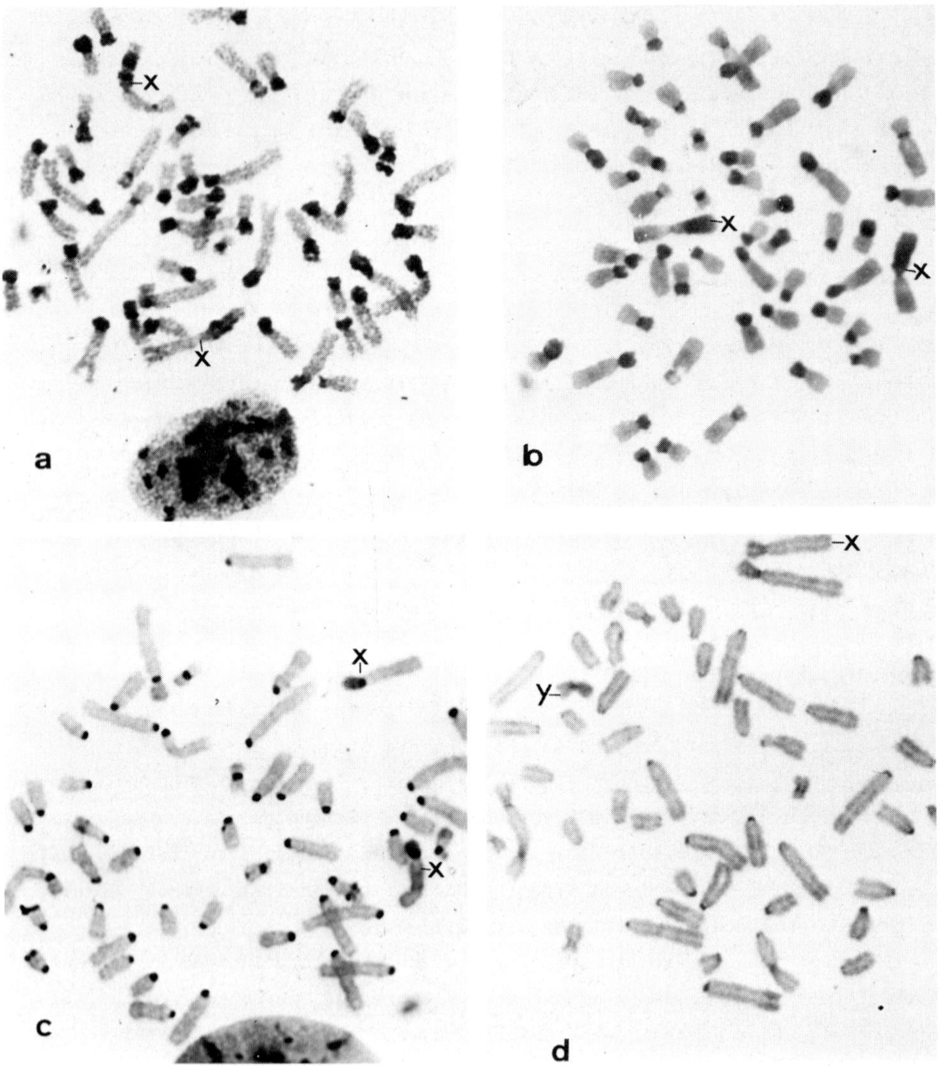

Fig. 1. Distribution of constitutive heterochromatin in (a) P. eremicus, (b) P. collatus, (c) P. crinitus and (d) P. stephani.

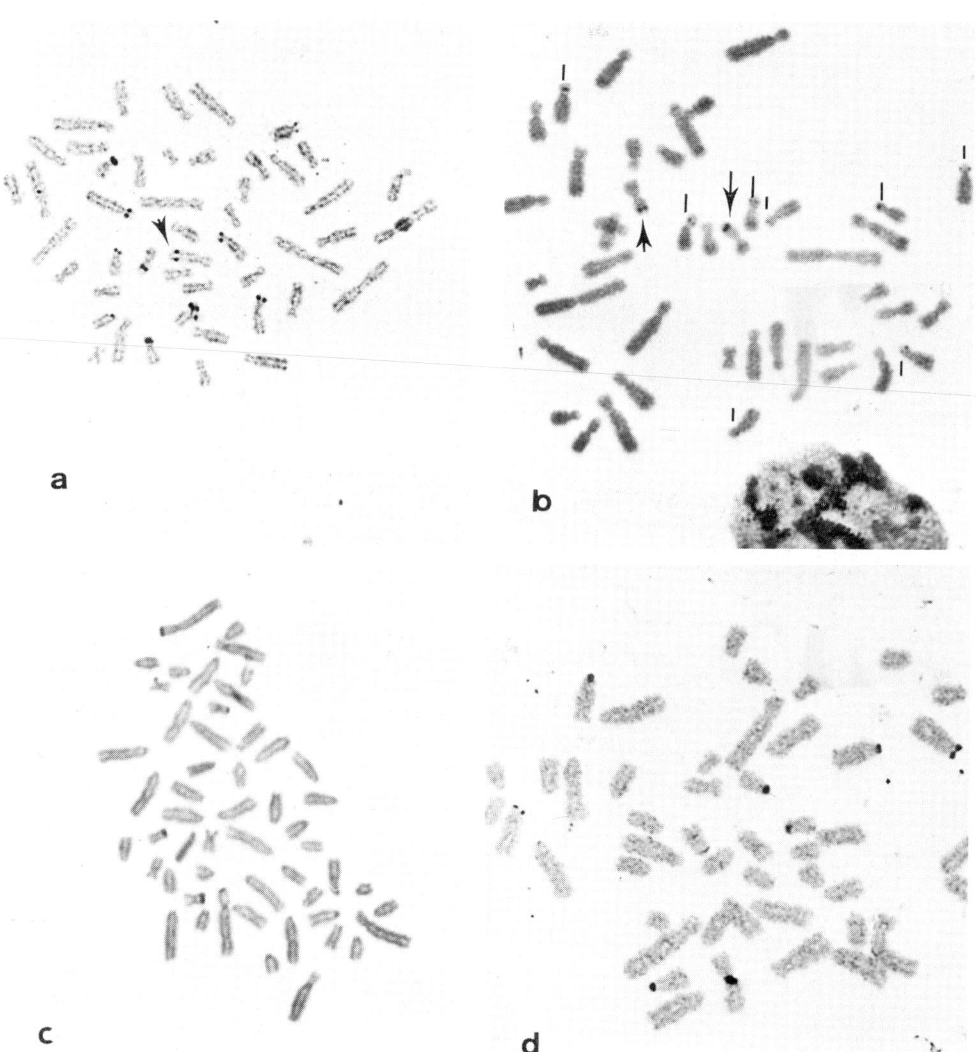

Fig. 2. Localization of nucleolus organizer regions (NORs), visualized as dark-staining areas by silver-staining in (a) P. eremicus, (b) P. collatus, (c) P. crinitus and (d) P. stephani. Note the interstitial localization of the silver deposit in the long arm of one chromosome of P. eremicus (a).

DNA density gradient profiles in neutral CsCl. After purified DNA samples from each species were centrifuged to equilibrium in neutral CsCl density gradients and photographed at 254 nm, the negatives were scanned as described to provide the profiles shown in Fig. 3. Both P. eremicus (3a) and P. collatus (3b) show a broad "main band" peak (ρ = 1.700 g/ml) with a hypersharp "satellite" peak on the heavy (GC-rich) side of the main band (ρ = 1.705 g/ml for P. eremicus and 1.704 g/ml for P. collatus). The density distribution of P. crinitus (3c) and P. stephani (3d) DNAs, however, contain only the main band peak (ρ = 1.700 g/ml) with no apparent satellite peaks.

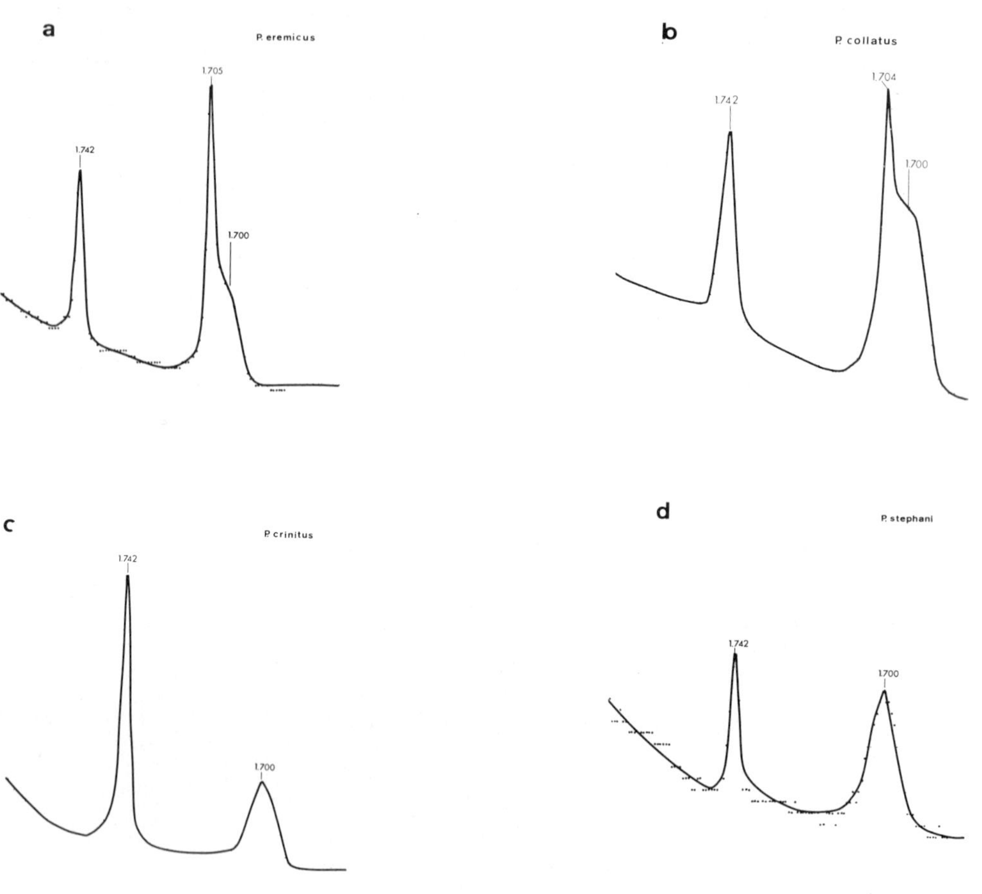

Fig. 3. Neutral CsCl density gradient profiles of purified DNA samples from (a) P. eremicus, (b) P. collatus, (c) P. crinitus and (d) P. stephani. Density values for main band and satellite DNAs are shown on each profile.

Localization of Peromyscus eremicus satellite DNA sequences. When metaphase cells of P. eremicus and P. collatus were fixed on slides, denatured and allowed to hybridize in situ with ^{3}H-cRNA transcribed from P. eremicus satellite DNA, the resulting autoradiographs showed localization of silver grains over many of the short arms of the chromosomes after 2-4 day exposures (Fig. 4a, 4b). After the same length of time, metaphase preparations of P. crinitus and P. stephani cells treated with the same ^{3}H-cRNA transcript showed no localization of grains, even in the C-band positive regions near the centromeres (Fig. 4c, 4d). The P. crinitus cell line used in these experiments was derived from a female animal, so we were unable to determine the in situ labeling pattern of the Y chromosome, although this experiment will be repeated on a male-derived P. crinitus cell line. No localization of silver grains is seen over the Y chromosome of P. stephani (Fig. 4d). Neither P. crinitus nor P. stephani show an accumulation of grains over the short arms of their X chromosomes.

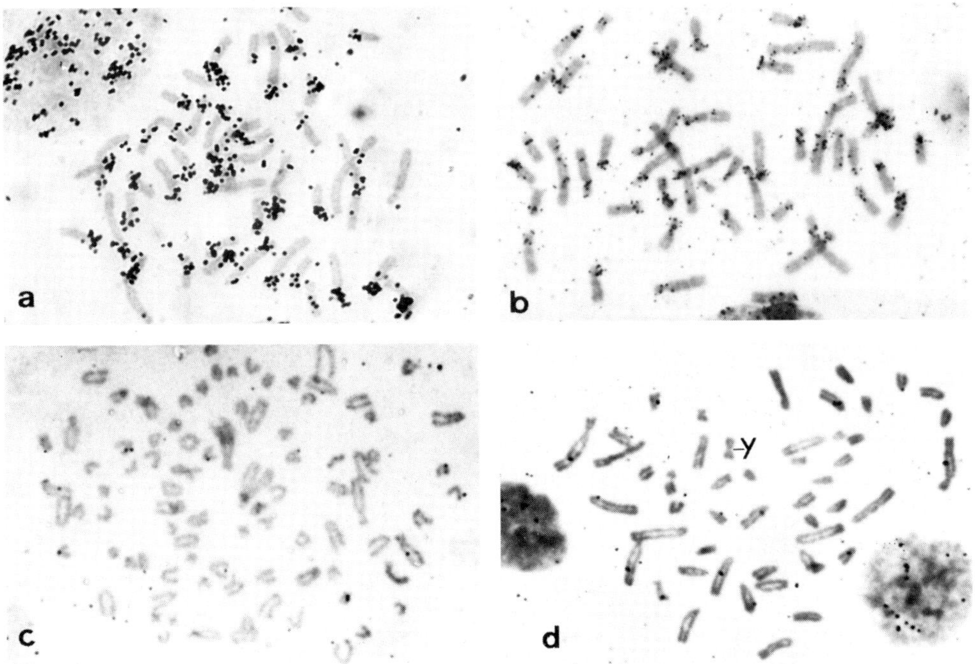

Fig. 4. Autoradiographs of in situ hybrids using ^{3}H-RNA complementary to P. eremicus satellite DNA as the mobile component: (a) P. eremicus, (b) P. collatus, (c) P. crinitus, and (d) P. stephani. Note grain cluster over the short arms of most metaphase chromosomes in (a) and (b), and lack of grain localization in (c) and (d).

DISCUSSION

The analysis of chromosomal and molecular variations among related species has been useful for quantitating and confirming the relationships between organisms whose taxonomy has been established on the basis of other morphological, ecological and behavioral characteristics. Recent cytological studies, for example, have demonstrated considerable conservation of G-banding patterns between related species[4,16,17] and even between related genera[18,19]. Some of these studies have been useful in determining the mechanisms which are responsible for the evolution of karyotypes of related organisms[4,19].

In a few cases, chromosome analyses have provided a significant contribution to the field of animal systematics. For example, in their survey of the chromosomes of the cotton rat, Sigmodon hispidus, Zimmerman and Lee[20] observed that animals from the southeastern United States had a diploid number of 52, while individuals from Arizona had a diploid number of 22. Certainly these two forms, with such drastic chromosomal differences, are not expected to produce fertile offspring even though they are morphologically similar. A similar phenomenon was observed by Gardner[21] in the species Didelphis marsupialis, which included all large American opossums from the United States, Mexico and Central America. Chromosome analysis showed that there were two distinct populations of animals, one with all acrocentrics and the other with many pairs of subtelocentrics. Subsequent morphological studies revealed subtle but consistent differences between the two, and two separate species have been identified.

In the genus Peromyscus, chromosome analysis supported the opinions of mammalogists who believed that the golden mouse, P. (Ochrotomys) nuttalli, should not be included in this genus. These mammalogists placed the golden mouse in the genus Ochrotomys. In studying the chromosomes of this genus, Patton and Hsu[22] found that the diploid number was 52, while all other Peromyscus species had diploid numbers of 48.

In this investigation, we have observed considerable variation in the neutral CsCl DNA buoyant density profiles, in the distribution of constitutive heterochromatin and in the locations of nucleolus organizer regions in 4 species of the genus Peromyscus. The taxonomic classification[23] of this genus places three of the species studied, namely P. eremicus, P. collatus and P. stephani, in the subgenus Haplomylomys and places P. crinitus in the crinitus group of the subgenus Peromyscus. Two of these species, P. eremicus and P. collatus, show very similar C-band and NOR distribution patterns, and both show similar profiles when their purified DNAs are centrifuged to equilibrium in neutral CsCl. In addition, we have seen that the ^{3}H-cRNA prepared from a P. eremicus satellite DNA template hybridizes in situ to the heterochromatic regions of both P. eremicus and P. collatus chromosomes.

Peromyscus crinitus and P. stephani, on the other hand, both show very little C-band positive material in their chromosomes, and their DNAs do not show obvious satellite peaks in neutral CsCl gradients. They both have karyotypes containing

40 acrocentric chromosomes, and only 8 chromosomes with easily-observable second arms. These characteristics tend to imply that P. stephani is more closely related to P. crinitus than it is to either P. eremicus or P. collatus. Neither P. crinitus nor P. stephani chromosomes show any DNA homology to the satellite sequences of P. eremicus.

Using conventional karyotype analysis, Hsu (personal communication) found that P. merriami and P. interparietalis (both members of subgenus Haplomylomys) possess short arms on all their chromosomes similar to P. eremicus and P. collatus. It is highly probable that members of Haplomylomys have a high content of C-band and that their satellite DNA sequences are related, if not identical. In this context, we believe that placing P. stephani in Haplomylomys is improper. Newer information on cytogenetics, biochemistry, and molecular biology should be incorporated into consideration of phylogenetic relationships.

ACKNOWLEDGEMENTS

We gratefully acknowledge Dr. Oscar G. Ward for providing us with the P. collatus and P. stephani animals used in this study, and Dr. T. C. Hsu for his encouragement and critical evaluation of the manuscript. We also thank Misses Mary Bea Cline, Dorothy A. Holitzke and Renee Halbach for their assistance in preparing the manuscript.

This work was supported in part by Research Grants DEB 76-10580 from the National Science Foundation, VC-21 from the American Cancer Society, CA-11430 from the National Cancer Institute, and G-373 from the Robert A. Welch Foundation. Marion Watt Hazen is presently supported by a Rosalie B. Hite predoctoral fellowship from The University of Texas M. D. Anderson Hospital and Tumor Institute, Houston, Texas 77030.

REFERENCES

1. Hsu, T. C. and Arrighi, F. E. (1968) Cytogenetics, 7, 417-446.
2. Bradshaw, W. N. and Hsu, T. C. (1972) Cytogenetics, 11, 436-451.
3. Duffey, P. A. (1972) Science, 176, 1333-1334.
4. Pathak, S., Hsu, T. C. and Arrighi, F. E. (1973) Cytogenet. Cell Genet., 12, 315-326.
5. Deaven, L. L., Vidal-Rioja, L., Jett, J. T. and Hsu, T. C., (in press) Cytogenet. Cell Genet.
6. Hsu, T. C. (1974) Annual Review Genet., 7, 153-176.
7. Bloom, S. E. and Goodpasture, C. (1976) Hum. Genet., 34, 199-206.
8. Lau, Y.-F. and Arrighi, F. E. (in press) in Monograph of Seminar-Workshop held in Montevideo, Uruguay, Feb. 1977.
9. Marmur, J. (1961) J. Mol. Biol., 3, 208-218.
10. Stefos, K. and Arrighi, F. E. (1974) Exptl. Cell Res., 83, 9-14.

11. Johnston, D. A., Cavnar, W. B., Marani, S., McLarty, J. W., Newton, L., Murdock, J., Hughes, J. P. and Smythe, A., An Interactive General Purpose Image Digitizing System — an Internal Document of the Department of Biomathematics, M. D. Anderson Hospital & Tumor Institute, July, 1977.

12. Flamm, W. G., Birnstiel, M. L. and Walker, P.M.B. (1972) in Subcellular Components Preparation and Fractionation, Birnie, G. D. ed., Butterworth & Co., London.

13. Hazen, M. W., Kuo, M. T. and Arrighi, F. E. (in press) Chromosoma.

14. Pardue, M. L., Gerbi, S. A., Eckhardt, R. A. and Gall, J. G. (1970) Chromosoma, 29, 268-290.

15. Arrighi, F. E., Stock, A. D. and Pathak, S. (1976) in Chromosomes Today, Vol. 5, Pearson, P. L. and Lewis, K. R. eds., John Wiley & Sons, New York, pp. 323-329.

16. Pathak, S., Hsu, T. C., Shirley, L. and Helm, J. D., III (1973) Chromosoma (Berl.), 42, 215-228.

17. Mascarello, J. T., Stock, A. D. and Pathak, S. (1974) J. Mammal., 55, 695-704.

18. Stock, A. D. and Mengden, G. A. (1975) Chromosoma (Berl.), 50, 69-77.

19. Stock, A. D. and Hsu, T. C. (1973) Chromosoma (Berl.), 43, 211-224.

20. Zimmerman, E. G. and Lee, M. R. (1968) Chromosoma (Berl.), 24, 243-250.

21. Gardner, A. (1973) Special Publication of the Museum, Texas Tech University.

22. Patton, J. L. and Hsu, T. C. (1967) J. Mammal., 48, 637-639.

23. Hooper, E. T. (1968) in Biology of Peromyscus (Rodentia), King, J. A. ed., Special Publ. No. 2, Amer. Soc. of Mammalogists, pp. 27-74.

Chromosomes Today Volume 6, A. de la Chapelle and M. Sorsa eds.
© 1977 Elsevier/North-Holland Biomedical Press, Amsterdam, The Netherlands

THE CONGRUENCE CODE

JEROME LEJEUNE
Institut de Progenèse, 15, rue de l'Ecole de Médécine, 75005 Paris (France)

The highly selective recognition of short segments of DNA by some specific proteins (restriction enzymes, regulation proteins, etc.) implies a quasi perfect coadaptation of the surfaces complementing each other topologically and electronically.

The purpose of this assay is to investigate if there is some general code ruling the congruence of these molecules.

The most obvious difficulty comes from the enormous number of possible combinations. A sequence of six bases of DNA can have 6^4 configurations (because of the four possible bases T, C, A and G). For each of them a possibly congruent polypeptide must be searched for among 20^3 or 20^6 candidates (depending upon the ratio: one amino acid per two bases or one amino acid per base).

An exhaustive analysis would thus require between 26 million and 26 billion comparisons. Such a formidable task being out of reach, a simplification seems quite desirable.

A heuristic approach

If we compare the molecular congruence to the relations between a statue and its mold, it is evident that if the cast is made of the repetition of an identical pattern, the replica must also exhibit a monotonous repetition of a complementary pattern. This means that a monotonous DNA would only accommodate a monotonous polypeptide.

The size of the repetitive pattern can vary from one to -n- bases and the most convenient seems to choose at the beginning a pattern of two adjacent bases.

Due to the complementarity of A with T, and G with C, in the Watson and Crick dextrogyre double helix configuration, there are only four DNA molecules of this semi-monotonous type, and two of the monotonous type, namely:

$$\overset{5'\rightarrow}{\begin{pmatrix} C & T \\ G & A \end{pmatrix}_n} ; \overset{5'\rightarrow}{\begin{pmatrix} G & T \\ C & A \end{pmatrix}_n} ; \overset{5'\rightarrow}{\begin{pmatrix} T & A \\ A & T \end{pmatrix}_n} ; \overset{5\rightarrow}{\begin{pmatrix} G & C \\ C & G \end{pmatrix}_n} \qquad \overset{5'\rightarrow}{\begin{pmatrix} T & T \\ A & A \end{pmatrix}_n} ; \overset{5'\rightarrow}{\begin{pmatrix} G & G \\ C & C \end{pmatrix}_n}$$

repetitive — monotonous

This leaves us with six DNA models and 20 monotonous polypeptides (poly ALA, poly ARG, poly ASN, poly ASP, etc.). Even if a polypeptide is congruent only to one thread of DNA, the maximum number of comparisons needed for an exhaustive search amounts to 200 (*). Although time consuming, such an approach is clearly feasible.

The second property of monotonous molecules is the automatic addition of small discrepancies, which could pass unnoticed for one pattern, but would produce intolerable defects after repetition. Thus the criteria of congruence will be at their utmost strictness.

The criteria used are threefold:

1) The topological adequation must be maximal in the sense that any other polypeptide should show an evident difficulty or impossibility. For small amino acids, like glycin, the congruence of the residue must discriminate down to one hydrogen atom.

2) The electronic adequation must by complete, including every charged atom of the residues as well as all the peptidic backbone. This requirement of complete saturation of all the potential anchoring points is absolute for the protein but not for the DNA which can also persist in its Watson-Crick configuration without any protein fixed to it.

3) The van der Waals forces, the hydrophobic interactions, and the conformation remolding by London effects must be entirely satisfied.

If, and only if, these three criteria are simultaneously fulfilled on indefinite repetitive sequences, the particular polypeptide will be reputed congruent to the particular DNA sequence which accommodates it.

Although a complete analysis of the Congruence has not yet been achieved, preliminary results can be presented.

Preliminary results

a) Monotonous DNA. As already presented in a short note to the Académie des Sciences de Paris (1), it can be shown that a poly T thread accommodates only a poly-PHE, a poly A thread accommodates only poly-LYS, a poly C accommodates poly-PRO, and poly G accommodates poly-GLY.

b) Semi-monotonous DNA. For these DNAs, as shown in table I, the congruent combinations are not univocal, but include two polypeptides for each thread. This is probably related to the fact that $(TC)^n$ is not different topologically from $(CT)^n$, although the anchoring of the two peptides is indeed different.

(*) Of the 12 DNA threads, only 10 are topologically distinguishable.

TABLE I

CONGRUENCE BETWEEN MONOTONOUS POLYPEPTIDES AND REPETITIVE DNA THREADS

	5'- DNA Thread - 3'	NH_2 congruent peptide - COOH
monotonous DNA	$(TT)^n$	poly PHE
	$(CC)^n$	poly PRO
	$(AA)^n$	poly LYS
	$(GG)^n$	poly GLY
semi-monotonous DNA	$(TC)^n$	poly LEU, poly ARG
	$(TA)^n$	poly ILE, poly TYR
	$(TG)^n$	poly VAL, poly MET
	$(CA)^n$	poly HIS, poly THR
	$(CG)^n$	poly (ALA-SER), poly ARG
	$(AG)^n$	poly GLU, poly ARG

Remark. Although not recorded as repetitively congruent, ASN, ASP, CYS and GLN are found locally congruent, for one pattern only: AA ← ASN; GA ← ASP; TG ← CYS and TRP; and CA ← GLN. Also, AG is locally congruent to SER.

General features of the Congruence Code

Although far from complete, the present observations delineate already some general features.

1) In the major groove of DNA, repetitive congruence is only achieved by amino acids of the natural - L - series.

2) The peptidic chain has always its NH_2 terminus oriented toward the 5' end. This orientation is similar to that observed during synthesis on the messenger RNA.

3) The sequence of the bases of a pattern, congruent to a given amino acid, is identical to that of the genetic code. Even the curious share of the AG pattern by ARG and SER, so remarkable in the genetic code, is found in the congruence code.

4) These similarities are blurred when the pattern includes a thymine. The methylation of uracil seems to introduce some transcoding effect.

5) Small amino acids (like GLY, ALA, SER) attach themselves to the banister, while others are anchored to the central pillar of the DNA spiral staircase. This introduces severe topological restriction in non-monotonous polymers.

Besides the preliminary "dictionary" presented here, the congruence code must be completed by rules of "syntaxis". It is already clear that some arrangements will pull apart the two threads of the DNA (opening the way to transcription or replication) while others will block the two threads together (repression).

Design of experiments

Molecular model building is not a final proof. The availability of monotonous and repetitive polymers of both DNA and protein should allow clear cut controls. If specific fixations occur, according to the congruence code, refined techniques such as separative ultra-filtration, Raman spectrum analysis, or neutron diffraction, to cite a few, would precisely detect them. Also a particular point concerning the electronic state of the N_7 nitrogen atom of guanine must be investigated. The present code implies that this N_7 nitrogen atom is protonated and, thus, is a hydrogen donor (and not a hydrogen acceptor, as commonly believed).

Such experimental verifications are so much the more needed that the Congruence Code, if fully established, would have ample theoretical and practical implications.

Theoretical implications

If the striking analogies between the Congruence Code and the Genetic Code are not rejected a priori as purely coincidental and insignificant, these deep similarities could have some bearing on the basic mechanisms of life.

In considering the syntactic restraints of the Congruence Code, it can be remarked that the "third base" of the genetic code, does relax them entirely. Hence the apparent statistical randomness of the amino acid sequences in ordinary proteins.

Also the transcoding effect of the methyl of thymine, together with the presence of the "third base", protects the DNA from being constantly "parasited" by every protein synthesized according to the structure of its corresponding messenger RNA.

Under this point of view, the two basic interplays between DNA and proteins, i.e. the protein synthesis according to the structure of DNA (via RNA), and the recognition of DNA sequences by the structural properties of particular proteins, would both be based upon the congruence properties between the two series of molecules.

If this view were accepted, the genetic code could no longer be considered an arbitrary convention (in the sense that any other convention would do just as well if once established). Neither could it be considered a "frozen solution" simply maintained by the monoclonal origin of all life and automatically controlled by natural selection.

On the contrary, if based upon congruence properties, the genetic code as we know it now, would not be a solution, but the solution. No other one would do, hence its universality.

Practical implications

As soon as the syntactic laws would be known, the congruence code would greatly help to unravel some basic mechanisms.

If really existing, the conformations and the selective affinities predicted by the congruence code would be the "most likely" and the "most exquisite" ones.

Their application to enzymatic proteins would eventually throw new light on the interplay between these molecules and the nucleotidic coenzymes like ATP for energy, GTP for proteins, CTP for lipids and UTP for glucides.

Also, the actual mode of binding of regulating proteins to their target DNA could be precisely analyzed. In this respect, the sequences of the DNA and of the repressor of the Lac operon will be a kind of "pierre de rosette" in deciphering the fundamental mechanisms of gene regulation.

Eventually also congruence models could profitably be applied to research in immunogenetics, concerning the interplay between genes and antigens.

Finally it could very well be that artificial regulation of gene functions by taylor made polypeptides, would become a major tool in genetic engineering. Among the most obvious possibilities are the compensation of activation failures due to genetic defects or the repression of supernumerary genes in trisomies.

Such achievements are still purely speculative nowadays, but they would lead molecular engineering to become the honest servant of curative medicine, aiming to help the disabled by alleviating their suffering.

1 - J. Lejeune - Congruence de polymères monotones ADN/proteine.
C.R. Acad. Sc. Paris 285, série D 249-252.

Chromosomes Today Volume 6, A. de la Chapelle and M. Sorsa eds.
© 1977 Elsevier/North-Holland Biomedical Press, Amsterdam, The Netherlands

COMPARATIVE BANDING AND GENE MAPPING IN THE PRIMATES. EVOLUTION OF CHROMOSOME 1 DURING FIFTY MILLION YEARS

JEAN DE GROUCHY, CATHERINE FINAZ, NGUYEN VAN CONG
Hôpital des Enfants-Malades 149 rue de Sèvres 75730 Paris Cedex 15 (France)

ABSTRACT

Equivalent chromosomes to chromosome 1 of man and the Pongidae were found in the baboon (PPP) and in the African green monkey (CAE). In PPP, chromosome 1 has a similar banding pattern and carries the ENO-1 and PGM-1 loci. In CAE, two telocentric chromosomes, n°4 and 13, are similar to 1p and 1q respectively. The former carries the ENO-1:PGM-1 synthenic group, and the latter Pep-C.

The identity of banding patterns and of gene contents shows that it is the same chromosome that has been transmitted during fifty million years from the common ancestor to the modern Catarrhini. Three simple evolutionary models are proposed to explain this evolution.

INTRODUCTION

The comparison of the karyotypes of man {*Homo sapiens* (HSA)}, the two varieties of chimpanzee {*Pan troglodytes* (PTR) and *Pan paniscus* (PPA)}, the gorilla {*Gorilla gorilla* (GGO)}, and the orangutan {*Pongo pygmaeus* (PPY)} has unravelled the chromosomal phylogeny of these species and allowed the reconstruction of the karyotype of their common ancestor, who lived some 25 to 30 million years ago [1,2,3,4,5,6].

By cellular hybridization it was shown that many genes with known localizations in man can be assigned to homologous chromosomes in the great apes [7,8,9,10].

The gene map of chromosome 1 is particularly rich in man and many genes have also been assigned to chromosome 1 of the Pongidae (table I).

When comparing this chromosome and apparently homologous chromosomes in two further related species, the baboon {*Papio papio* (PPP)} and the African green monkey {*Cercopithecus aethiops* (CAE)} it is now possible to trace the origin of chromosome 1 further back in time, some fifty million years ago, to the common ancestor of the Catarrhini [11,12] (fig. 1).

METHODS AND RESULTS

The Pongidae. Chromosome 1 is morphologically identical in man, the chimpanzee, the gorilla and the orangutan, except for the presence of a large secondary constriction in man only. Three genes have been localized in all four species : enolase-1

TABLE 1

COMPARATIVE GENE MAPPING OF CHROMOSOME 1 IN MAN AND THE PONGIDAE

Locus	Man	Chimpanzee	Gorilla	Orangutan
ENO-1	+	(P),+	(P),+	(P),+
AK-2	+	(P),+		
PGD	+		(P),+	
PGM-1	+	(P),+	(P),+	(P),+
UGP	+		(P),-	
Pep-C	+	(P),+	(P),+	(P),+
RN5S	+	(P),+	(P),+	(P),+
FH	+		(P),-	
GUK	+	(P),+		

P = provisional assignment

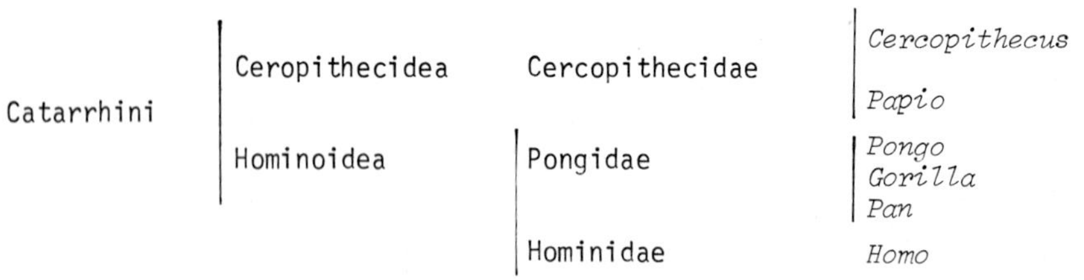

Catarrhini	Ceropithecidea	Cercopithecidae	*Cercopithecus* *Papio*
	Hominoidea	Pongidae	*Pongo* *Gorilla* *Pan*
		Hominidae	*Homo*

Fig. 1. Phylogeny of old world Primates.

(ENO-1), phosphoglucomutase-1 (PGM-1), and peptidase-C (Pep-C) (fig. 2). The RN5S gene(s) has also been assigned to this chromosome [13] but will not be considered further in this paper.

The baboon (PPP). Chromosome 1 in PPP, the largest of the complement, is apparently homologous to chromosome 1 of the Pongidae if it is inverted. Thus, the longest arm has a banding pattern identical to that of 1p in man and the short arm a pattern homologous to that of 1q in man, with the restriction that a large paracentric inversion is necessary and that the secondary constriction is nigh inexistant (fig. 3).

By cellular hybridization it was shown that ENO-1 and PGM-1 are localized on chromosome 1 in PPP (fig. 2). Unfortunatly, the electrophoretic pattern of Pep-C is identical to that of the mouse, thus masking its identification.

The African green monkey (CAE). The reference karyotype of CAE was established by Finaz et al. [14]. No chromosome is homologous to chromosome 1 of man or PPP but two telocentrics, n°s 4 and 13, which are homologous to 1p and 1q of PPP respectively. (Chromosome 13 of CAE therefore differs from 1q of man by the above mentioned paracentric inversion) (fig. 3).

Cellular hybridization showed that ENO-1 and PGM-1 are synthenic in CAE and can

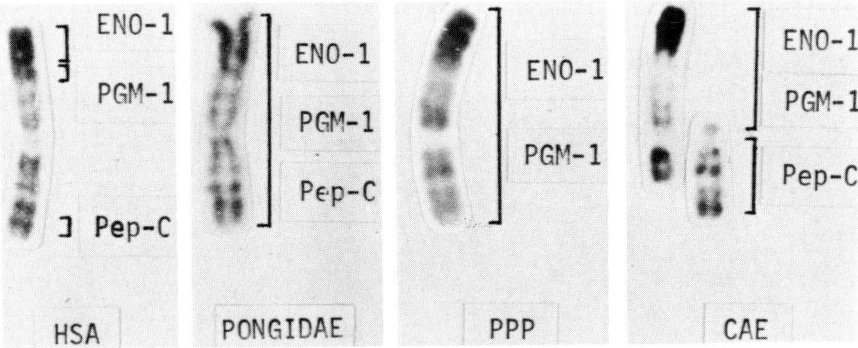

Fig. 2. Gene mapping of chromosome 1 in man (HSA), the Pongidae, the baboon (PPP), and the African green monkey (CAE).

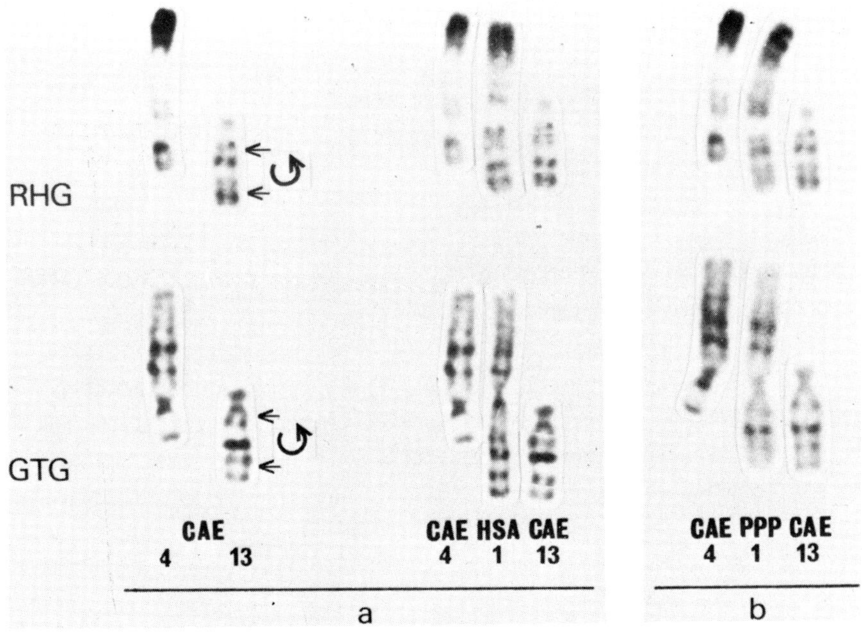

Fig. 3. Compared banding patterns of chromosome 1 in man (HSA) and of its equivalents in the African green monkey (CAE) and the baboon (PPP).RHG:R-bands.GTG:G-bands.

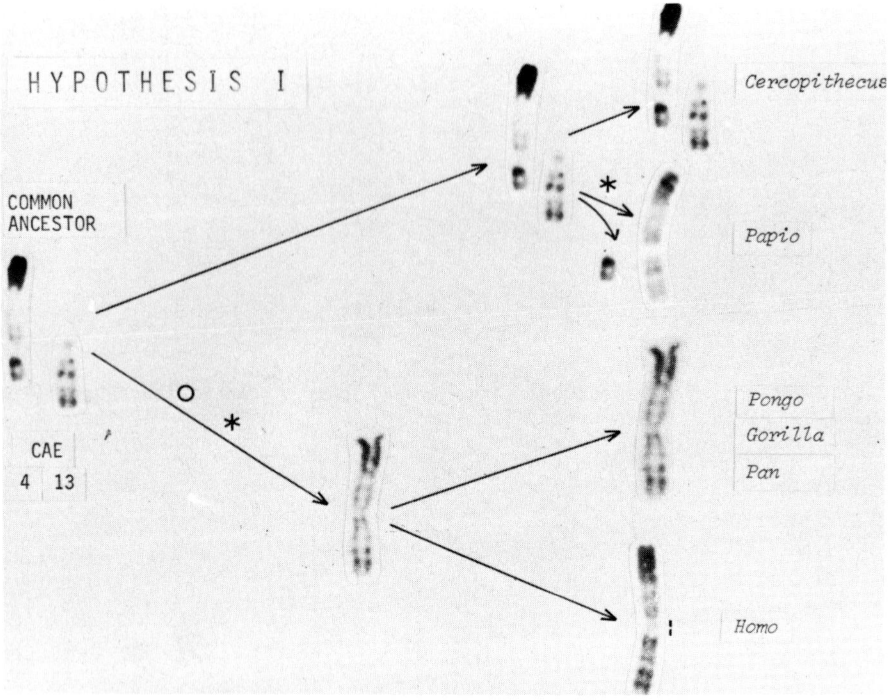

Fig. 4. Evolution of chromosome 1. Hypothesis I (see text).
O paracentric inversion
⁎ fusion
\+ fission
I acquisition of heterochromatic material.

be assigned to chromosome 4, and that Pep-C is not syntenic with the ENO-1:PGM-1 group and can be assigned to chromosome 13 (fig. 2).

DISCUSSION

The identity of banding patterns and of gene contents shows that it is the same chromosome that has been transmitted during fifty million years from the common ancestor to the modern Catarrhini. Three simple evolutionary models can explain this evolution.

Hypothesis I. The ancestral chromosome consisted of chromosomes 4 and 13 of CAE. Two independant fusions and one paracentric inversion are necessary. One fusion must have occurred early in the hominoidea phylum, and the second late in the *Papio* phylum after it separated from CAE (fig. 4).

Hypothesis II. The ancestral chromosome was chromosome 1 of PPP. Evolution involved a centromeric fission and the paracentric inversion of 1q. The fission (the

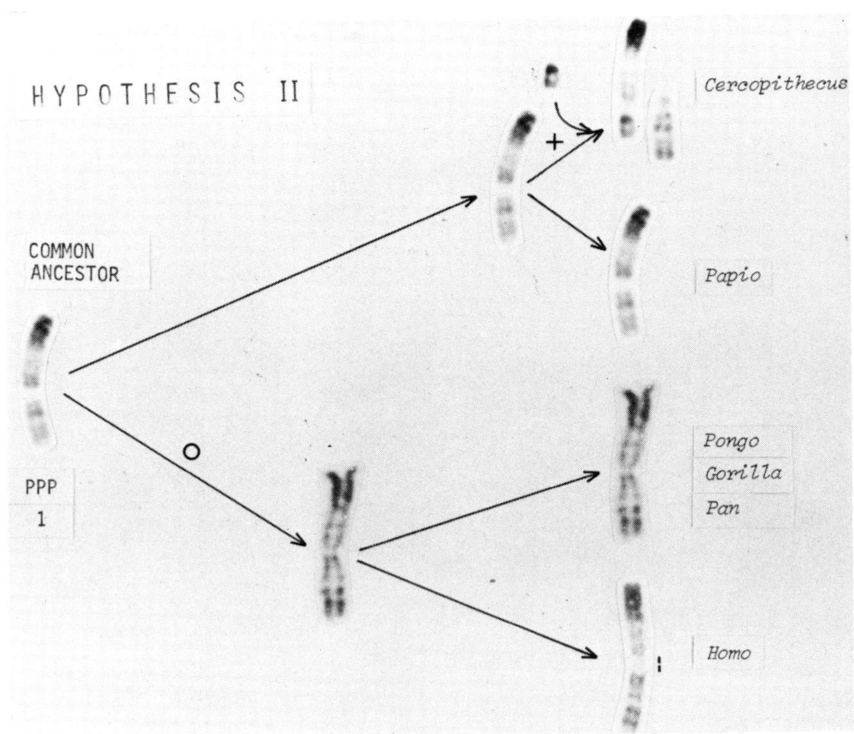

Fig. 5. Evolution of chromosome 1. Hypothesis II.

nature of which is discussed below) occurred in CAE and the inversion early in the hominoidea (fig. 5).

Hypothesis III. The ancestral chromosome was chromosome 1 of the Pongidae. The fission and the paracentric inversion both occurred in the Cercopithecidae, the former after CAE and PPP diverged and the latter before this divergence (fig. 6).

In each model, chromosome 1 of HSA differs from that of the Pongidae by the late acquisition of the juxtacentromeric heterochromatic material.

The nature of the fusion. Centromeric or Robertsonian fusions have been postulated for a long time to explain the reduction of chromosome numbers in many species. A fusion between acrocentrics occurred in man and was responsible for the reduction in the numbers of chromosomes from 48 in the Pongidae to 46 in man. At the origin of chromosome 2, this fusion was not however of the centromeric type but rather a telomeric fusion between the short arms of the two acrocentrics.This end-to-end rearrangement produced a "transiant" dicentric with secondary inactivation of one of the centromeres [15].

In hypothesis I, two fusions are postulated at different moments of evolution but between the same ancestral chromosomes. These fusions were however of the centromeric

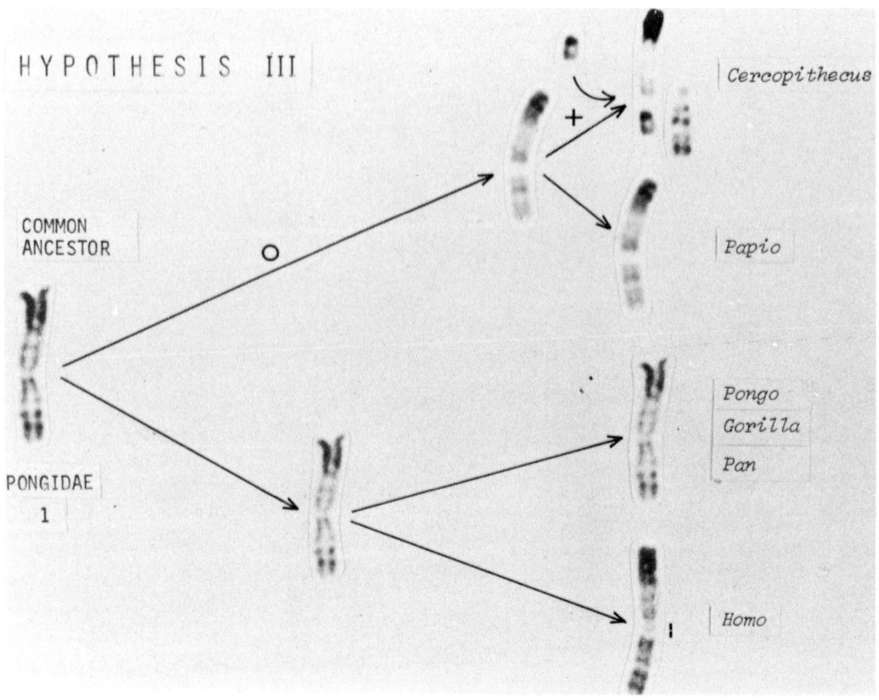

Fig. 6. Evolution of chromosome 1. Hypothesis III.

type as shown by banding. They do not account for the short arm of chromosome 4 in CAE, which could possibly have been translocated onto another chromosome.

Nature of the fission. Centromeric fissions producing two stable acrocentrics have been observed in many species as well as in man [16,17]. Their exact mechanism is far from being understood and one can imagine a variety of models :(1) A simple scission of the centromere produced two parts, both functional. (2) The fission could have been preceded by a duplication of the centromere. (3) The original chromosome may in fact have been the result of an ancient telomeric fusion having produced an active centromere and a "dormant" centromere. If a fission occurred between these two centromeres, the two new chromosomes may have utilized, one the active centromere, the other the resuscitated centromere. (4) The fission may also have been part of a more complex rearrangement involving, for instance, a pericentric inversion or a translocation onto another chromosome, thus accounting for the "loss" of the short arm of chromosome 4 of CAE.

Nature of the paracentric inversion. The paracentric inversion which has involved either chromosome 13 in CAE, or 1q in PPP, or 1q in the Pongidae, has occurred within very broad limits which correspond in fact to the entire chromosome arm.

From an evolutionary point of view, paracentric inversions are remarkable. When an inversion is relatively short, it is compatible with reproduction provided that no crossing-over occurs within its limits. Yet, when it involves practically all of a chromosome arm, at least one crossing-over is bound to occur. One or an odd number of crossovers result in lethal gametes. Only even numbers of crossovers are compatible with fertility. By contrast, as soon as it becomes homozygote, the paracentric inversion is no longer a cause of sterility. It thus behaves as a particularly efficient specifying factor.

CONCLUSIONS

From an economical point of view, hypotheses II and III, each involving a fission and a paracentric inversion, are the least prodigal since they imply only two events. Hypothesis I, which implicates three events - twice the same fusion and a pericentric inversion - may appear less likely. Yet, these economical arguments may not be valid. Nothing proves that identical events did not occur at different moments of evolution.

Whatever its exact evolution has been, chromosome 1 appears as a remarkable chromosome having little changed since fifty million years. One reason may be the nature of the genes the chromosome carries. We know already that its map is the richest of all autosome maps. But more important would be the order of these genes, an order that must not be disrupted by evolution.

Evolution now points to the fact that chromosome 1 is upside down in our modern karyotype. The long arm is in reality the short arm, and *vice versa*. Therefore, we propose that the Nomenclature Committee agrees to reverse chromosome 1.

ACKNOWLEDGEMENTS

The technical assistance of Mrs. Chantal COCHET is gratefully acknowledged.
This work was suported by : I.N.S.E.R.M., U.12 and C.N.R.S., ER 149.

REFERENCES

1. Turleau,C., Grouchy,J.de. (1972) C. R. Acad. Sc. Paris, 274, 2355-2357, serie D.

2. Pearson,P.L., Bobrow,M., Vosa,C.G., Barlow,P.W. (1971) Nature,231, 326-329.

3. Grouchy,J.de., Turleau,C., Roubin,M., Klein,M. (1972) Ann. Génét., 15, 79-84.

4. Turleau,C., Grouchy,J.de., Klein,M. (1972) Ann. Génét.,15, 225-240.

5. Grouchy,J.de., Turleau,C., Roubin,M., Chavin-Colin,F. (1973) Nobel Symposium 23, Stockholm, sept. 25-27, 1972, pp. 124-131.

6. Dutrillaux,B. (1975) Monographies des Annales de Génétique. Expansion Scientifique Ed. pp. 102.

7. Finaz,C., Turleau,C., Grouchy,J.de., Nguyen Van Cong., Rebourcet,R., Frézal,J. (1973) Biomedicine,19, 526-531.

8. Finaz,C., Cochet,C., Grouchy,J.de., Nguyen Van Cong., Rebourcet,T., Frézal,J. (1975) Ann. Génét., 18, 169-177.

9. Grouchy,J.de., Finaz,C., Cochet,C., Nguyen Van Cong., Rebourcet,R., Frézal,J. (1976) 3rd Human Gene Mapping, Conference Baltimore, 1975.

10. Baltimore Conference (1975) : Third International Workshop on Human Gene Mapping. Birth Defects : Original Article Series, XII, 7, 1976, The National Foundation, New York.

11. Finaz,C., Nguyen Van Cong., Cochet,C., Frézal,J., Grouchy,J.de. (1977) Cytogenet. Cell Genet., 18, 160-164.

12. Finaz,C., Nguyen Van Cong., Cochet,C., Frézal,J., Grouchy,J.de. (1977) Ann. Génét. 20, 85-92.

13. Henderson,A.S., Atwood,K.C., Yu,M.T., Warburton,D. (1976) Chromosoma, 56, 29-32.

14. Finaz,C., Dubois,M.F., Cochet,C., Vignal,M., Grouchy,J.de. (1976) Ann. Génét., 19, 213-216.

15. Lejeune,J., Dutrillaux,B., Rethoré,M.O., Prieur,M. (1973) Chromosoma, 43, 423-444.

16. Hansen,S. (1975) Humangenetik, 26, 257-259.

17. Dallapiccola,B., Mastroiacovo,P., Gandini,E. (1976) Hum. Genet., 31, 121-125.

Chromosomes Today Volume 6, A. de la Chapelle and M. Sorsa eds.
© 1977 Elsevier/North-Holland Biomedical Press, Amsterdam, The Netherlands

COMPARATIVE GENE MAPPING IN THE *PONGIDEA* AND *CERCOPITHECOIDEA*

GARVER, J.J., ESTOP, A., PEARSON, P.L., DIJKSMAN, T.M., WIJNEN, L.M.M. and MEERA KHAN, P.
Instituut voor Anthropogenetica, Universiteit van Leiden, Nederland

ABSTRACT

Interspecific cell hybrids from the chimpanzee, gorilla, orangutan, Rhesus monkey and African Green monkey were studied for gene mapping purposes. The segregation of enzyme markers homologous to human enzyme counterparts mapped to human chromosomes (HSA) 1, 6, 11, 12 and X examined were: 6PGD, PGM_1, PPH for HSA 1p, FH for HSA 1q, SOD_2 for HSA 6, LDH-A on HSA 11, LDH-B, Pep B, GAPD and TPI on HSA 12 and G6PD, αGal and PGK on the X. Our results show that the above syntenic groups tested are conserved and that the primate enzyme markers map to chromosomes believed to be, on the basis of banding patterns, homologous to those carrying the same enzymes in man.

The presence of identical syntenic groups and morphologically similar chromosomes in such widely divergent species as man and the *cercopithecoidea* indicates an ultra conservation of linked genes with chromosome banding morphology stretched over some tens of millions of years.

INTRODUCTION

Primate homologies to human chromosomes and their evolutionary consequences have long been speculated upon [1,2,3]. Since the introduction of cytogenetic banding techniques, this field has expanded considerably. Homologies to almost all of the human chromosomes have been proposed for the karyotypes of the chimpanzee, *Pan troglodytes* (PTR), the gorilla, *Gorilla gorilla* (GGO) and orangutan, *Pongo pygmaeus* (PPY) and many homologies have also been suggested in the Rhesus monkey, *Macaca mulatta* (MMA) and African Green monkey, *Cercopithecus aethiops* (CAE)[4,5].
These homologies are depicted in Fig. 1 in which the chromosomes are arranged from left to right as man, chimpanzee, gorilla, orangutan, rhesus and african green respectively. As yet there is no generally accepted nomenclature for the latter two species and in this article we use that previously proposed by us[5]. The chromosome identification for man and the great apes used is the same as that outlined in the supplement to the Paris nomenclature[6].

We have used the technique of somatic cell hybridization which has made possible the expansion of the human gene map as an additional tool for comparative chromosome evaluation. Our aim was to investigate whether primate chromosomes which look alike morphologically, contain similar linkage groups (syntenic groups). The combination of banding and gene mapping studies gives a more complete picture of the evolution of specific chromosomes with the divergence of species, and sheds more light on the possible mechanisms of chromosomal modification and their role in speciation.

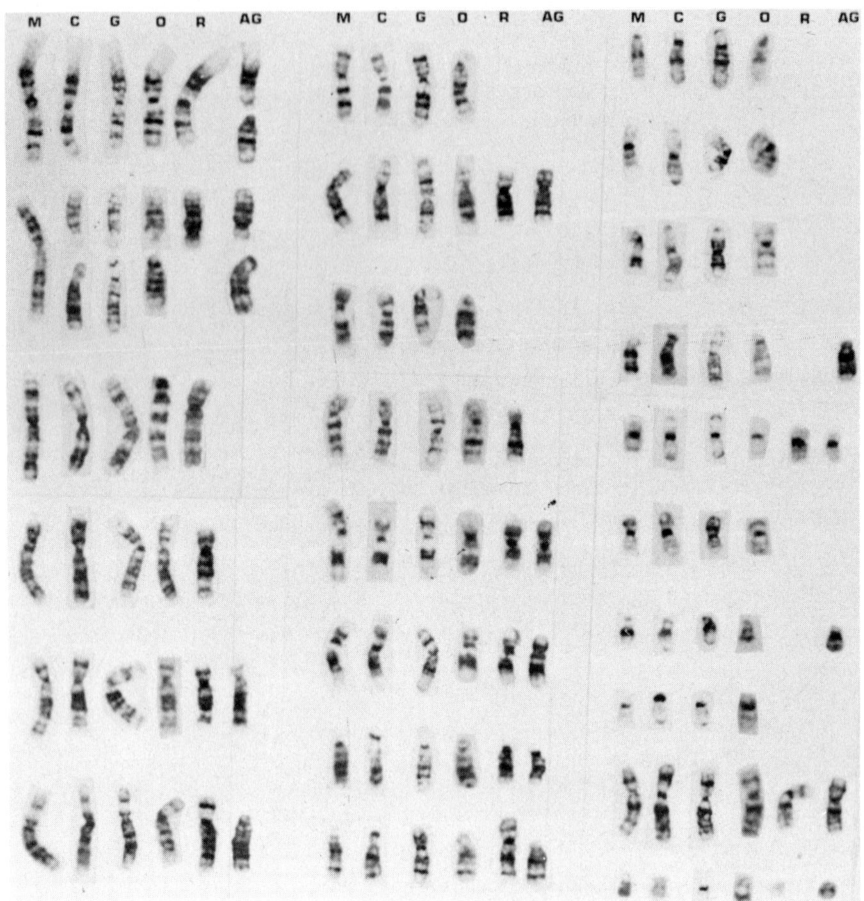

Fig. 1. The chromosomes of man (m), chimpanzee (c), gorilla (g), orangutan (o), Rhesus (R) and African Green (AG) arranged to show their supposed human homologies. The human chromosomes are ordinated conventionally from 1-22,X,Y. Spaces indicate absence of obvious banding homology.

MATERIALS AND METHODS

Peripheral lymphocytes from the before mentioned primates were fused with a thymidine kinase deficient Chinese hamster cell line by means of inactivated Sendai virus. Growth medium supplemented with hypoxanthine thymidine and amenopterine was used for selection (HAT medium) and isolation of interspecific hybrid clones as described by Westerveld[7]. Cellogel electrophoresis was performed on cell lysates to detect the following enzymes according to the systems prescribed by Meera Khan[8] and van Someren[9]: 6-phosphogluconate dehydrogenase (6PGD; E.C. 1.1.1.44), Phosphopyruvate hydratase (PPH; E.C. 4.2.1.11), Phosphoglucomutase (PGM; E.C. 2.7.5.1), Fumarate hydratase (FH; E.C. 4.2.1.2), Superoxide dismutase (SOD; E.C. 1.15.1.1), Lactate dehydrogenase (LDH; E.C. 1.1.1.27), Peptidase B (Pep B), Triose phosphate isomerase (TPI; E.C. 5.3.1.1), Glucose-6-phosphate dehydrogenase (G6PD; E.C. 1.1.1.49), α-Galactosidase (αGal; E.C. 3.2.1.22), 3-Phosphoglycerate kinase (PGK; E.C. 2.7.2.3). The system for glyceraldehyde 3 phosphate dehydrogenase (GAPD; E.C. 1.2.1.12) and citrate synthase (CS) was that of Herbschleb-Voogt et al. in preparation.

Chromosome analysis of the hybrid clones was performed using the alkaline-giemsa (G11) banding technique (Bobrow, 1973), followed by the trypsin giemsa technique (Seabright, 1971). In the case of the chimpanzee and gorilla hybrids, where fluorescent polymorphisms of the primate parents aided the identification of these particular chromosomes, atebrine stained preparations were also used for chromosome screening. About 20-30 metaphases for each hybrid cell line were scored for the primate chromosome complement.

RESULTS

The number of primary hybrid clones isolated and screened for each ape species studied was: 20 for the chimpanzee, 22 for the gorilla and 18 subclones, 18 for the orangutan, 11 for the Rhesus and 10 for the African Green. The segregation analysis results are given in table 1. The analysis pertaining to homologues of human 2, 9 and 14 are presented in the accompanying article[10]. Figures 2-6 show karyotypes of hybrid cells. It was found that the human 1p syntenic group was conserved through to the African Green monkey for the markers PGD, PPH and PGM_1 and carried on primate chromosomes homologous to the HSA 1, fig. 1 and in the case of the African Green to HSA 1p. FH was found to be syntenic to these markers in the great apes and the Rhesus but localized onto African Green 6[5] (HSA 1q). A single primary clone carrying a chromosome 1 deletion in the orangutan supports the relative HSA 1q and 1p distribution for these markers. We can therefore conclude that the HSA 1 syntenic group is preserved intact in all but the African Green, where two separate chromosomes (CAE 1 and 6) appears to represent this chromosome.

SOD-2 segregates concordently with the proposed HSA 6 homologues, namely PTR 5, GGO 5, PPY 5 and MMA 6 but not with the proposed homology for HSA 6q (CAE 2) in the African Green. We regard the lack of association reported in previous studies[2,11] as being due to either lability or detection of the enzyme.

Homologues for HSA 11 in all of the primates tested contain LDH-A. The syntenic group LDH-B, Pep-B, TPI is preserved and localized to a chromosome homologous to HSA 12 in all species tested. In the chimpanzee, orangutan, Rhesus and African Green monkey, GAPD was also found to belong to this syntenic group. It could not be assayed in our gorilla hybrids. In addition, although CS could only be separated in the Rhesus and African Green monkeys it also belonged to the HSA 12 syntenic group. The X-chromosome markers G6PD, αGal and PGK were syntenic in all of the primates except in the

TABLE 1

SEGREGATION ANALYSIS

Chromosome 1

PTR 1 (HSA 1)

	++	+-	-+	--
6PGD	9	0	0	11
PPH	9	0	0	11
PGM_1	9	0	0	11

GGO 1 (HSA 1)

	++	+-	-+	--
6PGD	9	0	0	12
PPH	9	0	0	10
PGM_1	9	0	0	13
FH	4	0	0	6
PEP-C	5	0	0	6

PPY 1 (HSA 1)

	++	+-	-+	--
6PGD	5	2	0	5
PPH	5	2	0	5
PGM_1	5	2	0	5
FH	4	1	0	6

MML 1 (HSA 1)

	++	+-	-+	--
6PGD	3	0	0	8
PGM_1	3	0	0	8
PPH	3	0	0	8
FH	3	0	0	8
AK_2	3	2	0	5
GUK	3	1	1	6

CAE 1 (HSA 1p)

	++	+-	-+	--
6PGD	5	0	0	5
PPH	5	0	0	5

CAE 6 (HSA 1q)

	++	+-	-+	--
FH	3	0	0	6

We found two clones on the orangutan hybrids which were positive for the enzymes 6PGD, PPH and PGM_1 but an intact PPY 1 could not be identified. One of the exceptional clones had a PPY 1p deletion and was FH negative and positve for the other markers supporting the human distribution. The other exceptional clone was positive for all the markers but carried no recognisable no. 1 chromosome.

Chromosome 6

PTR 5 (HSA 6)

	++	+-	-+	--
SOD_2	11	0	0	9

GGO 5 (HSA 6)

	++	+-	-+	--
SOD_2	4	0	0	6

PPY 5 (HSA 6)

	++	+-	-+	--
SOD_2	8	1	0	2

MML 6 (HSA 6)

	++	+-	-+	--
SOD_2	6	1	0	4

CAE 2 (HSA 6q)

	++	+-	-+	--
SOD_2	3	3	1	3

Table 1 cont.

Chromosome 11

	PTR 9 (HSA 11)			
	++	+-	-+	--
LDH-A	8	1	0	11

	GGO 9 (HSA 11)			
	++	+-		
LDH-A	21	2	1	10

	PPY 8 (HSA 11)			
	++	+-	-+	--
LDH-A	6	1	0	8

	MML 11 (HSA 11)			
	++	+-	-+	--
LDH-A	4	0	0	7

	CAE 12 (HSA 11)			
	++	+-	-+	--
LDH-A	6	0	0	4

Chromosome 12

	PTR 10 (HSA 12)			
	++	+-	-+	--
LDH-B	12	0	2	6
GAPD	9	0	0	1
PEP-B	11	0	3	6
TPI	12	0	2	6

	GGO 10 (HSA 12)			
	++	+-	-+	--
LDH-B	12	0	0	10
GAPD	3	0	0	4
PEP-B	11	0	1	9
TPI	12	0	0	9

	PPY 9 (HSA 12)			
	++	+-	-+	--
LDH-B	7	0	0	6
GAPD	4	0	1	6
PEP-B	4	1	1	6
TPI	4	0	1	6

	MML 12 (HSA 12)			
	++	+-	-+	--
LDH-B	9	0	0	2
GAPD	8	0	1	2
PEP-B	8	0	1	2
TPI	8	0	1	2
CS	7	0	1	1

	CAE 13 (HSA 12)			
	++	+-	-+	--
LDH-B	6	0	0	4
GAPD	6	0	0	4
PEP-B	6	0	0	4
TPI	6	0	0	4
CS	6	0	0	4

The X-chromosome

	PTR X			
	++	+-	-+	--
G6PD	6	0	0	14
PGK	6	0	0	14
αGAL	6	0	0	14

	GGO X			
	++	+-	-+	--
G6PD	2	0	0	8
PGK	2	0	0	8
αGAL	2	0	0	8

	PPY X			
	++	+-	-+	--
G6PD	7	0	0	6
αGAL	6	0	0	6

	MML X			
	++	+-	-+	--
G6PD	1	0	0	10
αGAL	1	0	0	10

	CAE X			
	++	+-	-+	--
G6PD	2	0	0	8
PGK	2	0	0	8
αGAL	2	0	0	8

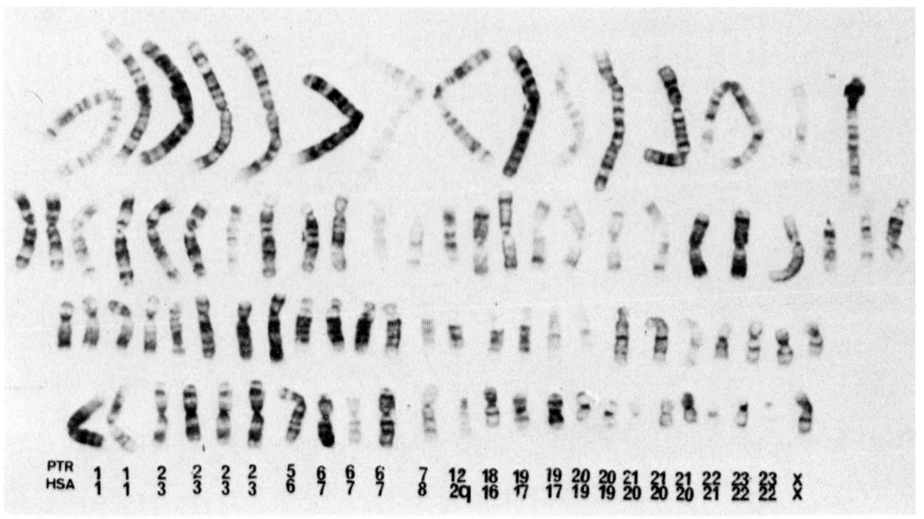

Fig. 2. Karyotype of the chimpanzee primary hybrid clone 7. The chimpanzee chromosomes are arranged below identified by both primate nomenclature and human chromosome homologies.

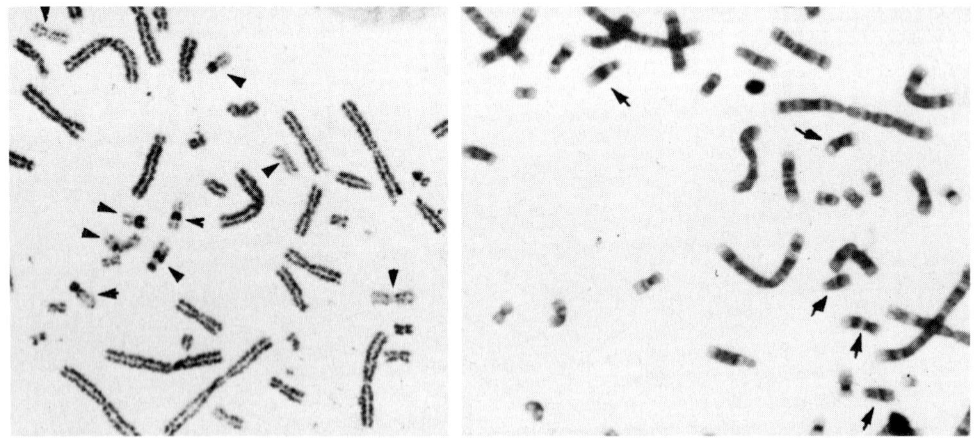

Fig. 3. G11 stained hybrid metaphase from gorilla primary hybrid clone 6. Indicated are GGO 2, 6, 17, 18 and 2d group, 2c group and 1f group chromosome.

Fig. 4. African Green hybrid metaphase containing the homologues to HSA 1q and 1p which are indicated.

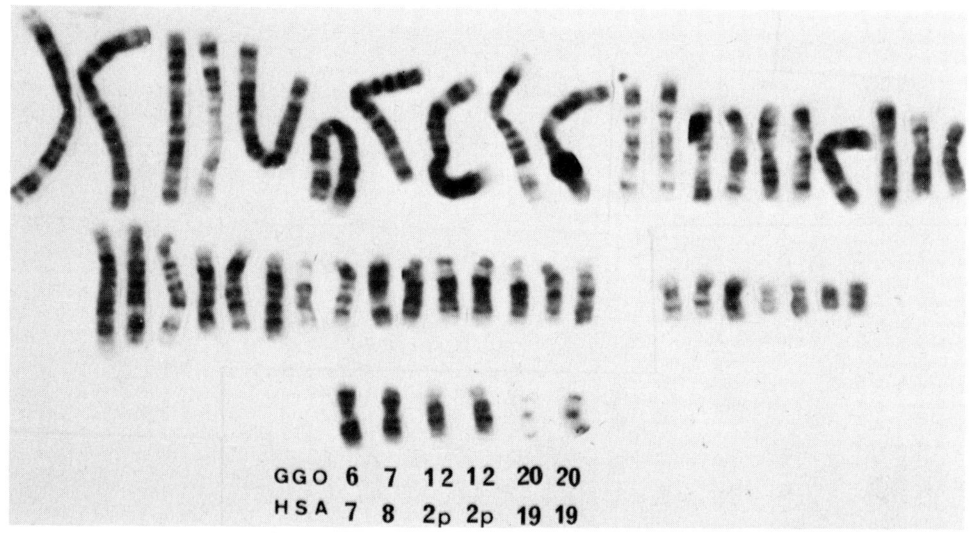

Fig. 5. Karyotyped gorilla hybrid metaphase from clone 3. The gorilla chromosomes are arranged below identified by gorilla nomenclature and human homology.

case of the orangutan and Rhesus where electrophoretic separation between the primate and Chinese hamster forms of PGK was not achieved.

DISCUSSION AND CONCLUSIONS

A common ancestor to the *cercopithecoidea* and *hominoidea* is believed to have existed during the early Eocene period. The earliest fossil evidence of a possible common cercopithecoid ancestor dates to about 35 million years ago[12]. Considerable chromosome homogeneity is to be found within the catarrhine apes, showing conservation of specific genetic material during evolution. Karyotypic similarities found within the catarrhines implies that these chromosomes would have existed in the karyotype of an ancient ancestor. Particularly interesting is the finding that certain ancestral chromosomes throughout evolution not only have retained similar banding patterns but have, as well retained the same syntenic enzyme groups.

General conclusions on the phylogeny of the chromosomes studied are:

1) The HSA 1 homologues have inherited and retained their q and p arm syntenic groups and banding morphology from an ancient common progenitor of the *cercopithecoidea* and *hominoidea*. The question of whether the Rhesus and Hominoid 1 arose from a centric fusion of two acrocentric chromosomes now present in the African Green or whether

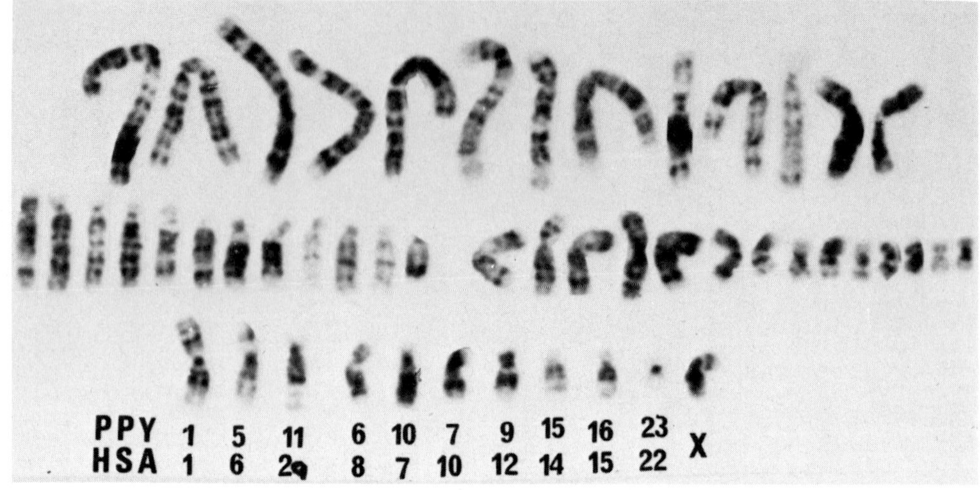

Fig. 6. A metaphase from the orangutan hybrid clone 60 is karyotyped above. The orangutan chromosomes are arranged below identified by both orangutan nomenclature and their human homologies.

these two acrocentrics arose from a fission of the present metacentric chromosome remains a matter of speculation.

2) Since the homologues to HSA X and 12 appear to be identical in both the *hominoidea* and *cercopithecoidea* studied, we can assume that these very ancient chromosomes have also inherited their banding pattern and gene content from a common progenitor.

3) However HSA 11 exhibits modest banding modifications whilst retaining an apparently similar gene content.

4) The homologues to HSA 6 also appear to be very similar in structure and gene content. This is of particular interest since the human 6 carries the genes for the major histocompatibility loci and although much evidence is available on their genetic variability in primates, especially in the chimpanzee and Rhesus monkey, as yet no direct chromosome localisation has been made. These studies pave the way to carrying out further investigations in localising this extremely important gene group.

We can conclude that a similarity in banding pattern in general implies a similarity in gene content. Individual exceptions to this are found in the lack of association of ITP to homologues of HSA 20, GPI to HSA 19 and SOD-2 to the African green HSA 6q. Discussion of these exceptions will form the basis of a further communication.

REFERENCES

1. Hamerton, J.L., Klinger, H.P., Mutton, D.E. and Lang, E.M. (1963) The somatic chromosomes of the Hominoidea. Cytogenetics (Basel) 2, 240-263

2. Pearson, P.L, (in press) Banding patterns, Chromosome polymorphism and primate evolution. ed. Yunis, J.

3. Dutrillaux, B. Sur la Nature et l'origine des chromosomes humains. Monographies des Annales de Génétique.

4. Stock, A.D. and Hsu, T.C. (1973) Evolutionary conservatism in arrangement of genetic material. Chromosoma (Berl.) 43, 211-224

5. Estop, A., Garver, J., Pearson, P.L., Wijnen, L.M.M. and Meera Khan, P. (in press) Gene mapping in the Cercopithecoidea. In Human Gene Mapping 4. Winnipeg Conference (1977) In Cytogenetics and Cell Genetics.

6. Paris Conference (1971) Supplement 1975. Standardization in Human Cytogenetics. In Birth Defects, Original Article Series, Vol. XI no. 9

7. Westerveld, A., Meera Khan, P., Visser, R.P. and Bootsma, D. (1971) Loss of human genetic markers in man-Chinese hamster somatic cell hybrids. Nature New Biol. 234, 20-24

8. Meera Khan, P. (1971) Enzyme electrophoresis on cellulose acetate gel: Zymogram patterns in man-mouse and man-Chinese hamster somatic cell hybrids. Arch. Biochem. Biophys. 145, 470-483

9. Van Someren, H., Beijersbergen van Henegouwen, H., Loss, W., Würzer-Figurelli, E., Doppert, B., Vervloet, M. and Meera Khan, P. (1974) Enzyme electrophoresis on cellulose acetate gel. II. Zymogram patterns in man-Chinese hamster somatic cell hybrids. Humangenetik 25, 189-201

10. Pearson, P.L., Estop, A., Garver, J.J., Wijnen, L.M.M. and Meera Khan, P. (in press) Evidence for the origin of human chromosomes 2, 9 and 14 from primate-rodent somatic cell hybrids. In Chromosomes Today Vol. 6. ed. de la Chapelle, Sorsa, M.

11. Warburton , D. and Pearson, P.L. (1975) Committee Report on comparative mapping. In Human Gene Mapping 3. Baltimore conference. Cytogenetics and Cell Genetics 14

12. Simons, E.L. (1967) The earliest Apes. Scientif American vol. 217, dec. 25-35

Chromosomes Today Volume 6, A. de la Chapelle and M. Sorsa eds.

EVIDENCE FOR THE ORIGIN OF HUMAN CHROMOSOMES 2, 9 AND 14 FROM PRIMATE-RODENT SOMATIC CELL HYBRIDS

PEARSON, P.L., ESTOP, A., GARVER, J.J., DIJKSMAN, T.M., WIJNEN, L.M.M. and MEERA KHAN, P.
Instituut voor Anthropogenetica, Rijksuniversiteit te Leiden, Nederland

ABSTRACT

A study of chromosome and enzyme segregation in primate-rodent somatic cell hybrids has shown that the genes for the enzymes IDH and MDH, believed to be carried on the long and short arms of human chromosome two respectively, dissociate in all species tested. In the chimpanzee and the gorilla these two enzymes are carried on the two chromosomes proposed to have undergone telomeric fusion to form chromosome 2 in man. A study of human chromosome 9 markers has shown that the same enzymes are coded by the proposed homologue in the chimpanzee and has helped to define which is the human 9 homologue in the gorilla. Although these genes are syntenic in the orangutan, African Green monkey and Rhesus monkey, there is no chromosome present identifiable as a homologue to the human 9 on the basis of banding pattern. The concordant segregation of the enzyme NP with a chromosome similar to the human 14 in all species studied indicates the ancient linkage of this particular chromosome.

INTRODUCTION

There has long been speculation over the possible ways in which the diploid chromosome no. of 48 present in the Pongidae could have been reduced to that of 46 in man. Hamerton et al.[1], and much more recently Chiarelli[2], proposed that a centric fusion involving two acrocentric chromosomes followed by establishment of homozygosity was the most probable way in which the reduction could have occurred. The latter author suggested, on the basis of chromosome measurement data, that the human no. 1 was the chromosome formed by the centric fusion. Following the introduction of chromosome banding techniques, de Grouchy et al.[3], proposed that the human 2 had been formed by a centric fusion of two chromosomes present in the Pongidae. This theory was subsequently modified by Lejeune et al.[4], who suggested a telomeric and not centric fusion origin of human chromosome 2 with suppression of one of the two centromeres. We have made use of the technique of somatic cell hybridisation to study the gene relationships of primate chromosomes to their human counterparts and to see whether any further information can be gained on the possible origin of human chromosome 2. This communication presents the results of that study and confirms the proposed two chromosome origin of human chromosome 2. See fig. 1.

In addition, the homologues to the human 9 and 14 have been further specified (figs. 2 and 3) by gene mapping, helping to erase some of the discrepancies and

differences of opinion over homologies based on banding patterns alone[5].

MATERIALS AND METHODS

The preparation and analysis of the hybrid cell material is as described by Garver et al.[6] elsewhere in this volume. The enzymes studied were MDH-1 (Malate dehydrogenase E.C. 1.1.1.37), IDH-1 (Isocitrate dehydrogenase, E.C. 1.1.1.42) and ACP-1 (Acid phosphotase-1, E.C. 3.1.3.2) for human chromosome 2 homologues; AK-1 (Adenylate-kinase -1, E.C. 2.7.4.3), AK-3 (Adenylate-kinase-3), and ACO-s (Soluble aconitase, E.C. 4.2.1.3) for human chromosome 9 homologues; NP (Nucleoside phosphorylase, E.C. 2.4.2.1) for human chromosome 14 homologues. The enzyme terminology is that proposed by Shows[7] and the number in parentheses refers to information on the enzyme given in McKusick's Mendelian Inheritance in Man[8].

RESULTS

Tables 1 to 3 give the hybrid cell segregation analyses for specified enzymes and the proposed homologues to human chromosomes 2, 9 and 14 respectively.
An excess in the ++ and -- segregation categories indicates a concordant segregation.

The trinomial followed by a number used in the table refers to the species concerned and a particular chromosome as defined in the supplement to the Paris Conference[5]. Figs. 4 and 5 show two hybrid cell karyotypes from clones thought to contain the gorilla and chimpanzee homologues to the human 2p and 2q arms.

DISCUSSION

From the segregation data we can conclude that the enzymes MDH and IDH, whilst being on a single chromosome in man[9], are carried on separate chromosomes in the primates studied. Further, in the chimpanzee and gorilla, they appear to segregate with the two chromosomes postulated to having been involved in the formation of human chromosome 2. This is in itself overwelming evidence on the origin of this chromosome. However in our study the gorilla chromosome thought to be homologous to 2p[5], was found to carry IDH and that thought to be homologous to 2q[5] the enzyme MDH. This is the reverse distribution to that found in man[9] and suggests a reconsideration of the 2p and 2q homologies of these gorilla chromosomes. The homologues to the human 2p and 2q could not be identified with certainty in the orangutan hybrid cell karyotype. Another variation to the human pattern was observed with the dissociation of ACP-1 with MDH in both the chimpanzee and orangutan, both of these enzymes being fairly closely located in the short arm of human chromosome 2[9].

Attempts to define a human-great ape chromosome nomenclature based upon banding homologies have been partially frustrated through difficulties in deciding which gorilla and orangutan chromosomes could be regarded as homologous to human chromosomes 9 and 14[5].

The present studies help to clarify the situation in that the chromosome 14 marker

Figs. 1-3 give the supposed chromosome homologies to human chromosomes 2, 9 and 14 respectively. The chromosomes are arranged from left to right in the order man, chimpanzee, gorilla, orangutan, Rhesus and African Green. Spaces indicate a lack of obvious homology.

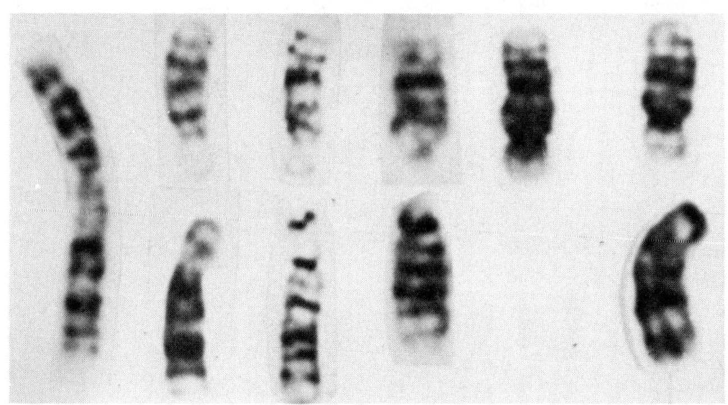

Fig. 1.
Homologies to human chromosome 2

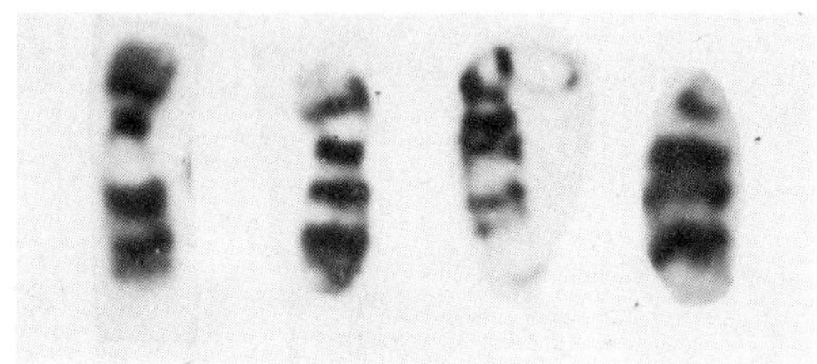

Fig. 2.
Homologies to human chromosome 9

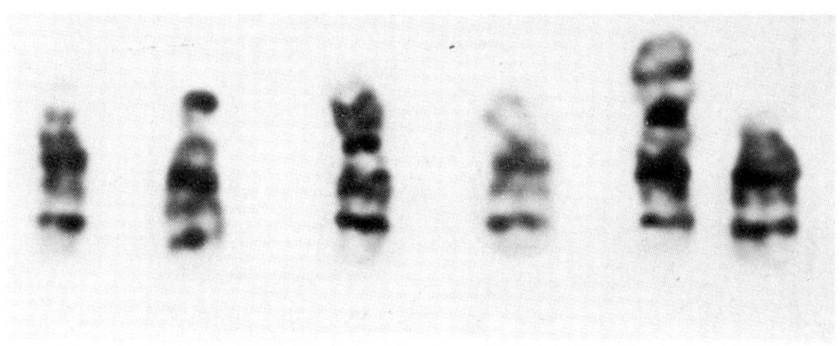

Fig. 3.
Homologies to human chromosome 14

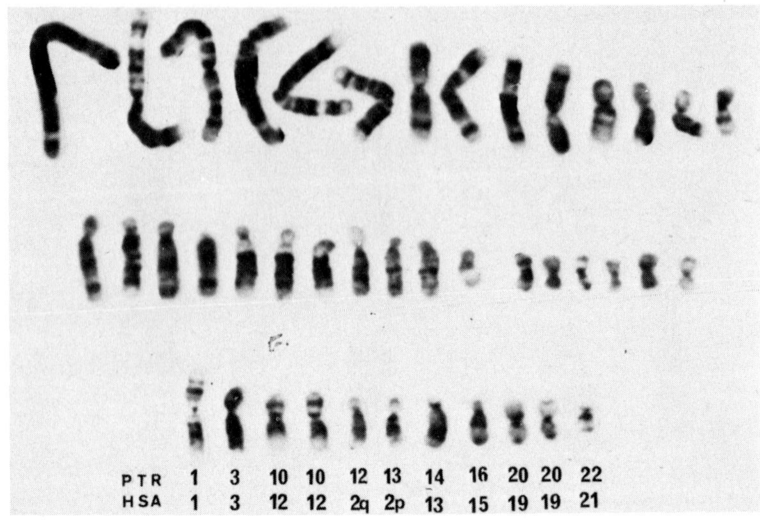

Fig. 4. G-banded karyotype of a chimpanzee/Chinese hamster hybrid cell showing the chimpanzee chromosomes on the bottom row. The supposed human homologies have been indicated below the chimpanzee numbering[5].

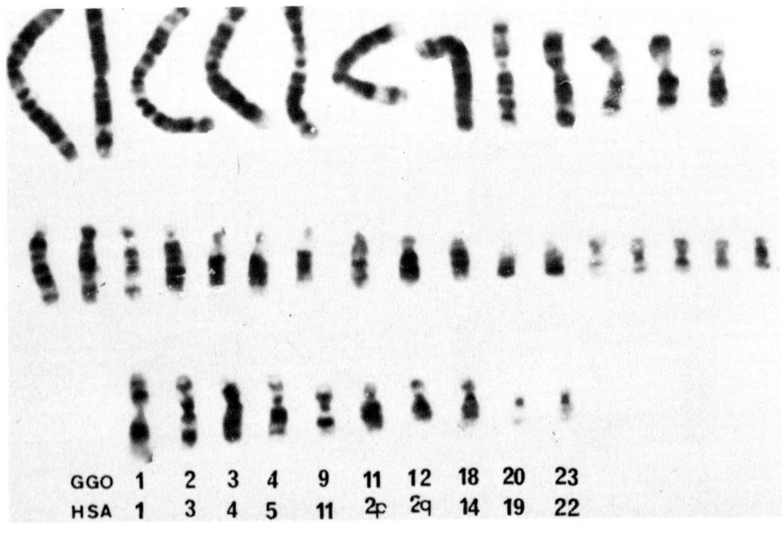

Fig. 5. G-banded karyotype of a gorilla/Chinese hamster hybrid cell.

TABLE 1

HUMAN CHROMOSOME 2 RELATIONSHIPS

Chimpanzee

	MDH			
	++	+-	-+	--
IDH	5	3	1	3

	ACP-1			
	++	+-	-+	--
MDH	4	0	2	2

	PTR 13			
	++	+-	-+	--
MDH	10	0	0	10

	PTR 12			
	++	+-	-+	--
IDH	8	0	0	4

Orangutan

	MDH			
	++	+-	-+	--
IDH	2	4	1	3

African Green

	MDH			
	++	+-	-+	--
IDH	2	1	3	3

Gorilla

	MDH			
	++	+-	-+	--
IDH	2	3	4	1

	GGO 11			
	++	+-	-+	--
MDH	5	0	0	3

	GGO 12			
	++	+-	-+	--
IDH	4	0	0	4

TABLE 2

HUMAN CHROMOSOME 9 RELATIONSHIPS

Chimpanzee

	PTR 11			
	++	+-	-+	--
AK-1	10	0	0	9
ACO-s	10	1	0	9
AK-3	5	0	0	4

Gorilla

	GGO 13			
	++	+-	-+	--
AK-3	4	0	0	6
ACO-s	4	0	0	6

Rhesus

	AK-3			
	++	+-	-+	--
ACO-s	4	0	0	7

African Green

	AK-1			
	++	+-	-+	--
ACO-s	5	0	0	5

TABLE 3

HUMAN CHROMOSOME 14 RELATIONSHIPS

Chimpanzee

	PTR 15			
	++	+-	-+	--
NP	14	0	0	4

Gorilla

	GGO 18			
	++	+-	-+	--
NP	5	0	0	5

Orangutan

	PPY 15			
	++	+-	-+	--
NP	3	4	2	1

African Green

	HSA 14			
	++	+-	-+	--
NP	3	1	1	4

NP segregates with gorilla chromosome 18 and the chromosome 9 markers with gorilla chromosome 13, thus supporting the homologies first proposed by Dutrillaux et al.[10]. Although the chromosome 9 syntenic group appears to be retained intact in the orang-utan, Rhesus monkey and African Green, as yet we have been unable to make any chromosomal assignment for it. Suffice it to say that the homologue to the human 9 in these species has been considerably modified in banding pattern to that seen in man suggesting a comparatively recent origin for this chromosome as opposed to the homologues of human 1, 6, 12 and X for example[6].

The finding that NP associates with a metacentric chromosome whose long arm resembles that of the human 14 in the African Green monkey hybrids (fig. 3) can be regarded as strong evidence that the human 14 also has a very long ancestry.

ACKNOWLEDGEMENTS

A. Estop has received a Dutch cultural exchange fellowship to participate in this work.

REFERENCES

1. Hamerton, J.L., Klinger, H.P., Mutton, D.E. and Lang, E.M. (1963) The somatic chromosomes of the Hominoidea. Cytogenetics (Basel) 2, 240-263

2. Chiarelli, B. (1972) Comparative chromosome analysis between man and chimpanzee. J. Hum. Evol. 1, 389-393

3. De Grouchy, J., Turleau, C., Roubin, M. and Klein, M. (1972) Evolution caryotypiques de l'homme et du chimpanzee. Etude comparative des topographies de bandes après dénaturation menagée. Ann. Genet. 15, 79-85

4. Lejeune, J., Dutrillaux, B., Rethoré, M. and Prieur, M. (1975) Comparison de la structure fine des chromatides d'Homo sapiens et de Pan troglodytes. Chromosoma 43, 423-444

5. Paris Conference (1971) Supplement 1975. Standardization in Human Cytogenetics. In Birth Defects, Original Article Series, Vol. XI, no. 9

6. Garver, J.J., Estop, A., Pearson, P.L., Dijksman, T.M., Wijnen, L.M.M. and Meera Khan, P. (in press) Comparative gene mapping in the Pongidea and Cercopithecoidea. In Chromosomes Today vol. 6. ed. de la Chapelle, A. and Sorsa, M.

7. Shows, T.B. (1975) Gene markers for mapping the human genome. The 1974 listing. In Human Gene Mapping 2. Cytogenetics and Cell Genetics 14, 29-37

8. McKusick, V.A. (1971) Mendelian inheritance in man, 3rd Ed. (John Hopkins Univ. Press, Baltimore)

9. Hamerton, J.L. (1976) Report of the committee on the genetic constitution of chromosomes 1 and 2. In Human Gene Mapping 3. Cytogenetics and Cell Genetics 16, 7-23

10. Dutrillaux, B., Rethoré, M.O., Prieur, M. and Lejeune, J. (1973) Analyse de la structure fine des chromosomes du Gorilla (Gorilla gorilla) comparison avec Homo sapiens et Pan troglodytes. Humangenetik 20, 343-354

FUNCTION OF CHROMOSOMES

Chromosomes Today Volume 6, A. de la Chapelle and M. Sorsa eds.
© 1977 Elsevier/North-Holland Biomedical Press, Amsterdam, The Netherlands

FUNCTION OF CHROMOSOMES - AN INTRODUCTION

NILS R RINGERTZ
Institute for Medical Cell Research and Genetics, Medical Nobel Institute,
Karolinska Institutet, S-104 01 Stockholm 60, Sweden

"Function of chromosomes" is a very broad topic and one could never hope to cover this topic fully at any single scientific meeting. The contributions presented in this section of the Helsinki Chromosome Conference do, however, illustrate some of the new experimental techniques which make it possible to analyze the function of chromosomes. It is also worth noting that some of the "old" biological systems continue to yield important information when combined with the modern methods of molecular biology. One example of this are the well-known polytene chromosomes of insects. Ashburner very appropriately has named his paper "Happy Birthday-Puffs". With this title he celebrates the fact that it is now 25 years since Beerman, Pavan and Breur suggested that the puffs of Dipteran polytene chromosomes represent active genes. That this experimental system is still highly productive is illustrated by Ashburner´s own report on the molecular basis of "puffing" and the correlation between puffing and gene expression. In a subsequent paper Edström and his colleagues discuss transcription in the giant puffs of *Chironomus tentans* known as Balbiani rings. This work examines a question which has been the subject of much controversy in mammalian cells; the intranuclear degradation of messenger RNA. The data presented show that for at least one type of gene, there is very little degradation of the transcription product in the nucleus. Further information about the accumulation of proteins in puffs is presented by Tanguay and coworkers. They find different protein patterns in different subnuclear structures. Activation of genes is associated with the accumulation of specific nonhistone proteins at chromosome regions undergoing puffing. Although much remains to be done this study brings us closer to an understanding of how specific proteins influence the activity of specific genes. Another contribution to this field is the paper by Jamrich *et al* which shows that it is possible to localize specific histones and RNA polymerases in salivary gland polytene chromosomes using immunofluorescence technique. Dr Steffensen, on the other hand, studies the intranuclear localization of specific genes in *Drosophila* interphase cells and the possible role of gene positioning for gene expression and the process of determination. The interphase nucleus is far from being a bag of randomly oriented chromosomes. Using nucleic acid hybridization he shows that 5s rRNA genes take a precise position at "2PM" between the nucleolus and the nuclear membrane. Dr. Stahl and his colleagues present another study pertaining to intranuclear anatomy. This report examines the fine localization of rDNA genes during meiotic prophase in mouse oocytes, and presents evidence that rDNA genes are present in the fibrillar centers of the nucleoli. Dr. Traut, in his presentation, discusses the transcriptional activity of chromosomes during oogenesis in a moth. During this process the chromosomes change from the lampbrush type to a stage where the loops have been retracted and the chromosomes have assumed a beaded appearance.

Plant cell chromosomes have been extensively studied. Yet there have been relatively few studies of chromosome structure in relation to cell differentiation. This situation may soon change. Dr Kessler in his paper reports on chromosomal DNA changes affecting nucleotide sequence redundancy and the size of the single gene fraction during developmental shifts in plant cells. Such observations and the recent success of plasmid DNA research in defining the genetic basis for plant tumors make it likely that there will soon be a rapid surge of interest in plant cell chromosomes in relation to cell differentiation.

Dr Lima de Faria´s contribution points to another important method in the analysis of chromosome function, namely somatic cell hybridization. Genetics has so far mainly been based on sexual reproduction and the analysis of the progeny from sexual crosses. Fusion of cells by Sendai virus or polyethylene glycol has changed this situation and made it possible to obtain hybrid cells from crosses involving different species and phyla. It remains to be seen what can be done with exotic crosses like fusions of plant protoplasts and human cells.

Chromosomes Today Volume 6, A. de la Chapelle and M. Sorsa eds.
© 1977 Elsevier/North-Holland Biomedical Press, Amsterdam, The Netherlands

HAPPY BIRTHDAY - PUFFS!

MICHAEL ASHBURNER

Department of Genetics, University of Cambridge, Cambridge, England.

ABSTRACT

After a brief historical review of the puffs of Dipteran polytene chromosomes the criteria required for the formal proof that puffs are indeed active genes are considered. Two experimental systems are discussed in some detail, the heat shock puffs and their proteins and the puffs coding for salivary gland secretion proteins, both in *Drosophila melanogaster*. A new model, proposing a two step control of transcription of polytene chromosome bands. is considered in the light of the behaviour of puff mutants.

DEDICATION

The birthday, of course, is the twenty fifth since the now classic papers of Beermann[1], Pavan and Breuer[2] and Mechelke[3] who laid the foundations for what has turned out to be one of the most fruitful analyses of the control of gene activity in eukaryotes. I cannot let this occassion pass without remembering the names of four friends who did much good work in this field and who died too young: I refer to Hans Berendes, Karlheinz Bier, Ulrich Clever and Joan Whitten.

> Time present and time past
> Are both perhaps present in time future
> And time future in time past.
> If all time is eternally present
> All time is unredeemable.
>
> T.S. Eliot: Burnt Norton.

HISTORICAL

The detailed studies of Beermann, and of Pavan and Breuer which led to the realisation of the importance of puffs were motivated by a hypothesis, which they discredited, that the banding patterns of the polytene chromosomes were tissue specific. It is perhaps ironic that this hypothesis, strongly favoured by Kosswig and Sengun[4], may not be as generally incorrect as has been thought since 1952. A comparison of the banding patterns of the polytene chromosomes of the bristle forming epidermal cells in the blowfly *Calliphora erythrocephala* with those of the ovarian nurse cells of the same species fails to show <u>any</u> extensive homology (Ribbert[5]). If these observations are confirmed and cannot be explained trivially they have important implications for our ideas of polytene chromosome structure and genetic organisation.

Nevertheless the analysis of the polytene chromosomes of different larval tissues of *Chironomus*, by Bermann [1] and of *Rhynchosciara* by Pavan and Breuer[2] showed that the apparent differences in banding patterns between larval tissues were not real and that many of these differences were due to reversible and transient modifications of the form of individual bands into structures named, almost 20 years earlier by Bridges, puffs. It was the discovery that puffs, and the larger Balbiani Rings, were reversible structural modifications of, perhaps, single bands and that the pattern of puffs was not only tissue specific but also changed in a regular manner as development proceeded that led Beermann, in particular, to the proposal that the puffs are visible manifestations of gene activity: a view most forcibly put by Beermann in his paper to the 1956 Cold Spring Harbor Symposium[6].

Subsequently, the tissue and developmental specificity of puffs was amply shown in several species of Diptera and correlations drawn between different patterns of puffing activity and different cell functions. Coincidentally progress was made in understanding puffs at the biochemical level, the single most important observation being that puffs and Balbiani Rings are sites of active RNA synthesis (Pelling[7]). In 1959 Becker[8] implied that the patterns of puffing activity seen during development may be under hormonal control, a prediction verified by Clever and Karlson[9] one year later. With *Chironomus tentans* these authors showed that the changes in puffing activity seen during normal larval and prepupal development were controlled by the steroid hormone ecdysone, and that certain puffs could be induced within minutes of exposure of the salivary glands to the hormone. This was one, if not the very first, experimental indications that hormones could act by changing gene activity.

What has, until very recently, been missing is the rigorous demonstration that puffs are active genes. This demonstration requires more than a correlation between the activity of certain puffs and certain cellular functions, even though such a correlation be supported by the close genetic linkage of the puff and function as in Beermann's[10] and Grossbach's[11] elegant studies of the Sonderzellen secretion in *Chironomus*. In addition to genetic linkage, what is required is the proof that the puff contains sequences being transcribed which code for the correlated function and, preferably, mutational data showing that loss of the puff results in loss of the function. The purpose of this review is to consider some recent data on puffs and proteins from studies with *Drosophila melanogaster*, studies which very strongly support, but which still do not quite prove beyond any shadow of doubt, Beermann's hypothesis. Particular attention will be given to just two systems, the heat shock puffs and the puffs coding for salivary gland secretion components.

HEAT SHOCK

It is to Ritossa[12] that we owe the discovery that a brief heat shock, for example from 25°C to 37°C for 10 minutes, given to *Drosophila* larvae or isolated tissues dramatically affects the pattern of puffing activity. In *D. melanogaster* heat shock results in the very rapid induction of nine new puffs. Despite an occasional excursion (e.g. Ashburner[13]) this phenomenon was only of parochial interest until Tissieres et al[14] showed a parallel induction of the synthesis of a group of new proteins in heat shocked salivary glands. This important observation was rapidly confirmed, by McKenzie et al[15], Lewis et al[16] and Koninkx[17] and was shown to occur in almost any tissue including, importantly for subsequent work, tissue culture cells[15]. For reasons yet to be elucidated a very rapid response to heat shock is the cessation of transcription at all but the induced puff sites and a cessation of translation of most preexisting messenger RNA's leading to a dissociation of the polysomes (the histone and mitochondrial messenger RNA's and the histone proteins appear to be exceptional in this respect) (McKenzie et al[15], Spradling et al[18], Mirault et al[19]). It is as if the cells are devoting all of their transcriptional and translational efforts to the synthesis of a small number of heat shock proteins and we have suggested that this is a homeostatic response of the cells (Lewis et al[16]). Whatever the mechanism of, or the reasons for, this devotion to the synthesis of heat shock mRNA's and heat shock proteins it is experimentally useful since it allows the newly synthesised messenger RNA's to be obtained without too much difficulty. It was originally shown that newly synthesised mRNA's from heat shocked tissue culture cells would hybridise in situ to precisely those sites on the polytene chromosomes that puffed on heat shock (McKenzie et al[15], Spradling et al[18,20], Lubsen and Sondermeijer[21], Sondermeijer and Lubsen[22]). Partially purified preparations of messenger RNA's from heat shocked cells will predominantly hybridise to single puff sites in many instances[20,21,22]. To complete the circle it has also been found that these messenger preparations can be translated, in a cell free system, and translate to give products of the same molecular weight and with the same tryptic digest patterns as the heat shock proteins from whole cells (Moran et al[23], Mirault et al[19]).

The messenger RNA's from heat shocked tissue culture cells have also been used, in several laboratories, to isolate plasmids containing complementary DNA sequences, thus allowing a physical analysis of the heat shock 'genes' (J. Liss, personal communication, M. Meselson, personal communication). These cloned sequences, isolated from total *Drosophila* DNA, hybridise to particular bands that form heat shock puffs.

If the heat shock puffs are indeed coding for the heat shock proteins, a conclusion strongly suggested by the results just mentioned, then an absence of one puff should result in the absence of a single heat shock protein. In fact matters are not quite so simple. Ish-Horowitz et al[24], have recently made a series of deletions covering either the 87C heat shock puff alone or both the 87C puff and the closely linked 87A heat shock puff. These deletions are many bands long and, not surprisingly, are homozygous lethal. However the deletion homozygotes do develop far enough for their pattern of protein synthesis after heat shock to be studied and they are morphologically distinguishable from their heterozygous sibs. Surprisingly embryos homozygously deficient for 87C alone synthesise all of the heat shock proteins after heat treatment. On the other hand homozygotes for a deletion that includes both the 87C and the 87A puffs do lack one of the major heat shock proteins. Unfortunately stocks deleted for the 87A puff alone are not yet available.

Why it is necessary to delete both 87A and 87C to see the absence of a single protein is not yet clear. The most likely explanation is that 87A and 87C are duplicate loci. This hypothesis is supported by two other observations: firstly even 'purified' preparations of heat shock messenger RNA's will hybridise to both 87A and 87C (though to different extents) (Spradling et al[20], Henikoff and Meselson[25]), secondly the studies of the sequence organisation of plasmids containing sequences complementary to heat shock messenger RNA suggest that these two sites share certain sequences (J. Liss, personal communication).

SALIVARY GLAND SECRETION

The larval salivary gland of *Drosophila* synthesises (Korge[26]), during the greater part of the third larval instar, a secretion whose function is, at puparium formation, to affix the animal to its substrate. This function accounts for the secretion being known as 'glue'. The synthesis of the secretion begins a few hours after the moult from the second instar (Beckendorf and Kafatos[27], Korge[26]), a response, I suggest, to a fall in ecdysone titre at that time. Synthesis continues until the time of puparium formation and then abruptly ceases. For most of the instar the secretion is stored within the cells in large paracrystalline granules (Rizki[28], Lane et al[29]) only to be released into the lumen a few hours before puparium formation. The release of the secretion into the lumen is a response to the increase in ecdysone titre before puparium formation (Poels[30]).

The puffing patterns of the salivary gland chromosomes change very dramatically as a consequence of the release of ecdysone before puparium formation (Ashburner[31]). Prior to this time, however, there are few large puffs active (e.g. Zhimulev[32]). Of the half dozen or so of these large puffs at least two probably code for components of the glue.

After suitable chemical treatment the secreted glue can be resolved into several different proteins on either acid-urea or SDS-acrylamide gels (Korge[33], Beckendorf and Kafatos[27]). For example, on urea gels glue is found to be composed of four major proteins. These proteins are very polymorphic: in fact two types of genetic variations are found. On the one hand different stocks may differ in the relative electrophoretic mobility of certain proteins and, on the other hand, certain stocks may either lack certain proteins or show additional protein bands. The electrophoretic variants of "protein-4" and of "protein-3" have allowed the genetic loci responsible for this variation to be mapped.

The locus for protein-4 was mapped genetically by Korge[33,34] close to white on the X-chromosome and cytologically, by the use of deletions, to between bands 3C10 and 3D1. The locus responsible for the protein-3 variants was mapped to chromosome arm 3L by Korge[33] and to between bands 67F3.4 and 68C8 by Akam et al[35]. In both cases the amount of the protein made was shown to be a function of the genetic dosage of the respective region. The critical observation is that both 3C and 68C are the sites of two of the very largest puffs active at the time that the glue is synthesised by the salivary gland. Both puffs regress at the end of larval development, probably a response to the increase in ecdysone titre at that time (Ashburner[36]) and both puffs are absent from other tissues with polytene chromosomes, such as the larval fat body and ring gland (Richards and Ashburner, unpublished).

These data strongly suggest that the 3C and 68C puffs are the visible manifestations of the genes coding for protein-4 and protein-3 respectively. In the case of 3C and protein-4 this conclusion is strengthened bythe fact that at least two stocks are known which lack protein-4 in their glue: both also lack activity of the 3C puff (Korge[34], Ashburner, unpublished). Most of the glue proteins are extensively glycosilated and it remains possible that what has been mapped in the experiments of Korge and of Akam et al are loci coding for enzymes concerned with the glycosilation. If this possibility is to be seriously entertained then it will be necessary to propose the extra hypothesis that the abnormal protein-4 in those stocks which lack the 3C puff, a protein abnormal not in its polypeptide chain but in its sugar side chains, is either not stored by the cells or is invisible with the analytical techniques used. This question can only be considered to be formally settled when the polypeptides of these proteins have been chemically characterised, and their messenger RNA's isolated, translated in vitro and shown to map at the correct puff sites.

ON THE CONTROL OF PUFFING

Korge[34] has discovered that the 'mutants' lacking the 3C puff (which I will call $3C^{0}$) and the corresponding protein-4 show, with respect to both puff and protein phenotype, unusual behaviour in heterozygotes with 'wild type' (i.e. $3C^{+}$) strains.

It is important to recall that, normally, the maternally and paternally derived homologues are intimately synapsed in the polytene chromosomes though occasionally such synapsis may fail (asynapsis). Asynapsis may be promoted by making heterozygotes of a standard sequence homologue and one carrying inversions. In a standard sequence $3C^{0}/3C^{+}$ heterozygote the wild type homologue puffs at 3C irrespective of chromosome synapsis. This is to be expected. What is interesting is that the $3C^{0}$ homologue also puffs at 3C *if*, and only if, its homologue is synapsed in the 3C region with the wild type homologue. Precisely the same situation was discovered some years ago for strains differing in puffing activity at the 64C site (Ashburner[37] for review). As a control the case of the mutant for the 46A puff can be cited: in $46A^{0}/46^{+}$ heterozygotes the $46A^{0}$ homologue does *not* puff despite synapsis with the wild type homologue (Ashburner[37]).

Korge[34] has published evidence that, when activated to form a puff as a consequence of synapsis with a $3C^{+}$ homologue, the $3C^{0}$ homologue is *truly* active, that is to say codes for a protein (protein-4h) which, luckily, is electrophoretically distinguishable from that coded for by the $3C^{+}$ homologue. If $3C^{0}/3C^{+}$ heterozygotes are made in which the wild type chromosome carries so many inversions that synapsis of the 3C region is a very rare event then protein-4h is *not* present. That is to say, not only is the puffing of the $3C^{0}$ chromosome dependent upon intimate synapsis with a $3C^{+}$ chromosome but also its phenotypic expression at the protein level. Again one must enter a slight caveat: the homology between protein-4 (from $3C^{+}$) and protein-4h (from the activated $3C^{0}$) has yet to be confirmed chemically.

At face value these experimental data would rule out an explanation of the synapsis dependent non-autonomy of puff mutants suggested to me by Beermann: This proposed that the mutant chromosome was not really active but appeared puffed because RNA transcribed from the wild type chromosome accumulated on it. To explain the behaviour of the 64C puff mutant I earlier[37] considered that a chromosome limited diffusion of a regulatory substance, perhaps an RNA, could occur between synapsed homologues but not between homologues separated by nuclear sap. In this view, the mutant homologue was mutant in a cis-acting, or synapsis dependent trans-acting regulatory element whose product was diffusible within, but not between, chromosomes.

Recent considerations of chromosome structure in general, and of polytene chromosome structure in particular, provide an interesting alternative model. The bands of the polytene chromosomes would appear to be highly ordered structures - this is witnessed by their birefringence in the polarizing microscope (Sedat and Manuelidis[38]). Moreover, theoretical considerations of chromatin organisation and chromosome structure have led Alberts *et al*[39] to propose that the bands are microcrystalline arrays of nucleosomes. There is little doubt that the transition from a compact band to decondensed puff involves extensive changes in the supramolecular structure of the banded chromatin which might, rather loosely, be viewed as a disruption of its ordered state. If so, then this transition may be a cooperative process which can be initiated at any point on a band and then propogate across the entire band width.

It follows that this process must be triggered by the interaction of 'something' with specific chromosomal receptors and that if these receptors are lost, by mutation, then the band could not unfold to form a puff. However if such a mutant homologue is synapsed with a wild type homologue the mutant band may unfold as a consequence of the propogation of the cooperative loss of the crystalline structure of the wild type band region. Viewed in this light puff mutants, such as $3C^0$ and $64C^0$, which show synapsis dependent nonautonomy, are mutant at the level of the unfolding of the banded chromatin into such an extended configuration that transcription can occur.

It is possible that this is only the first level of control of transcription of sequences contained within single bands. A second level of control may exist which determines, from amongst the available (i.e. unfolded) sequences just those which are to be transcribed. If this is so then it follows that genes packed together within a single band (if there is more than 'one gene' per band) share the first, but not necessarily the second, control step. Thus puffs which are judged, by cytological criteria to be the same, seen for example in different tissues or within a tissue at different stages of development, may be nonidentical with respect to the sequences transcribed.

The attractiveness of these considerations is that they suggest a solution to a very long standing problem: That of the raison d'etre of polyteny versus, for example, a non—polytene endopolyploid nuclear organisation. Except for rather vague statements concerning control the only obvious advantage of polyteny hitherto has been the advantage it gives to the scientist: It may be that the true significance of polyteny is that it achieves a striking economy of control.

ACKNOWLEDEMENTS

The author's own studies have been supported by grants from the Science Research Council, London. Thanks are due to several friends and colleagues for pre-publication copies of their papers and to Bruce Alberts, Steve Beckendorf, Bill Gelbart, Horst Kress and Abe Worcel for many hours of an enjoyable discussion.

REFERENCES

1. Beermann, W., (1952) Chromerenkonstanz und Spezifische Modifikationen der Chromosomenstrukter in der Entwicklung und Organdifferenzierung von *Chironomus tentans*. Chromosoma, 5, 139-198.

2. Pavan, C., and Breuer, M., (1952) Polytene chromosomes in different tissues of *Rhynchosciara*. J. Heredity, 43, 150-157.

3. Mechelke, F., (1953) Reversible Strukturmodifikationen der Speicheldrüsenchromosomen von *Acricotopus lucidus*. Chromosoma, 5, 511-543.

4. Kosswig, C., and Sengün, A., (1947) Vergleichende Untersuchungen uber den Bau der Riesenchromosomen der Diptera. C.R. Soc. Turque Sci. Phys. Nat., 13, 94-101.

5. Ribbert, D., (1977) Nonhomologizable chromomere patterns in experimentally induced polytene chromosomes in the germ line and that of true somatic cells in the fly *Calliphora erythrocephala*. Helsinki Chromosome Conference. Abstracts, p.152.

6. Beermann, W., (1956) Nuclear differentiation and functional morphology of chromosomes. Cold Spring Harbor Symp. Quant. Biol., 21, 217-232.

7. Pelling, C., (1959) Chromosomal synthesis of ribonucleic acid as shown by incorporation of uridince labelled with tritium. Nature, 184, 655-656.

8. Becker, H.J., (1959) Die Puffs der Speicheldrüsenchromosomen von *Drosophila melanogaster*. I. Beobachtungen zum Verhalten des Puffmusters im Normalstamm und bei zwei Mutanten, giant and lethal-giant-larvae. Chromosoma, 10, 654-678.

9. Clever, U., and Karlson, P., (1960) Induktion von Puff-veranderungen in den Speicheldrüsenchromosomen von *Chironomus tentans* durch Ecdyson. Expt. Cell Res., 20, 623-626.

10. Beermann, W., (1960) Ein Balbiani-Ring als Locus einer Speicheldrüsenmutation. Chromosoma, 12, 1-25.

11. Grossbach, U., (1968) Chromosome-Aktivitat und biochemische Zelldifferenzierung in der Speicheldrüsen von *Camptochironomus*. Chromosoma, 28, 136-187.

12. Ritossa, F., (1962) A new puffing pattern induced by temperature shock and DNP in *Drosophila*. Experientia, 18, 571-573.

13. Ashburner, M., (1970) Patterns of puffing activity in the salivary gland chromosomes of *Drosophila*. V. Responses to environmental treatments. Chromosoma, 31, 356-376.

14. Tissieres, A., Mitchell, H.K., and Tracey, U., (1974). Protein synthesis in salivary glands of *Drosophila melanogaster*: Relation to chromosome puffs. J. Mol. Biol., 84, 389-398.

15. McKenzie, S.L., Henikoff, S., and Meselson, M., (1975) Localization of RNA from heat-induced polysomes at puff sites in *Drosophila melanogaster*. Proc. Nat. Acad. Sci. U.S.A., 72, 1117-1121.

16. Lewis, M., Helmsing, P., and Ashburner, M., (1975) Parallel changes in puffing activity and patterns of protein synthesis in salivary glands of *Drosophila*. Proc. Nat. Acad. Sci. U.S.A., 72, 3604-3608.

17. Koninkx, J.F.J.G., (1976) Protein synthesis in salivary glands of *Drosophila hydei* after experimental gene induction. Biochem. J., 158, 623-628.

18. Spradling, A., Penman, S., and Pardue, M.L., (1975) Analysis of *Drosophila* mRNA by in situ hybridization: Sequences transcribed in normal and heat-shocked cultured cells. Cell, 4, 395-404.

19. Mirault, M.E., Goldschmidt-Chermont, M., Moran, L., Arrigo, A.P., and Tissieres, A., (1978) The effect of heat shock on gene expression in *Drosophila melanogaster*. Cold Spring Harbor Symposium, 42 (In Press).

20. Spradling, A., Pardue, M.L., and Penman, S., (1977) Messenger RNA in heat-shocked *Drosophila* cells. J. Mol. Biol. 109, 559-587.

21. Lubsen, N., and Sondermeijer, P.J.A., (1977) The products of the "heat shock" loci of *Drosophila hydei*: Correlation between locus 2-36A and the 70,000 molecular weight "heat shock" peptide. (In Press).

22. Sondermeijer, P.J.A., and Lubsen, N., (1977). Heat shock peptides in *Drosophila hydei* and their in vitro synthesis. (In Press).

23. Moran, L., Mirault, M-E., Arrigo, A.P., Goldschmidt-Clermont, M., and Tissieres, A., (1977) Heat shock of *Drosophila melanogaster* induces the synthesis of new messenger RNAs and protein. Phil. Trans. Roy. Soc. London. B. (In Press).

24. Ish-Horowitz, D., Holden, J., and Gehring, W., (1977) Deletions of two heat activated loci in *Drosophila melanogaster* and their effects on heat induced protein synthesis. Cell, (In Press).

25. Henikoff, S., and Meselson, M., (1977) Transcription at two heat shock loci in *Drosophila*. Cell, (In Press).

26. Korge, G., (1977) Larval saliva in *Drosophila melanogaster*: Production, composition and relationship to chromosome puffs. Develop. Biol., 58, 339-355.

27. Beckendorf, S.K., and Kafatos, F.C., (1976) Differentiation in the salivary gland of *Drosophila melanogaster*: Characterization of the glue proteins and their developmental appearance. Cell, 9, 365-373.

28. Rizki, T., (1967) Ultrastructure of the secretory inclusions of the salivary gland cell in *Drosophila*. J. Cell Biol., 32, 531-534.

29. Lane, N.J., Carter, Y.R., and Ashburner, M., (1972) Puffs and salivary gland function: The fine structure of the larval and prepupal salivary glands of *Drosophila melanogaster*. Wilhelm Roux' Archiv., 169, 216-238.

30. Poels, C.L.M., (1970) Time sequence in the expression of various developmental events induced by ecdysterone in *Drosophila hydei*. Develop. Biol. 23, 210-225.

31. Ashburner, M., (1972) Patterns of puffing activity in the salivary gland chromosomes of *Drosophila*. VI. Induction by ecdysone in salivary glands of *D. melanogaster* cultured in vitro. Chromosoma, 38, 255-281.

32. Zhimulev, I., (1974) Comparative study of the function of polytene chromosomes in a laboratory stocks of *Drosophila melanogaster* and the l(3)tl mutant (lethal tumorous larvae). Chromosoma, 46, 59-76.

33. Korge, G., (1977) Direct correlation between a chromosome puff and the synthesis of a larval saliva protein in *Drosophila melanogaster*. Chromosoma, 62, 155-174.

34. Korge, G., (1975) Changing puff activity and protein synthesis in larval salivary glands of *Drosophila melanogaster*. Proc. Nat. Acad. Sci. U.S.A., 72, 4550-4554.

35. Akam, M., Ashburner, M., Richards, G., and Roberts, M., (1977) *Drosophila*: The genetics of two major larval proteins. (In preparation).

36. Ashburner, M., (1973 Sequential gene activation by ecdysone in polytene chromosomes of *Drosophila melanogaster*. I. Dependence upon ecdysone concentration. Develop. Biol., 35, 47-61.

37. Ashburner, M., (1970) The genetic analysis of puffing in polytene chromosomes of *Drosophila*. Proc. Roy. Soc. London. B., 197, 319-327.

38. Sedat, J., and Mauelides, L., (1977) A direct approach to the structure of eukaryotic hromosomes. Cold Spring Harbor Symp. 42, (In Press).

39. Alberts, B., Worcel, A., and Weintraub, H., (1977). On the biological implications of chromatin structure. In: "The Organisation and expression of the eukaryotic genome" Eds. Bradbury, E.M., and Jovaherian, K., Academic Press, London, pp 165-191.

Chromosomes Today Volume 6, A. de la Chapelle and M. Sorsa eds.

KINETICS OF RNA EXPORT FROM THE BALBIANI RINGS IN VIVO

J.-E. EDSTRÖM, S. LINDGREN, U. LÖNN and L. RYDLANDER
Department of Histology, Karolinska Institutet, S-104 01 STOCKHOLM, Sweden

ABSTRACT

Balbiani ring (BR) RNA in salivary gland cells of *Chironomus tentans* was studied with regard to production and export rate *in vivo* and half life in nucleus and cytoplasm. After a transcription time of about 20 min it enters the nuclear sap where it spends on the average 60 min before most or all of it is delivered to the cytoplasm where its half life is about 20 hours. Because of the large amounts of BR-RNA in the cytoplasm the total amount degraded per time unit is high, explaining the need for an intense synthetic activity in the BR.

INTRODUCTION

Little is known about the export kinetics of defined gene products in fully differentiated cells in the intact living animal. This may be one of the reasons why the nuclear RNA metabolism remains unclear in many essential respects.

The RNA produced in the Balbiani rings (BR) of the salivary glands of *Chironomus tentans* (75 S RNA) is suitable for a study of export kinetics because it can be isolated and analyzed both at the site of production and in the cytoplasm[1,2] and because it has properties of messenger RNA, containing poly(A)[3] and being present in polysomes[4]. We have estimated the RNA production rate from the RNA content in the isolated BR 1 and BR 2 and the time it takes to transcribe this RNA[5]. Export was determined from the export rate of ribosomal RNA (rRNA) and the ratio of labelled BR-RNA to rRNA after precursor administration, extrapolated to zero time. The amounts exported are similar to the amounts produced and there is no evidence for a quantitatively dominating intranuclear degradation *in vivo*.

From the content of BR 1+2 RNA in the cytoplasm and data on amounts exported we could determine the half life of this RNA[6]. It turned out to be relatively short, surprising in view of the reported stability of the protein synthesis in the salivary glands after administration of actinomycin D and the probable dependence of most of the protein synthesis on the BR[7]. Determinations of half life with this drug showed that the half life of BR-RNA was about the same as in untreated animals, in both cases around 20 hrs[8].

The content of BR-RNA could also be determined in nuclear sap[6]. From the knowledge that the production is similar to the export it could be decided that it lies *en route* to the cytoplasm rather than in a separate pathway. The lag in its cytoplasmic appearance[3] may be a consequence of a dilution by the pool. The content corresponds to the degradation occurring in the cytoplasm during 90 min and is thus quantitatively a rather small reserve.

PRODUCTION OF RNA FROM BR 1+2

RNA in the process of being transcribed is present in the BR. If there is only such RNA and the transcribed molecules form a continuous array of growing molecules it is possible to determine the rate of RNA production from the RNA content and the transcription time. The formula is $s=2 \times m \times \frac{60}{t}$, where s is the amount produced per hour, m the mass of RNA in the transcribing site, and t the transcription time in minutes, i.e. the time it takes for polymerases to travel from one end of the unit to the other one. The formula applies whether there are one or several transcription units at the site as long as the transcription rate is similar in all units. There have been several determinations of transcription time for explanted glands[9,10,11] and it has been found that the time *in vivo* is similar[5] and likely to be about 20 min, although somewhat shorter, i.e. 15 min, or longer times, i.e. 25 min, cannot be excluded.

The RNA content can be determined in isolated BR and also the degree to which the BR are contaminated by nuclear sap derived RNA and RNA from non-BR puffs[6]. There are two additional possible errors. Finished BR-RNA molecules may accumulate for some time after transcription. This is unlikely because it would lead to a peaked electrophoretic distribution in the 75 S range, so far not observed. Another possible error is that all molecules do not reach finished size. In both cases the application of the formula leads to an overestimate of the RNA production. The production rate has been calculated on the basis of different transcription times. For 20 min the production rate is about 2500 pg per day whereas 15 min gives 3300 pg per day. The latter value can probably be considered as an upper limit for the production rate.

EXPORT OF RNA FROM BR 1+2

In principle export could be determined from the BR-RNA content in the cytoplasm and the rate of degradation in a system in equilibrium. The latter parameter, however, cannot easily be determined *in vivo* because of label reutilization and the extended duration of the labelling pulse. An indirect procedure was therefore chosen[5]. The rate of export for rRNA is more easily determined because of its slow degradation and the possibilities of using methyl groups for labelling. The ratio between label in 75 S RNA and rRNA at different times after precursor administration was determined and the ratio extrapolated to zero time. The labelling ratio is then converted to a mass ratio with the aid of data for base composition and specific activity for RNA in microisolated nucleoli and BR. Thus the export at zero time for 75 S RNA can be expressed as a fraction of that of rRNA. The export rate of rRNA was found to be about 3000 pg per day, about 1000 pg being required for growth and 2000 pg for turnover. The mass ratio at zero time 75 S RNA/rRNA was about 0.7. The export rate of BR-RNA is thus about 2000 pg per day which is only slightly less than the rate of production with a 20 min transcription time. Even if 15 min transcription time is assumed with 3300 pg as an upper limit for the production it is clear that a considerable fraction of the RNA produced is also exported.

HALF LIFE OF BR-RNA

The half life of BR 1+2 RNA, or an integrated value if this RNA is heterogeneous with respect to half life, can be determined from the rate of export and the content of 75 S RNA in the cytoplasm. The cells contain of the order of 2200-2300 pg of BR 1+2 RNA[6]. Since 1—2% of the export is required for growth, almost all of the 2000 pg of RNA exported is required to balance degradation. This necessitates a half life of about 19 hrs for BR-RNA.

If the decay of labelled RNA is followed in animals given actinomycin D a similar value is obtained, i.e. 21 hrs[8]. Both values are surprisingly short in view of earlier reports on a stability of protein synthesis in these glands after exposure to this drug[12,13]. Probably most of the protein synthesis is determined by the BR[7]. The discrepancy may have several explanations without invalidating a role for the BR in the genetic determination of the secretory proteins.

THE NUCLEAR POOL

The amount of BR-RNA in the nucleus is only 5 % of that of the whole cell. In the nucleus about 90 % of this RNA is in the nuclear sap[6]. There are two possible roles for the BR-RNA in the sap. One is that it largely represents a separate pathway; the other one that it consists of RNA *en route* to the cytoplasm. Since we know that the production and export rates are similar, it is possible to distinguish between these two alternatives. If the sap RNA represents a separate pool and is largely shut off from the pathway of BR-RNA to the cytoplasm it follows that the exported BR-RNA has to enter the cytoplasm as soon as it is synthesized. If on the other hand it is in equilibrium with a pool of sap RNA one would expect a delay in its cytoplasmic appearance. This is also what happens and it takes as a rule at least two hours for 75 S RNA to appear in the cytoplasm[3], or about the same time as for the large ribosomal RNA precursor. On the assumption that the sap RNA represents future cytoplasmic RNA and that it is delivered stochastically one can calculate a half life for its nuclear residence time of about 60 min. The further delay observed in the cytoplasmic appearance *in vivo* is probably due to the additional effects of transcription time, initial low specific activity in precursor pools and the fact that the appearance was measured in the cytoplasm at some distance from the nuclear envelope.

DISCUSSION

In several systems most of the non-ribosomal high molecular weight nuclear RNA, hnRNA, degrades in the nucleus[14]. The *in vivo* results for salivary gland cells of *Chironomus tentans* clearly show that intranuclear degradation of BR-RNA, if it occurs at all, cannot be quantitatively dominating, since the data for production rate are likely to be maximal values. If the export rate was lower it would lead to a longer half life for BR-RNA. The values are, however, strikingly similar to the determinations based on actinomycin D. In an earlier report a high extent of intranuclear degradation was found for BR-RNA in explanted glands[11]. The animals used, were, however, considerably younger than in the present investigations and it cannot be stated at present whether the difference in results is due to animal age or to different techniques.

Intranuclear degradation of hnRNA could in principle be due to degradation of nuclear-specific parts of messenger precursors, molecules entirely without a direct cytoplasmic function or to degradation of the whole messenger precursors, as well as to combinations between these alternatives. If the results for BR-RNA are typical also for differentiated mammalian cells *in vivo* they would tend to suggest that the extensive hnRNA degradation does not concern messenger sequences.

ACKNOWLEDGEMENTS

Supported by a grant from the Swedish Cancer Society.

REFERENCES

1. Edström, J.-E. and Beermann, W. (1962) J. Cell Biol. 14, 371-380.
2. Daneholt, B. and Hosick, H. (1973) Proc. Nat. Acad. Sci. 70, 442-446.
3. Edström, J.-E. and Tanguay, R. (1974) J. Mol. Biol. 84, 569-583.
4. Daneholt, B., Anderson, K. and Fagerlind, M. (1977) J. Cell Biol. 73, 149-160.
5. Edström, J.-E., Ericson, E., Lindgren, S., Lönn, U. and Rydlander, L. (1977) Cold Spring Harbor Symp. Quant. Biol. 42, in press.
6. Edström, J.-E., Lindgren, S., Lönn, U. and Rydlander, L. (1977) Manuscript in preparation.
7. Grossbach, U. (1973) Cold Spring Harbor Symp. Quant. Biol. 38, 619-627.
8. Rydlander, L. and Edström, J.-E. (1977) Manuscript in preparation.
9. Daneholt, B., Edström, J.-E., Egyházi, E., Lambert, B. and Ringborg, U. (1969) Chromosoma 28, 399-417.
10. Egyházi, E. (1975) Proc. Nat. Acad. Sci. 72, 947-950.
11. Egyházi, E. (1976) Cell 7, 507-515.
12. Clever, U., Storbeck, I. and Romball, C.G. (1969) Exp. Cell Res. 6, 195-201.
13. Doyle, D. and Laufer, H. (1969) Exp. Cell Res. 57, 205-210.
14. Darnell, J.E. (1975) Harvey Lectures, Series 69, 1-47.

Chromosomes Today Volume 6, A. de la Chapelle and M. Sorsa eds.

SYNTHESIS AND INTRANUCLEAR DISTRIBUTION OF NON-HISTONE PROTEINS IN MICRODISSECTED NUCLEI OF *Chironomus tentans*

ROBERT M. TANGUAY and LOUIS M. NICOLE
Génétique Humaine, Centre Hospitalier de l'Université Laval,
2705 Boul. Laurier, Ste Foy, Québec (Canada)

ABSTRACT

Highly labeled proteins from microdissected nuclei of salivary glands incubated *in vitro* with ^{3}H-leucine were analysed by SDS-polyacrylamide gel electrophoresis. Electrophoretic profiles of nuclear proteins from glands incubated for various periods of time show a limited differential rate of appearance and/or turnover of some of these proteins in the nucleus. Nuclei were further microdissected into four fractions, chromosomes I-III, chromosome IV, nucleoli and nuclear sap. Analysis of the proteins from the different intranuclear compartments reveals qualitative and quantitative differences in some of these. The nucleoli contain numerous specific bands. The nuclear sap has many bands in common with the chromosomal fraction. There are no detectable differences between the labeled proteins of chromosome I-III and chromosome IV which contains the large Balbiani rings.

INTRODUCTION

The chromosome puffs of Diptera have been shown to represent a morphological manifestation of gene activity [1-3]. These puffs are actively engaged in RNA synthesis and have been considered to represent units of transcription [4,5]. It has previously been shown that during development and hormone or temperature induced puffing, gene activation is accompanied by a local increase of non-histone proteins in the puffed region [6,7]. However little is known about the nature and the function of the proteins accumulating at the puff loci. Non-histone proteins have been suggested for many years to have an important role in the regulation of gene activity ([8-10]for reviews). The polytene chromosomes of *Chironomus tentans* with their large characteristic puffs, the Balbiani rings, represent a potentially useful system for studying specific protein changes presumably accompanying gene activation or repression. The elegant techniques developed by Edström's group [11-14] for studying RNA synthesis and maturation in microisolated puffs as well as in the other nuclear or cytoplasmic compartments and the possibility of experimentally inducing activation or repression by physical [15] or chemical means[16-17] offer an interesting approach for studying protein changes in single puffs during these processes. As a first step, we have analysed the proteins from microdissected nuclei and cytoplasm by high resolution SDS-polyacrylamide gel electrophoresis[18]. Here we report our results on the synthesis and intranuclear accumulation of proteins from glands labeled *in vitro* with ^{3}H-leucine. Furthermore the highly labeled proteins extracted from various microdissected intranuclear compartments, chromosome I-III, chromosome IV, nuclear sap and nucleoli were analysed and compared by gel electrophoresis.

MATERIALS AND METHODS

Animals and labeling procedures. *Chironomus tentans* larvae originally obtained from Prof. Edström (Stockholm) and Prof. Laufer (Storrs, Conn.) were raised in this laboratory at 18-20°C by a modified procedure [18]. Salivary glands from fourth Instar larvae were incubated at 20°C in 50μl of Cannon's modified medium[19] (without leucine) supplemented with 100μCi of ^{3}H-leucine (New England Nuclear, 110Ci/mmole).

Fixation and microdissection. After incubation, glands were usually fixed for 30 minutes at 4°C in ethanol: acetic acid (3:1 v/v), washed with 70% ethanol and stored in ethanol: glycerol (1:1) as described by Lambert and Daneholt[19]. In certain experiments, glands were fixed in ethanol: acetone (1:1) instead of ethanol: acetic acid. The intracellular compartments were microdissected with a De Fonbrune micromanipulator according to the procedure of Edström[11].

Extraction and electrophoresis of proteins. The proteins from the pooled microdissected components were extracted in a small volume of extraction medium (1.0% SDS, 0.14M β-mercaptoethanol, 10mM sodium phosphate buffer, pH 7.2) and analysed on a 8.75% acrylamide-SDS slab gel with a 3% acrylamide stacking gel, in the discontinuous buffer system of Laemmli[20] as modified by LeStourgeon and Wray [21]. The position of the molecular weight markers used, myosin heavy chain, β-galactosidase, γ-globulin heavy chain, tropomyosin or α-chymotrypsinogen A are indicated on the figures.

Radioactivity measurements and fluorography. Half of the gel was sliced and the slices were digested at 37°C in 5% Protosol supplemented with toluene cocktail and counted in a Beckman liquid scintillation counter. The other half of the gel was prepared for fluorography by the procedure of Bonner and Laskey[22] and exposed at $-70^{\circ}C$ to Kodak XR-2 films for the time indicated in the figures. The developed film was scanned with a Vitatron TLD 100 densitometer.

A full description of the various methods is given elsewhere[18].

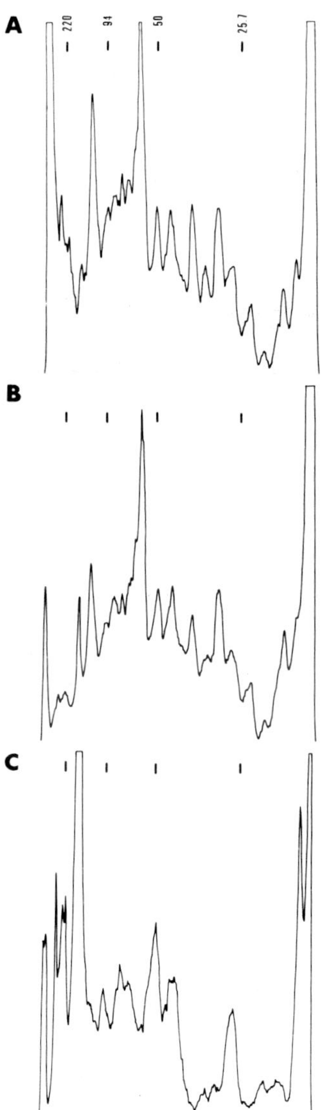

Fig. 1. Densitometric scans of fluorograms of labeled nuclear and cytoplasmic proteins extracted from glands fixed in different media. Glands were incubated for 180 minutes with ^{3}H-leucine, fixed, microdissected and the proteins extracted and analysed as described in Materials and Methods. (Exposure time, 7 days). Nuclear proteins from glands fixed in ethanol: acetone (A) and nuclear (B) and cytoplasmic (C) proteins from glands fixed in ethanol: acetic acid.

RESULTS

Figure 1 compares the fluorogram scans of labeled proteins extracted from nuclei and cytoplasm of glands fixed either in the usual Carnoy mixture or in ethanol: acetone. Between 20 and 25 polypeptide bands can be resolved in this short SDS denaturing gel. Both nuclear proteins profiles (Fig 1A, B) are quite comparable. The acidic fixative does not seem to extract certain proteins selectively at least at the level of resolution achieved here. In fact one extra band (approximate M.W. 150000) which is absent from the nuclear profile of ethanol: acetone fixed glands (Fig 1A) is present in the acidic ethanol fixed nuclear extract (Fig 1B). This band could represent a cytoplasmic contaminant since a protein of similar molecular weight can be seen to be a major labeled protein in the cytoplasmic extract (Fig 1C). It could also be a membrane protein since in the acid fixed glands, the nucleus is usually isolated with its nuclear membrane while in the ethanol: acetone fixed glands it is isolated as a compact nuclear mass without the nuclear membrane.

The electrophoretic stained profiles of total cellular proteins extracted from acid fixed gland is also undistinguishable from that of non-fixed glands[18]. The recovery of total labeled proteins extracted from nuclei or cytoplasm of glands fixed by the two methods is also comparable. We have no data yet on the complete or partial removal of histones from the nuclei of acid-fixed glands.

In the presence of cycloheximide (10μg/ml), the inhibition of labeling of nuclear and cytoplasmic proteins was respectively

TABLE I

TIME COURSE OF THE INCORPORATION OF ^{3}H-LEUCINE IN NUCLEI AND CYTOPLASMS FROM GLANDS INCUBATED IN VITRO

Time of incorporation (min)	cpm/nucleus	cpm/cytoplasm	$\frac{\text{cmp/nucleus}}{\text{cpm/cytoplasm}} \times 100$
45	193	23230	0.83
90	465	25250	1.81
180	2580	89800	2.87
360	3400	96370	3.53

Four groups of salivary glands were incubated with ^{3}H-leucine in Cannon's modified medium for 45, 90, 180 and 360 minutes respectively. Nuclei and cytoplasms were isolated by microdissection and dissolved in extraction medium as described in Materials and Methods. The radioactivity of small aliquots was determined by dilution in a scintillation cocktail containing 5% Protosol.

99% and 92%. Table I illustrates the gradual appearance of radioactivity in the nucleus and the increasing nuclear/cytoplasmic cpm ratio after different periods of *in vitro* labeling. This result and the cycloheximide sensitivity of the labeling suggest that we are dealing with a class of newly synthesized proteins which migrate rapidly to the various intranuclear compartments.

Figure 2 shows the electrophoretic profiles of nuclear proteins extracted from glands labeled *in vitro* from 45 to 360 minutes. The number and positions of the bands are similar at the level of resolution achieved by the gel slicing technique. However after longer incubation periods (180 and 360 minutes) there is a preferential accumulation of label in the higher molecular weight region of the gels (Fig 2C, D). At least 2 bands (approximate M.W. 95000 and 60000) which seem absent after 45 and 90 minutes (Fig 2A, B) start to accumulate in the nucleus after 180 minutes (arrows, Fig 2C, D). In

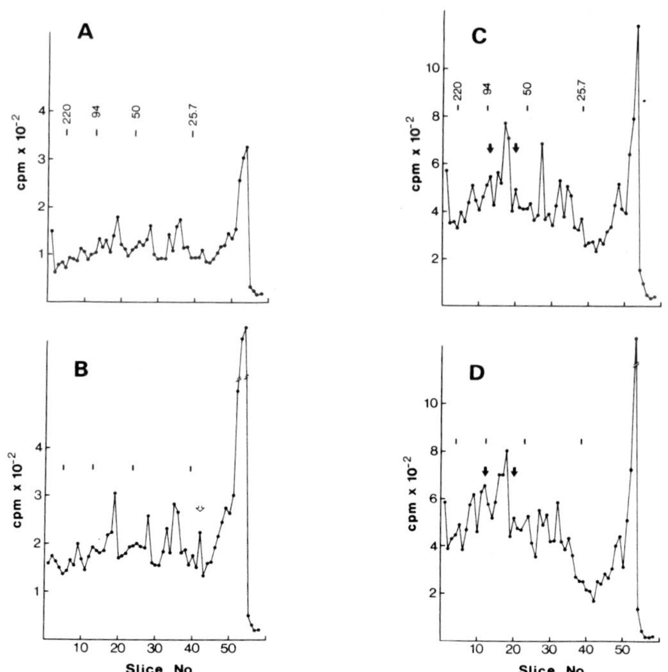

Fig 2. Electrophoretic profiles of nuclear proteins from glands labeled for (A), 45, (B), 90, (C), 180 and (D), 360 minutes. The incubation of the glands and the microdissection of the nuclei are described in Table 1. After solubilisation of the nuclei in extraction medium, the proteins were analysed by SDS-polyacrylamide gel electrophoresis as described in Materials and Methods. The radioactivity profiles were plotted after digestion of the sliced gel in 5% Protosol. Number of nuclei used in this experiment: (A), 270, (B), 140, (C), 28 and (D), 21.

the lower molecular weight region, one band (M.W. 23000) which is particularly prominent after 90 minutes (open arrow, Fig 2B) is practically absent at 360 minutes (Fig 2D). The fluorograms of these preparations show that apart from these bands others not resolved in the sliced gel accumulate and/or turn-over at a differential rate (data not shown). The suggestion of a limited differential turnover is substantiated by studies of the labeled proteins after pulse and chase experiments (with cold leucine or cycloheximide)[18].

After labeling for 180 minutes, nuclei were further microdissected into 4 fractions: chromosomes I, II and III, chromosome IV, nucleoli and nuclear sap. The fluorogram banding pattern of nucleolar proteins (Fig 3, B) is quite distinct from those of the other nuclear fractions; many specific highly labeled protein bands, mainly in the 19 to 34000 daltons range, can be seen. The nuclear sap proteins profile (Fig 3 D) has many bands in common with the chromosome proteins (Fig 3, A ,C). Apart from the quantitative varia-

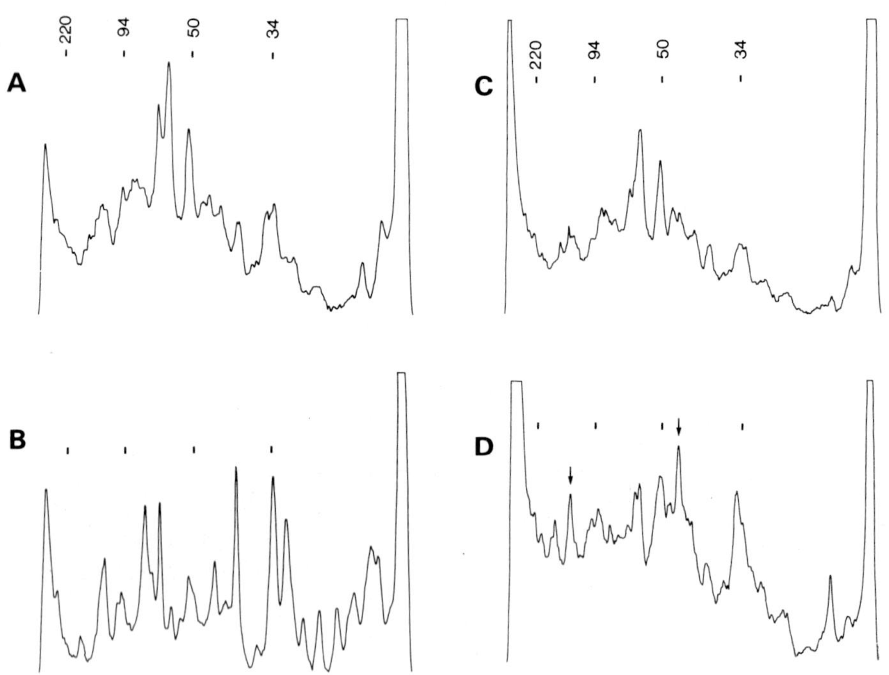

Fig 3. Fluorographic profiles of labeled proteins from (A), chromosomes I-III, (B), nucleolus, (C), chromosome IV and (D), nuclear sap. Glands were incubated *in vitro* with ^{3}H-leucine for 180 minutes. Nuclear compartments were dissected and their proteins were analysed by SDS-polyacrylamide gel electrophoresis as described in Materials and Methods. The gel was treated for fluorography. Scans were obtained after a six week exposure.

tions between those profiles, at least two major labeled bands (M.W. 123000 and 49000) seem specific to the nuclear sap (arrows Fig 3 D). The two prominent peaks (M.W. 68000 and 63000 daltons) of chromosome I-III or chromosome IV are underrepresented in the nuclear sap and nucleolar protein profiles. Finally at the level of resolution achieved here, there is no significant difference between the fluorographic profiles of chromosome I-III and chromosome IV, the small chromosome which harbors the large Balbiani rings.

DISCUSSION

This system offers an alternative approach to the study of proteins from different nuclear compartments. Instead of using different nuclear washing procedures [23], fixed nuclei are microdissected into chromosomes, nucleoli and nuclear sap. During the procedure, there is a constant visual control of the purity of the separated intranuclear fractions. The acidic ethanol fixative has been shown to extract histones [24,25]. We do not know if histones are extracted from our material but the profiles of labeled non-histone proteins are quite similar whether the glands are fixed in the acidic ethanol or the ethanol: acetone fixative. The presence of specific bands in the various fractions seems to indicate that there is no major redistribution of proteins during fixation as has been shown to be the case for RNA in microdissected cells [13-19].

Results presented here and elsewhere [18] seem to indicate a differential turnover of some of these nuclear proteins. Preliminary results also show that many of these highly labeled protein bands are different from the steady state nuclear proteins extracted from unlabeled nuclei. Therefore we are probably looking at a population of metabolically labile proteins.

While there are at least two bands which are specific to the nuclear sap, our results indicate that most labeled proteins found in the nuclear sap are also present in the chromosomes. This is not unexpected since in a dynamic state, many proteins become associated with RNA during the process of transcription and are released in the nuclear sap as RNP particles [26,27]. We do not think that the similarities result from contamination since the purity of the fractions is visually controlled and some proteins are specific. The similarities probably reflect the *in vivo* situation where there is a constant exchange of macromolecules between the different intranuclear compartments. Using a different approach, Comings and Harris [23] have also reported that there are no clear differences between many nucleoplasmic and chromatin non-histone proteins.

The situation in chromosome IV is also of interest. While most of the RNA produced can be accounted by the transcription of BR-1 and BR-2 [28], there is no indication of accumulation of specific labeled proteins different from those in the much more transcriptionally heterogeneous sites on chromosomes I-III. Therefore, these proteins do not seem to show specificity with regards to transcriptional loci.

We do not know the function of these highly labeled proteins. They could be involved in the regular maintenance and housekeeping of the cell nucleus. For example, the proteins in chromosome I-III and IV could perform common functions such as transcription, processing or transport of RNA. We do not think that they are regulatory proteins, at least specific ones, since most are common to the chromosomes and nuclear sap. We are presently using 2-dimensional gels in order to search for specific regulatory proteins.

Finally we hope that this approach which is complementary to the immunofluorescent technique of Jamrich and others [24,28-31], will yield useful information on the function of non-histone proteins in the control of gene expression.

ACKNOWLEDGEMENTS

This work was supported by the Medical Research Council of Canada and by a Special Research grant for equipment from Université Laval. We also thank Mrs F. Biron for skillful and patient assistance in the animal culture.

REFERENCES

1. Beerman, W. (1952) Chromosoma, 5, 139-198.
2. Beerman, W. (1972) in Results and Problems in Cell Differentiation Vol 4 Beerman, W. ed., Springer Verlag, Berlin pp 1-33.
3. Berendes, H. (1973) Int. Rev. Cytol., 35, 61-116.
4. Pelling, C. (1976) Cold Spring Harbor Symp. Quant. Biol. 35, 521-531.
5. Grossbach, U. (1973) Cold Spring Harbor Symp. Quant. Biol. 38, 619-627.
6. Berendes, H. (1974) in Acidic Proteins of the Nucleus, Cameron, I.L. and Jeter, J.R. eds, Academic Press, New York pp. 191-212.
7. Holt, T.K.H. (1971) Chromosoma, 32, 428-435.
8. Elgin, S.C.R. and Weintraub, H. (1975) Ann. Rev. Biochem., 44, 726-774.
9. Stein, G.S. and Kleinsmith, L.J. eds (1975) Chromosomal Proteins and their role in the Regulation of Gene expression, Academic Press, New York, pp. 1-307.

10. Cameron, I.L. and Jeter, J.R. eds (1974) Acidic Proteins of the Nucleus, Academic Press, New York, pp 1-343.

11. Edström, J.E. (1964) in Methods in Cell Physiology Vol 1, Presscott, D., ed. Academic Press, New York, pp. 417-447.

12. Daneholt, B. (1975) Cell, 4, 1-9.

13. Edström, J.E. and Lönn, U. (1976) J. Cell Biol., 70, 562-572.

14. Edström, J.E. and Tanguay, R. (1973) Cold Spring Harbor Symp. Quant. Biol., 38, 693-699.

15. Yamamoto, H. (1970) Chromosoma, 32, 171-190.

16. Beerman, W. (1973) Chromosoma, 41, 297-326.

17. Lezzi, M. and Gilbert, L.I. (1969) Proc. Nat. Acad. Sci., USA, 64, 498-503.

18. Tanguay, R.M. and Nicole, L.M. Submitted.

19. Lambert, B. and Daneholt, B. (1975) in Methods in Cell Biology, Vol. X, Presscott, D.M. ed., Academic Press, New York, pp. 17-47.

20. Laemmli, U.K. (1970) Nature, 227, 680-685.

21. LeStourgeon, W.M. and Wray, W. (1974) in Acidic Proteins of the Nucleus, Cameron, I.L. and Jeter, J.R. eds, Academic Press New York, pp 59-102.

22. Bonner, W.M. and Laskey, R.A. (1974) Eur. J. Biochem., 46, 83-88.

23. Comings, D.E. and Harris, D.C. (1976) J. Cell Biol., 70, 440-452.

24. Silver, L.M. and Elgin, S.C.R. (1976) Proc. Nat. Acad. Sci. USA, 73, 423-427.

25. Retieff, A.E. and Rüchel, R. (1977) Exptl Cell Res., 106, 233-237.

26. Malcolm, D.B. and Sommerville, J. (1977) J. Cell Sci., 24, 143-165.

27. Miller, O.L. and Bakken, A.H. (1972) in Gene Transcription in Reproductive tissue, Diczfalusy, E. ed., Karolinska Institutet, Stockholm, pp. 155-173.

28. Egyhazi, E. (1975) Proc. Nat. Acad. Sci. USA, 72, 947-950.

29. Jamrich, M., Greenleaf, A.L. and Bautz, E.K.F. (1977) Proc. Nat. Acad. Sci. USA, 74, 2079-2083.

30. Alfageme, C.R., Rudkin, G.T. and Cohen, L.H. (1976) Proc. Nat Acad. Sci. USA, 73, 2038-2042.

31. Plagens, U., Greenleaf, A. and Bautz, E.K.F. (1976) Chromosoma, 59, 157-165.

Chromosomes Today Volume 6, A. de la Chapelle and M. Sorsa eds.

STRUCTURAL AND FUNCTIONAL CHANGES OF CHROMOSOMAL DNA DURING AGING AND PHASE CHANGE IN PLANTS

BEZALEL KESSLER and SHIMEON RECHES
ARO, The Volcani Center, Bet Dagan, and Department of Life Sciences, Bar-Ilan University, Ramat Gan (Israel)

ABSTRACT

Chromosomal plant DNA is heterogenous and its nucleotide sequence redundancy is not constant during the life cycle. It differs in different organs, changes with age and environmental conditions, and is modulated by gibberellin in an organ-, age- and phase-specific mode. In *Hedera helix*, the size of repeated and single-gene fractions and the renaturation profiles of chromosomal DNA change with the transition from the juvenile to the adult phase. The primer activity of DNA in a plant DNA-polymerase system is modified by gibberellin and is age- and phase-dependent. Our data support the hypothesis that the chromosomes themselves are the site for primary changes in volving a phase change from juvenile to adult growth.

INTRODUCTION

In the course of their life cycle, organisms pass through a series of minor and often quite imperceptible changes which gradually result in the ultimate acquirement of new developmental stages. In some cases, however, the shift from one developmental phase to another occurs suddenly. Such developmental transitions, which are recognizable by a sudden transformation of physiological and morphological characteristics, are termed phase changes[1].

Phase changes are rather common in the development of animals and plants, including phenomena such as the switch from larva to adult in insects, morphological and physiological changes in parasitic animals and plants upon change of the host, the switch of the shoot meristem from the production of vegetative to the production of reproductive organs, the transitions between gametophyte to sporophyte, etc.

Of particular theoretical and practical interest is the case of phase change in woody perennials when the plant is transformed from juvenile to the mature phase[2,3]. For many years, juvenility in plants has attracted the attention of plant biologists[4]. The juvenile phase is readily observed in plants that persist in this phase for some time. A classic case of juvenility is that of the common ivy, *Hedera helix* L.[5], which may retain its juvenile stage of a creepy vine and palmate leaves for an indefinite time and only occasionally switch to its mature form of an arborescent upright plant with entire leaves, which then proceeds to reproduce and fructify. These changes in growth pattern upon transition from the juvenile to the mature stage are well known for many plant species, in some of which the juvenile phase may continue over many decades. The differences arising upon transition in the cells are distinct and intrinsic and are retained even in cell cultures established from juvenile and adult shoots[6,7] or when their environment is changed drastically.

Developmental events are ultimately controlled by the state and differential expression of genes at a given time. However, no single concept has yet been developed to explain the physiological and genetic basis of the juvenile and mature stages. It was the purpose of this study to learn more details about molecular differences between the juvenile and adult phases and possible relationships between them and aging phenomena.

MATERIALS AND METHODS

Plant material. *Hedera helix* L. leaves were sampled from mature and juvenile plants growing in Nahal Amud in the Upper Galilee.

Cell number determinations. The issues were digested for 18 h in 10% chromic acid[8] to disrupt the tissue to single cells which were counted in a hemactyometer.

Experimental methods

Determinations of DNA, RNA and protein. Leaves were ground to a fine powder in liquid nitrogen before extraction. The tissues were homogenized and then left overnight in cold methanol. The precipitate was extracted with methanol containing 0.05M formic acid. After centrifugation the precipitate was suspended in 10% TCA. Then, after centrifugation, the precipitate was successively extracted with warm ethanol, ethanol: ether (2:1 v/v), and ether. The dry residue was then extracted with 5% perchloric acid for 10 min. at 90°C. DNA[9] and RNA[10] were determined in the supernatant. The precipitate was dissolved in 0.5 M NaOH and served for the assay of proteins[11].

Preparation of DNA. DNA was prepared by a method modified from Marmur[12] and Saito and Miura[13], as described previously[14]. The average molecular weight of native DNA was determined by sedimentation analysis with a Spinco model E analytical ultracentrifuge, using the equations of Studier[15]. DNA buoyant density was determined by isopycnic CsCl centrifugation.

Shearing of DNA and sizing of DNA fragments. Native DNA was sheared by sonication. $S^{o}_{20,w}$ values of the sheared DNA were obtained by alkaline banding in a Spinco model E analytical ultracentrifuge. The relationships between $S^{o}_{20,w}$ and the single strand nucleotide length were calculated according to Studier[15].

Optical melting of DNA. Thermal denaturation profiles of DNA were determined with a Zeiss spectrophotometer in water-jacketed cuvettes attached to a Haake circulating water bath. Temperatures were recorded by a thermistor probe connected to a telethermometer.

DNA/DNA reassociation. After denaturation of DNA at 100°C, DNA samples were reassociated in 0.12 M sodium phosphate at 60°C to the desired Cot values. The rate of reassociation was measured by hydroxyapatite chromatography[16].

Preparation and assay of DNA polymerase. DNA polymerase was prepared after Stout and Arens[17], as modified in our laboratory. DNA polymerase activity was assayed in the presence of primer DNA prepared from the experimental plants.

RESULTS

Characteristics of juvenile and mature leaves of *Hedera*. Tables 1 and 2 summarize some characteristics of mature and juvenile *Hedera* leaves. The RNA and DNA content tend to be higher in juvenile leaves on a dry weight basis, while the protein level is lower. The juvenile leaves are larger than mature leaves, but the average number of cells per leaf is similar. Physical properties of the DNA and its GC content are presented in Table 3 and Fig. 1.

TABLE 1

GENERAL CHARACTERISTICS OF AVERAGE LEAVES OF *HEDERA HELIX*

Parameter	Phase of Development	
	Juvenile	Adult
Size (cm^2/leaf)	23.9	17.3
Fresh wt (mg/leaf)	480.7	448.8
Dry wt (mg/leaf)	181.2	182.2
No. of cells/leaf	2.5×10^7	2.3×10^7
Volume (nL/cell)	232	220

TABLE 2

DIFFERENCES IN RNA AND DNA CONTENT OF *HEDERA HELIX* LEAVES

Leaf Nucleic Acid Origin	Basis of Calculation					
	mg/g dry matter	mg/leaf	pg/cell	pg/genome	RNA/DNA	ploidy
Juvenile						
DNA	0.935	0.145	8.0	5.1	10.5	2N
RNA	9.81	1.780	83.2	—		
Mature						
DNA	0.710	0.122	6.7	1.83	10.4	4N
RNA	7.440	1.370	70.2			

TABLE 3
SOME PROPERTIES OF *HEDERA HELIX* LEAF DNA

	DNA						
	Native				Unique		
Phase of Development	T_m [a] °C	Density gm cm^{-1}	GC [a] %	Hyper-chromicity %	ΔT_m	Hyper-chromicity %	Mismatch %
Juvenile	81.4	1.688	30	37	10.0	22.0	10.0
Mature	83.6	1.692	35	40	8.5	9.4	8.5

[a] The values are means calculated from the thermal denaturation profiles and the buoyant densities.

On a per cell basis, the content of DNA is higher in the leaf cells of juvenile plants (8.0 pg/cell) than in adult cells (6.7 pg/cell). The chromosomes of *Hedera* cells have not yet been examined to determine their ploidy. An estimation of the haploid genome size can be obtained by comparing the reassociation kinetics of *Hedera* DNA with that of a bacterial DNA of known genome size. At 350 nucleotide fragment length, the $Cot_{\frac{1}{2}}$ (pure) of non-repetitive mature *Hedera helix* DNA is 1.79 x 10^3 (Table 4). The proportion of mature *Hedera* DNA that is non-repetitive is 56% (Fig. 2). The $Cot_{\frac{1}{2}}$ of *E. coli* DNA under our conditions is 8 and its genome size is 2.8 x 10^9. Thus, mature *Hedera* DNA is 225 times more complex than *E. coli* DNA and the whole mature *Hedera* DNA contains 400 times more nucleotides than does *E. coli*. Its minimum haploid DNA content (genome mass) is, therefore, 1.4 x 10^{12} or 1.85 pg per haploid genome. The chemically determined DNA content is 6.7 pg per mature cell, indicating a ploidy of 4 N for the mature cells. A similar computation for the DNA of juvenile cells shows a genome size of 2.3 x 10^{12} or 3.62 pg. The cellular DNA content is 8.0 pg, indicating a ploidy of 2 N.

DNA/DNA reassociation kinetics. The reassociation kinetics of *Hedera* DNA, measured by hydroxyapatite chromatography, is shown in Fig. 2. The average length of the sheared DNA was about 350 base pairs. Fig. 2 indicates an apparent presence of four components of different reassociation frequencies in DNA of both mature and juvenile origin. An analysis of these data reveals notable differences in the sequence frequency classes of juvenile and mature DNA. If we define "highly repeated" sequences as those sequences reassociating at Cot = 10^{-1} and "repeated" sequences as those binding to hydroxyapatite between Cot = 10^{-1} and Cot = 10^{0}, then 6% of juvenile DNA is composed of "highly repeated" sequences and 12% of "repeated" sequences. The distribution of these fractions is different in mature DNA, where 21% is highly repeated and only 4% contain the less repeated sequences. The component of highly

TABLE 4

KINETIC ANALYSIS OF MATURE AND JUVENILE DNA OF *H. HELIX*

Category of DNA[a]	Fraction of Genome	Mass of Genome[b] (pg)	$Cot_{\frac{1}{2}}$ observed	$Cot_{\frac{1}{2}}$ pure[c]	$Cot_{\frac{1}{2}}$ expected[d]	Redundancy[e]	Kinetic Complexity[f]
			Juvenile				
H. Repetitive	0.06	0.31	8.5×10^{-3}	5.1×10^{-4}	6.5×10^{2}	7.6×10^{4}	4.66×10^{2}
M. Repetitive	0.12	0.62	2.5×10^{-1}	3.0×10^{-2}	1.3×10^{3}	5.2×10^{3}	2.67×10^{4}
Intermediate	0.04	0.21	1.6×10^{0}	6.4×10^{-2}	4.3×10^{2}	2.7×10^{2}	5.92×10^{4}
Slow	0.78	3.97	7.9×10^{3}	6.2×10^{3}	8.4×10^{3}	1.0×10^{0}	1.80×10^{10}
			Mature				
H. Repetitive	0.21	0.38	1.2×10^{-1}	2.5×10^{-2}	8.0×10^{2}	6.7×10^{3}	2.33×10^{4}
M. Repetitive	0.04	0.07	4.5×10^{-1}	1.8×10^{-2}	1.5×10^{2}	3.3×10^{2}	1.66×10^{4}
Intermediate	0.19	0.35	1.1×10^{1}	2.1×10^{0}	7.4×10^{2}	6.7×10^{1}	1.93×10^{6}
Slow	0.56	1.03	3.2×10^{3}	1.8×10^{3}	2.2×10^{3}	0.7×10^{0}	1.65×10^{9}

[a] Definition of reassociation of each category of DNA: Highly repetitive: Cot = 0,1; medium repetitive: Cot between 0.1 and 1.0; intermediate: Cot between 1.0 and 100; slow: Cot > 100.

[b] The haploid mass of the genome was computed on the basis of the cellular mass of the genome (Table 1).

[c] $Cot_{\frac{1}{2}}$pure describes the reassociation of a homogenous kinetic component of which 100% of the reassociation DNA binds to DNA. $Cot_{\frac{1}{2}}$pure = $Cot_{\frac{1}{2}}$ observed x fraction of genome.

[d] Genome size relative to *E. coli* (3.8×10^{-15}g; $Cot_{\frac{1}{2}}$ = 8 under our experimental condition); Cot expected if all the sequences were unique.

[e] Redundancy = $Cot_{\frac{1}{2}}$ expected/$Cot_{\frac{1}{2}}$ observed.

[f] Kinetic complexity in nucleotide pairs was calculated relative to the complexity of *E. coli* DNA (4.2×10^{6} nucleotide pairs) and the rate constant for the reassociation of 350 nucleotide fragments of *E. coli* DNA ($0.22\ M^{-1}\ S^{-1}$).

repetitive DNA contains probably an estimated 2-3% of zero-binding DNA, i.e., that portion of DNA that binds to hydroxyapatite at a Cot < 10^{-4} because of an apparent formation of intramolecular duplexes between palindrome sequences[16]. The third kinetic component of juvenile DNA, reassociating between Cot 1 and 100, shows very little reassociation over the next two decades of Cot, comprising only 4% of the genome. The situation is different in the mature DNA. In this case there is no clear separation between the repeated components reassociating at Cot = 1 and those reassociating from Cot > 100. The intermediate fraction, reassociating between Cot 1 and 100, comprises 19% of the genome while the fourth unique component constitutes 56%.

Transition of phases. Phase changes are probably derived from cellular differences of the shoot meristem. The adult meristem of *H. helix* consists of smaller cells than the juvenile meristem[18]. Also, the cells of adult leaves are smaller than juvenile leaves (Table 1) and contain less RNA and DNA, both on a weight[19] and cellular (Table 2) basis. Furthermore, DNA-RNA hybridization studies indicate that phase change may involve differences in the rate of transcription in the adult phase[20]. It seems, therefore, that the differences between the juvenile and adult phases of *H. helix* arise from intrinsic differences of their cells, differences which are expressed in the sequence organization of their genome DNA (Table 4). These differences in DNA sequence organization may explain why each developmental phase may persist for long periods of development and why factors controlling selective gene activity may be transmitted through many cell generations. On the other hand, DNA sequence organization is not stable and does change during ontogenesis[21], maturation[22] and under the influence of gibberellin[22].

Although normally the juvenile and adult phases are rather persistent, considerable information exists that a reversion of the phases can be experimentally manipulated. Thus, gibberellins can induce phase changes from juvenile to adult[23] or reverse the adult phase to juvenile[24], as in the case of *Hedera helix*. From the fact that the adult organ contains less cellular DNA, we may expect it to be more responsive to gibberellins[25].

The latter case, being typified by an increase in cellular DNA (Table 2) and a change in its sequence organization,should probably be accompanied by a gibberellin-induced DNA synthesis. Therefore, we conducted studies of possible effects of gibberellins on DNA polymerases of plants (I. Snir and B. Kessler, to be published). We isolated and purified DNA polymerase from some plants. Under *in vitro* conditions the activity of this enzyme was entirely dependent on the presence of primer DNA and the four deoxyribonucleotides in the reaction mixture. Before the start of the reaction the various components of this system were selectively pre-incubated with gibberellins and the following results were obtained:

(a) The purified enzyme, even after a prolonged pre-incubation with gibberellin, does not respond to the hormone (Table 5).

(b) A pre-incubation of primer DNA with gibberellin resulted in an activation of the DNA polymerizing reaction (Table 5).

(c) DNA's of different sources and synthetic polymers served as good primers, such as DNA from cucumber, peas, as well as dGdT. An AT polymer, on the other hand, namely, dAT, had no primer activity.

These data indicate that gibberellins might be directly involved in the extent and specificity of DNA polymerase activity. Gibberellin has indeed been shown to bind to DNA[14], causing changes in the rate of RNA synthesis within 1-2 hours of application[26], leading to the initiation of gene activity[27].

TABLE 5

THE EFFECT OF PRE-INCUBATION WITH GIBBERELLINS ON THE ACTIVITY OF DNA POLYMERASE

Pre-incubated Reaction Component	% Activation			
	Gibberellin A_3		Gibberellin$_{4/7}$	
	2×10^{-5}M	2×10^{-6}M	2×10^{-5}M	2×10^{-6}M
DNA polymerase (6)[a]	0	0	0	0
DNA primer (6)	25	11	18	23

[a] The numbers in parentheses indicate the number of experiments

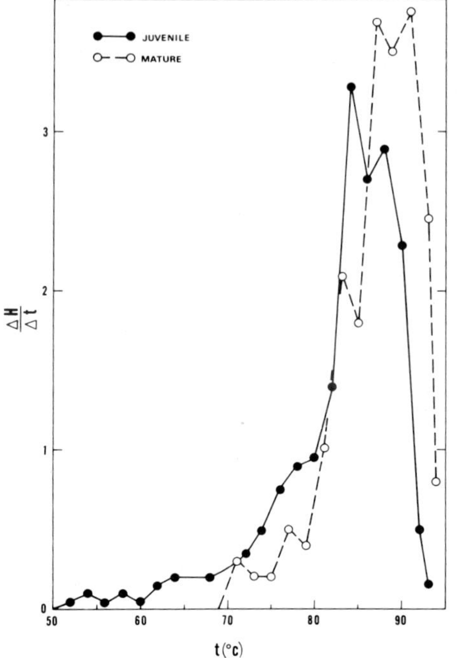

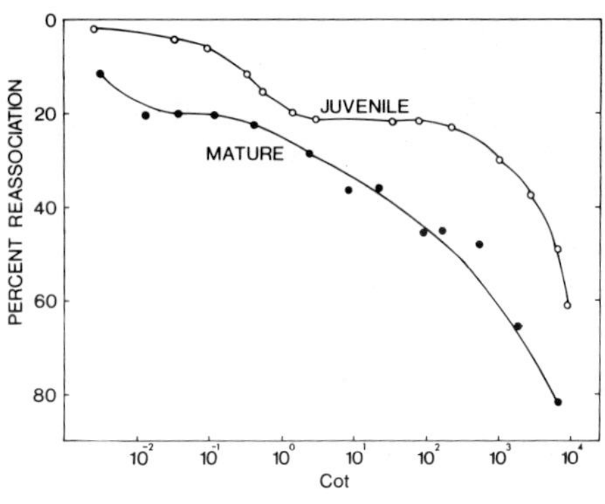

Fig. 1. Derivative absorption-melting curves of DNA from juvenile and mature leaves (left).

Fig. 2. Renaturation profile of DNA from juvenile and mature *H. helix* (right).

DISCUSSION

It is quite generally accepted that the differences between the juvenile and adult phases are derived from changes in the shoot apices, but very little is known about factors that trigger the phase transition. A possible candidate to act as such trigger is gibberellin, which induces phase change[24] and activates DNA polymerase activity (Table 5) via binding[14] to primer DNA.

A model for a genetic regulation of phase change has been proposed by Brink[1,28]. Brink characterized chromosomes as paragenetic and genetic. Paragenetic chromosomes are primarily concerned with the ontogenetic development and the genetic ones with heredity. Brink suggests that chromosomes embody an apparatus for the primary regulation of gene actions during development and that the chromosomes themselves are the site for primary changes leading to phase change.

According to Brink, the expression of a structural gene can be altered in a specific way when linked with a heterochromatic segment, and can switch back to its original state when the association is terminated. This phenomenon was termed "paramutation." The heterochromatic repressor segment consists of repeating subunits, the number of which determines the expression of the structural gene. Changes in the number of the subunits may result from overreplication or nonreplication during the cell cycle. Paramutation thus constitutes a kind of genetic variability which represents a regular heritable change in function. Paramutable genes are inherently metastable and paramutational alterations may persist over long periods.

In this account we described experiments which indicate that major changes in the developmental pattern of *H. helix* are distinctly expressed and probably located within the chromosomes. Similar to cucumber[25], *H.helix* has a relatively small genome size and a relatively high proportion of the unique DNA fraction, as compared with other higher plant genomes. Juvenile leaves contain more DNA per cell and per haploid genome and their ploidy seems to be 2N. Upon transition to the mature phase the cellular as well as the genomic DNA level decreases, but their ploidy becomes 4N. It seems that the polyploidization upon phase transition results from a selective amplification of sequences rather than from endoduplication of the entire genome as evidenced by (i) the shift of $Cot_{\frac{1}{2}}$ upon transition, and (ii) the change in the relative proportions of the different DNA classes.

On the basis of our results we propose tentatively that paramutation is involved in some way in phase changes as it might account for (i) gibberellin-induced phase changes, (ii) the gibberellin-induced activation of primer DNA, (iii) the higher level of DNA in juvenile *H. helix*, (iv) the larger and more redundant genome mass of the juvenile cells which could result from a formation of repressor subunits (sequences), and (v) the longer persistent juvenile phase during which polyploidization is suppressed.

Studies are now in progress to test in detail the above working hypothesis.

SUMMARY

1. We studied cellular changes during phase change in *Hedera helix*.
2. Juvenile leaves contain a higher level of cellular DNA and RNA than do mature leaves.
3. The size and mass of the juvenile genome is larger than that of the mature genome.
4. Calculations indicated that the ploidy of juvenile cells is 2N and of mature cells 4N.
5. Gibberellin, which induces phase transition, activates the DNA primer in plant DNA polymerase systems.
6. The results are interpreted in the light of the paramutation theory.

REFERENCES

1. Brink, R.A. (1962) Quart. Rev. Biol., 37, 1-22.
2. Wareing, P.F. (1959) J. Linn. Soc. Lond. (bot.), 56, 282-289.
3. Zimmerman, R.H. (1972) Hort. Science 7, 447-455.
4. Goebel, K. (1900) Organography of Plants. Clarendon Press, Oxford.
5. Stoutemyer, V.T. and Britt, O.K. (1965) Am. J. Bot., 52, 805-810.
6. Stoutemyer, V.T. and Britt, O.K. (1969) Am. J. Bot., 56, 222-226.
7. Robbins, W.J. and Hervey, A. (1970) Am. J. Bot., 57, 452-457.
8. Brown, R. and Rickless, P.P. (1949) Proc. Roy. Soc. ser. B, 136, 110-125.
9. Keck, K. (1956) Arch. Biochem. Biophys., 63, 446-451.
10. Mejbaum, W. (1939) Z. Physiol. Chem., 258, 117-122.
11. Lowry, O.H., Rosebrough, N.J., Farr, A.L. and Randell, R.J. (1951) J. Biol. Chem., 193, 265-275.
12. Marmur, J.A. (1961) J. Mol. Biol., 3, 208-218.
13. Saito, H. and Miura, K.I. (1963) Biochim. Biophys. Acta, 72, 619-629.
14. Kessler, B. and Snir, I. (1969) Biochim. Biophys. Acta, 195, 207-218.
15. Studier, F.W. (1975) J. Mol. Biol., 11, 373-390.
16. Britten, R.J., Graham, D.E. and Nenfeld, B.R. (1974) in Methods in Enzymology 29E, Grossman, L. and Moldave, K., eds., Academic Press, N.Y.
17. Stout, E.R. and Arens, M.Q. (1970) Biochim. Biophys. Acta, 213, 90-100.
18. Stein, O.L. and Fosket, E.B. (1969) Am. J. Bot., 56, 546-561.
19. Millikan, D.F. and Ghosh, B.N. (1971) Pl. Physiol., 24, 10-13.
20. Rogler, E.R. and Dahmus, M.E. (1974) Pl. Physiol., 54, 88-94.
21. Miksche, J.P. and Hotta, Y. (1973) Chromosoma, 41, 29-36.
22. Kessler, B. (1973) in Biochemistry of Gene Expression in Higher Organisms, Australia and New Zealand Book Comp., Sydney.

23. Pharis, R.P. and Morf, W. (1961) Can. J. Bot., 45, 1519-1524.
24. Robbins, W.J. (1957) Am. J. Bot., 44, 743-746.
25. Snir, I. and Kessler, B. (1975) Pl. Sci. Letters, 5, 163-170.
26. Sussex, I., Clutter, M. and Walbot, V. (1975) Pl. Physiol., 56, 575-578.
27. Varner, J. (1964) Pl. Physiol., 39, 413-415.
28. Brink, R.A., Styles, D.A. and Axtell, J.D. (1968) Science 159, 163-170.

Chromosomes Today Volume 6, A. de la Chapelle and M. Sorsa eds.

CHROMOSOME ARCHITECTURE AND THE INTERPHASE NUCLEUS: DATA AND THEORY ON THE MECHANISMS OF DIFFERENTIATION AND DETERMINATION

DALE M. STEFFENSEN
Department of Genetics and Development, University of Illinois, Urbana, Il. 61801, U.S.A.

ABSTRACT

With the advent of in situ RNA:DNA hybridization to chromosomes and nuclei one can examine the three-dimensional structure of genes at interphase. Although we have data on human cells and several primates, the main research effort has been with *Drosophila melanogaster*. Using whole mount preparations of tissue with interphase nuclei the 5S rRNA genes were shown to take a precise position at 2 PM between the nucleolus and the nuclear membrane. The 5S genes are on 2R near the end at 56F, while the nucleolar organizer (NO) is in the middle of the centric heterochromatin on the X chromosome and on the short arm of the Y chromosome. The 2 PM position has been established with nuclei from salivary glands, Malphighian tubules, rectal cells and the large nuclei brain cells. Although not confirmed by cell lineage study, it appears that genes change their position in some nuclei during terminal differentiation. A number of other RNAs are being examined, singly and in combination with 5S RNA. There are histone DNA, a few tRNA species and several DNA sequences obtained from Drosophila DNA grown in plasmids. The purpose is to put together a complete interphase map for several differentiated cell types. At least one highly repeated DNA (DNA satellite IV) has been examined in interphase. This DNA sequence and the others appear to control chromosome arm orientation and coordinate gene position by connecting various segments to the chromocenter.

Several lines of evidence suggest that specific gene positioning and nuclear architecture are fundamental and are prerequisites for determination and development to occur. Simply, gene placement is nuclear determination. A theory has been developed to explain how the egg and sperm proceed to produce the differentiated embryo. The spacial symmetry of the egg nucleus specifies the radially related symmetry in the egg cortex.

INTRODUCTION

Almost every biologist would concede that the egg and sperm nuclei possess the ability to control differentiation in the developing zygote. The agreement on how development is regulated and controlled ends there. Through a series of events involving both theoretical and experimental discoveries, I believe that a simple theory can be proposed to explain the modes and patterns of development in plants and animals. It is called a theory rather than a hypothesis because the ideas were

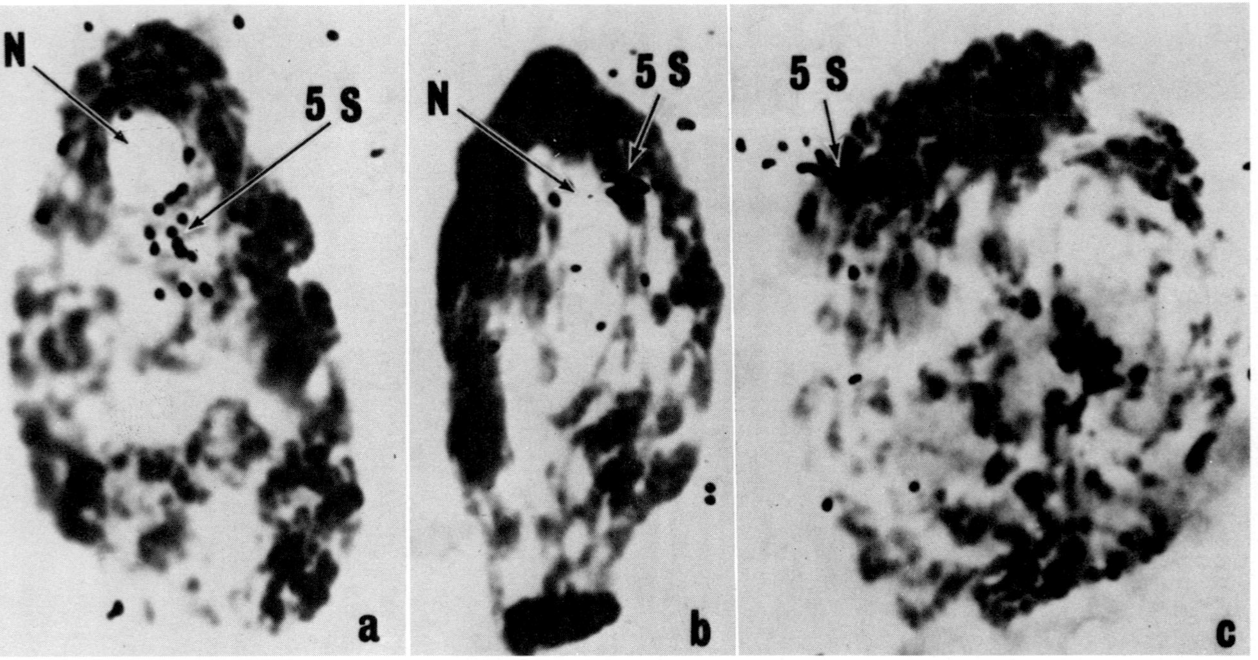

Fig. 1. Salivary gland nuclei of Drosophila labeled in situ with ^{125}I-5S rRNA to illustrate nucleolar association. N indicates the space occupied by the nucleolus after ribonuclease digestion. a and b have the 5S loci (5S rRNA) in the upper middle part of the cell and if rotated 90° either way the genes would be at 10 AM or 2 PM. In C the 5S genes are at the edge at 10 AM. The nucleolar area is obscured by overlapping chromatin. All three cells have a well defined chromocenter at noon.

intuitively embedded in the works of T. Boveri and E.B. Wilson[1]. Time and space do not permit the development of the evidence in any depth; thus, only the barest essentials of this theory will be outlined.

The data come from an extension of mapping genes by in situ hybridization on chromosomes. As long as 6 or 7 years ago, it had become apparent that certain genes, such as 5S rRNA, were not located randomly in the interphase nuclei but were always associated near nucleoli. Evidence will be presented from Drosophila. A number of technical and biochemical advances were required in order to develop convincing evidence for these observations. In particular, three pure radioactive probes were needed for triangulation and establishment of gene positions.

The theory revolves around one simple statement: genes, chromosome segments and the genome as a whole are arranged in prescribed three-dimensional configurations and associations which position certain gene products within the nucleus and regulate the location of their transfer from the nuclear membrane to symmetrically related sites in the cytoplasm.

MATERIALS AND METHODS

The wild type of Drosophila melanogaster, Canton S, was used. Procedures for in situ hybridization have been described elsewhere, as have the methods for purifying and iodinating 5S rRNA and 18 + 28S rRNA[2]. The ^{3}H-C RNA copied from the histone genes of the sea urchin have been used in new experiments in collaboration with Drs. R.H. Cohn and L.H. Kedes. The histone gene work is described in one of their papers[3].

The cells of the various organs used were dissected in insect Ringers solution, then teased into a thin layer while they were monitored under the phase microscope. Next they were frozen in liquid nitrogen, fixed in ethanol : acetic acid (3:1), given three changes of ethanol, and air-dried in preparation for in situ methods.

RESULTS

The salivary gland nuclei in Fig. 1 illustrate a basic observation in somatic nuclei. The 5S rRNA genes near the end of 2R (56F) associate specifically with nucleoli. When the cell is prepared so that the 5S gene falls at 2 PM, the nucleolus is always immediately toward the left, as shown in the drawings of autoradiographs of Malphighian tubules, Fig. 2.

Conclusions from annealing RNA in situ with three different probes (histone cRNA, 5S rRNA and 18 + 28S rRNA) can be summarized by simple geometrical analyses or the clock analogy. When the end of chromosome 2R is located between the nuclear membrane and the nucleolus at 2 PM, the histone genes at the proximal part of 2L (39DE) are always in the AM quadrant at 8 AM or 11 AM. This triangulation suggests that the arms 2R and 2L assume the same left-right position relative to the nucleolar rDNA in females having two X chromosomes or in males with single X and Y chromosomes. The rDNA on the X chromosome is in the middle of the X heterochromatin and the rDNA of

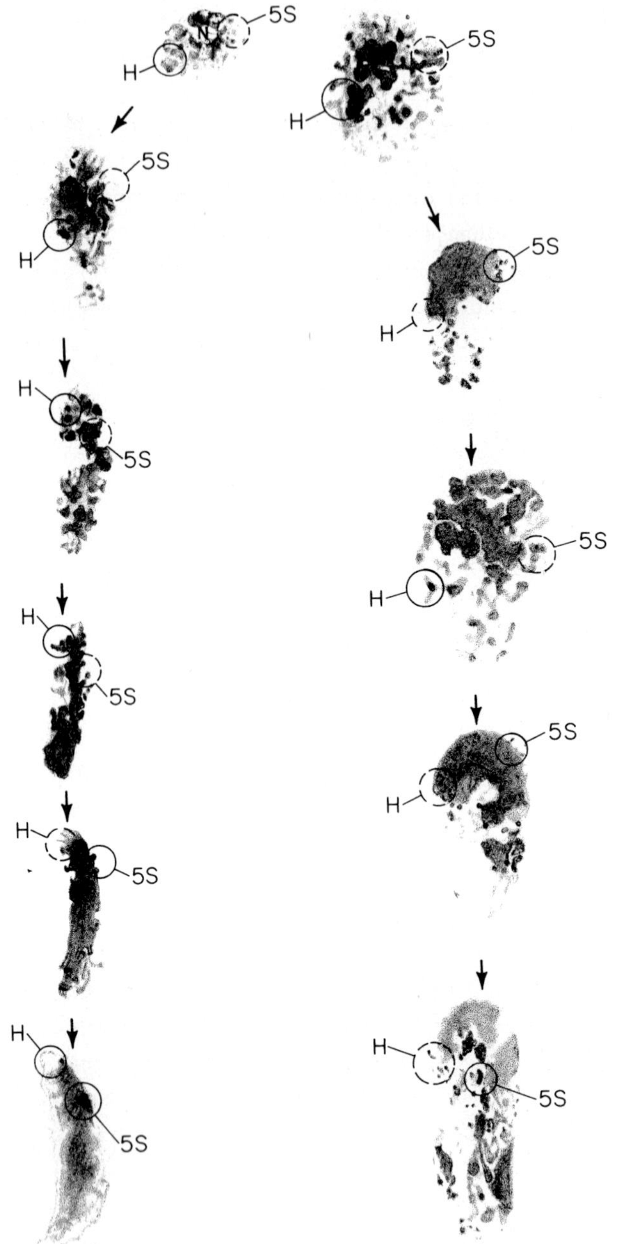

Fig.2. A drawing of two nuclear types from Malphighian tubules of Drosophila hybridized with histone cRNA-^{3}H and/or 5S rRNA-^{125}I. The kidney-shaped nuclei on the left always have the 5S rRNA genes (5S) on the convex side at 2 PM, while the histone genes (H) are always on the concave side at 8 AM or 11 AM, depending on the developmental stage. The stage switch from 8 AM to 11 AM appears to be a specific unfolding of 2L toward the top of the nucleus. The nuclear type on the right is at a higher polytene level than the one on the left. Here the chromosome arm of 2L remains at 8 AM. These nuclei remain in the "jack-knife" position and never unfold completely as do the "kidney-shaped" nuclei on the left. Solid circles indicate observed heavy labeling in autoradiographs, while dotted circles indicate the inferred sites of each gene. The nucleus at the bottom left had both histone and 5S RNA used in the *in situ* hybridization.

the Y is in the middle of the Y^S.

Diploid Malphighian nuclei or undifferentiated nuclei at low polytene stages have prominent chromocenters. The 9 or 10 sites of DNA satellite 1.705 (IV) on all chromosomes form a single "dumbbell" cluster of label (Peacock and Steffensen, 1975[4] and unpublished), occupying about 1/4 of the chromocenter. Since the Y chromosomes have four large blocks of DNA 1.705, including the one near the end of Y^L, the Y^L arm must fold back to the chromocenter in order to place all of the sequences together. Male nuclei exhibit a large fold-back loop on the 5S gene side at 3-4 PM, probably representing both 2R and Y^L. (This assumption is being tested directly.) These and many other observations by other Drosophila workers, including position effect phenomenon, force one to the conclusion that the highly repeated DNA sequences have a primary role in initiating the early phases of nuclear differentiation.

SUMMARY AND CONCLUSIONS

1. In the egg, the nucleolus and the nucleolar organizer chromosomes are positioned at 11 AM or the mirror image position of 1 PM during early egg development in every plant or animal observed thus far. Other chromosomes assume their "bouquet" configurations at zygotene and pachytene, so that the distal chromosome ends will be fixed at the top of the egg and proximal segments at the basal end. The basal end will eventually give rise to the vegetable pole of the animal embryo.
2. Most of the homoeotic mutants of *Drosophila melanogaster* are clustered principally in the proximal euchromatic region of chromosome 3 but also occur in the same position in 2 (Fig. 3). In *Zea mays*, most of the macromutations and developmental mutants are also located mainly on the proximal segment of the long arm of chromosome 3 but also occur on chromosome 1 long according to a recent map[7]. These genes cluster at the basal region of the egg nucleus, where gene products can be transported to the future vegetable pole or germ cell end. A few developmental genes at chromosomal ends are positioned to control the anterior embryonic development.
3. The individual highly repeated DNA sequences aggregate and initiate the basic gene program of positional information at both early and late stages.
4. Chromosome associations between certain DNA segments and the nuclear pore membrane complexes provide a mechanism for moving nuclear pores into the annulate lamella. Gene products are then moved outward to the cortical cytoplasm. Centriole, microtubules and actin-containing microfilaments coordinate this precise transfer.
5. Positionally located determinants often observed as cortical granules or polar granules near the cell membrane enter embryonic nuclei[8] and determine the cell. Actual nuclear determination probably occurs when the nucleus changes from the diffuse state to an uneven heterogeneous state with membrane attachments and active nucleoli at or about the blastula stage. In many vertebrates, one X chromosome becomes late-labeling and becomes associated with the nuclear membrane.
6. The differentiated state is propagated from one cell generation to the next as

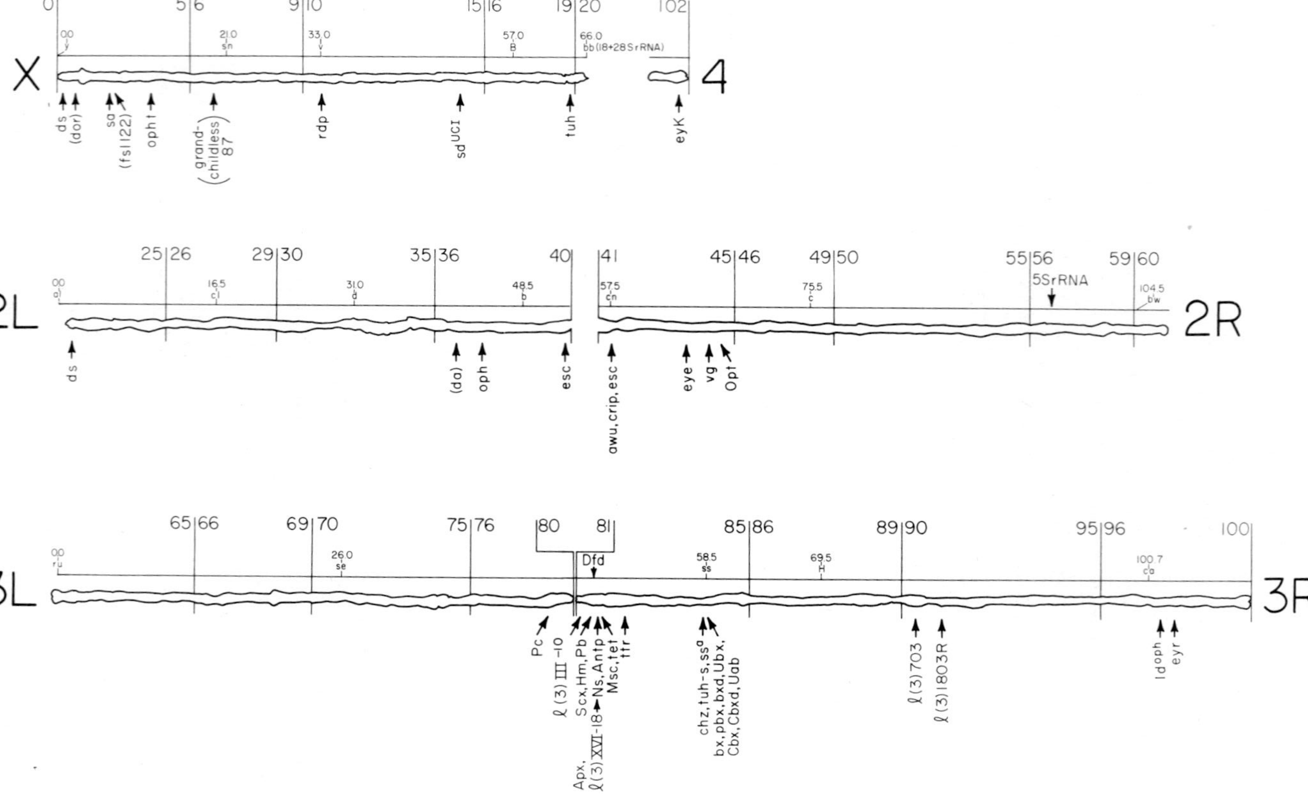

Fig. 3. Homoeotic mutants of Drosophila melanogaster (Ouweneel, 1976)[5]. Maternal effect mutations are indicated in parentheses. One gene (da) is located in section 31 rather than 37 according to Sandler (1977)[6]. Many of the genes have not been mapped exactly. Pc is on 3L in 79 according to T. Kaufman (unpublished).

clones. The cell biology literature on nuclear membrane and nuclear pore attachment during mitosis is extensive[9].

REFERENCES

1. Wilson, E.B. (1925) The Cell in Development and Heredity, Macmillan Co., New York.
2. Steffensen, D.M. (1977) Human gene localization by RNA:DNA hybridization in situ. In: Molecular Structure of Human Chromosomes, Chromosomes in Biology and Medicine 1 (Editor J.J. Yunis). Academic Press, Inc., New York.
3. Cohn, R.H., Lowry, J.C. and Kedes, L.H. (1976) Histone genes of the sea urchin (S. purpuratus) cloned in E. coli: Order, polarity and strandness of the five histone-coding and spacer regions. Cell 9, 147-161.
4. Peacock, W.J. and Steffensen, D.M. (1975) Mapping the highly repeated DNA sequences 1.705 of Drosophila melanogaster. J. Cell Biol. 328a.
5. Ouweneel, W.J. (1976) Developmental genetics of homoeosis. Adv. Genetics 18, 179-248.
6. Sandler, L. (1977) Evidence for a set of closely linked autosomal genes that interact with sex chromosome heterochromatin in Drosophila melanogaster. Genetics 86, 567-582.
7. Neuffer, M.G., Jones, L. and Zuber, M.S. (1968) The mutants of maize, Published by Crop Sci. Soc. Amer., Madison, Wisc.
8. Mahowald, A.P., Illmensee, K. and Turner, F.R. (1976) Interspecific transplantation of polar plasm between Drosophila embryos, J. Cell Biol. 70, 338-373.
9. Maul, G.G. (1977) Nuclear pore complexes, J. Cell Biol. 74, 492-500.

Chromosomes Today Volume 6, A. de la Chapelle and M. Sorsa eds.

LOCALIZATION, STRUCTURE AND ACTIVITY OF RIBOSOMAL CISTRONS IN THE MOUSE OOCYTE DURING MEIOTIC PROPHASE I

A. STAHL, C. MIRRE, M. HARTUNG, B. KNIBIEHLER and A. NAVARRO
Laboratoire d'Histologie-Embryologie II, Faculté de Médecine, 13385 Marseille Cedex 4, France.

ABSTRACT

The mouse oocyte is the site of nucleolar synthesis during pachytene. The chromosomes containing a nucleolar organizer are attached to the nuclear envelope by their paracentromeric heterochromatin. The nucleolus appears at the junction between the paracentromeric heterochromatin and the euchromatic portion of the bivalent. In this region, which corresponds to the secondary constriction, 30 Å diameter fibers extend from the lateral element of the synaptonemal complex up to the nucleolar fibrillar center in which they penetrate. In situ hybridization demonstrates that the ribosomal cistrons are localized in the secondary constriction and in the fibrillar center of the nucleolus. Following brief incorporation of ^{3}H-uridine, labeling is localized over the dense fibrillar component which is always adjacent to the fibrillar center. These observations might indicate that the nucleolar rDNA is localized in both the fibrillar center and its associated dense fibrillar component and that rDNA transcription occurs in the latter.

INTRODUCTION

Ribosomal cistrons have been studied for a long time using purely morphological techniques, and were known as nucleolar organizers. The latter were rapidly localized in the secondary constriction of metaphase chromosomes[1,2,3]. In situ hybridization confirmed the presence of the ribosomal cistrons in the secondary constriction of the nucleolar chromosomes[4]. In the mouse, several autosomes present a secondary constriction immediately distal to the centromere[5,6]. Polymorphism in the number and distribution of secondary constrictions has been noted among inbred strains[7]. Using in situ hybridization the ribosomal DNA was localized in the centromeric region of chromosomes n° 12, 15, 16, 18 and 19 in the mouse, and variations in the number of genes present on a given chromosome were demonstrated depending on strain. It appears that a large secondary constriction indicates a large gene number, whereas absence of the secondary constriction does not exclude the presence of ribosomal cistrons, but suggests the presence of a reduced amount[8,9,10].

Localization of the ribosomal cistrons in the interphase nucleus is more difficult to determine. The presence of labeled nucleoli in the human lymphocyte nucleus was reported by Henderson et al[11]. In Xenopus, Pardue[4] indicated that the genes coding for ribosomal RNA are situated within or in close proximity to the nucleolus.

Until recently, no studies have been performed in order to establish the correlation between gene localization as revealed by in situ hybridization, and the various components of the nucleolus as observed in the electron microscope. This situation can be explained by the lack of precision of in situ hybridization which presently can only be used in light microscope studies. The difficulty might be overcome by comparing the results of in situ hybridization to the ultrastructural data from large nucleoli whose components are widely separated from each other, as seen in the case of natural nucleolar segregation. This study was performed in our laboratory using quail oocytes during meiotic prophase I. During pachytene, we observed de novo synthesis of nucleoli which present characteristic relations with the microchromosomes containing the nucleolar organizer. Chromatin fibers emanating from certain bivalents penetrate into the fibrillar center of the newly formed nucleolus. The fibrillar center is surrounded by a layer of dense fibrils. Granules appear later on the outer side of the dense fibrillar region, in such a way that at a given moment the nucleolus presents 3 distinct zones[12]. The region which becomes labeled after in situ hybridization corresponds to the fibrillar center of the nucleolus[13]. The same technique was applied to human skin fibroblasts containing a nucleolus without a fibrillar center. In this case, the distribution of the ribosomal cistrons is more diffuse and essentially located on the nucleolar periphery[14].

In the mouse oocyte, de novo nucleolar synthesis occurs at pachytene[15,16]. Successive formation of the fibrillar center, dense fibrils and granules, which at the onset of synthesis are well-separated from each other, allows useful comparison with data from in situ hybridization.

MATERIAL AND METHODS

In situ hybridization

Ovaries from 18,19 and 20 day-old mouse embryos were removed and treated according to the technique of Luciani et al[17]. Following fixation in methanol-acetic acid (3:1), the gonads were placed in 45 % acetic acid. Cellular dissociation was obtained by repeated pipetting. The resulting suspension was spread on precooled slides which were then dried by hot air.

In situ hybridization was performed with tritium labeled ribosomal RNA obtained from HeLa cells according to the technique described by Ascione and Arlinghaus[18]. Tritiated uridine (24 Ci/mM, CEA, France) was added to cultures (1mCi per 75cm^2 Falcon dish). Extraction of ribosomal RNA was followed by isolation of 18 S and 28 S RNA on a sucrose gradient and purification on a Sephadex G25 column. The specific radioactivity of the tritium labeled ribosomal RNA was 1.4×10^6 cpm per μg. Hybridization was performed according to the technique described by Gall and Pardue[19]. The hybrid preparations were covered by Kodak NTB2 emulsion, exposed for 90 days at 4 $^\circ$C, and developed using Kodak D 19b developer. The autoradiographs were stained with Giemsa solution.

Electron microscopy

The gonads of female mouse embryos aged 13 through 20 days were removed at daily intervals. Ovaries of 1,2 and 3 day-old mice were also removed. The ovaries were fixed for 15 min at 4°C in 3 % glutaraldehyde in 0.1 M phosphate buffer, pH7.2, containing 0.1 M sucrose. After washing in the buffer solution, the specimens were postfixed in 2 % osmium tetroxyde in identical buffer for 20 min. After dehydration, the ovaries were embedded in Epon and sectionned with a diamond knife on a Reichert OMU2 ultramicrotome. The sections were contrasted with uranyl acetate and lead citrate.

High resolution autoradiography

The entire ovaries were incubated 30 min. or 45 min. at 37°C in 1 ml of 80 % Eagle and 20 % calf serum medium containing 100 μCi/ml of ^{3}H-uridine (sp. act. 24 Ci/mM, CEA, France). After the above described fixation and before postfixation, the specimens were repeatedly washed during 30 min. in the buffer in order to remove unincorporated radioactive uridine. After dehydration and embedding in epon, sections were realized for the following technique according to Granboulan[20] and Salpeter and Bachmann[21]. The sections were transfered to a collodion-coated slide. Before coating with emulsion, the sections on the slides were stained with uranyl acetate and lead citrate. A thin carbon layer was vacuum evaporated over the stained sections. Slides were then coated with a diluted Ilford L4 emulsion by the dipping method. After exposure for 4 months at 4°C the preparations were developed in Microdol-X(Kodak) for 4 min. at 18°C and fixed with Kodak Rapid Fixer for 3 min. The specimen was stripped off onto a surface of double-distilled water and copper grids were placed over the sections. The specimen on the grids was then picked up with a filter paper. After drying, the grids were carefully released from the remnant film. The collodion film was thinned for 2 min. in iso-amylacetate. Preparations were examined with a Siemens Elmiskop 101 microscope at 80 KV.

OBSERVATIONS

In situ hybridization

At pachytene several chromocenters formed by association of the centromeric heterochromatin of 3 to 5 bivalents can be seen. In situ hybridization reveals the presence of a maximum of 3 sites corresponding to the ribosomal cistrons. These sites are located on certain bivalents in close proximity to the chromocenter(fig. 1 and 2). This zone presents a particular appearance since it is less stained and frequently thinner than the remaining regions of the bivalent. Such an aspect exactly corresponds to the secondary constriction of mitotic bivalents. Labeling grains are often situated on either side of the secondary constriction rather than over the chromosome.

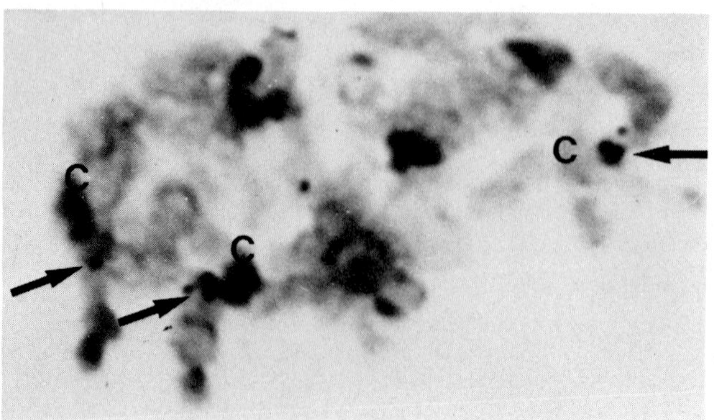

Fig.1. Pachytene stage of mouse oocyte after in situ hybridization with ^{3}H-rRNA. Arrows indicate 3 labeled nucleolar organizer regions (secondary constrictions) closely located to paracentromeric heterochromatin. C = chromocenters formed by association of paracentromeric heterochromatin of several bivalents.

Electron microscopy

Electron microscopy reveals that the chromocenters formed by association of the centromeric heterochromatin of several bivalents are in contact with the nuclear envelope. The newly synthesized nucleoli most often appear in contact with one of the bivalents whose heterochromatin is contained in the chromocenter (fig.3). The nucleoli may also be observed in contact with an apparently isolated bivalent whose centromeric end is attached to the nuclear envelope (fig.4,5, and 6). A synaptonemal complex is present in the axial region of the bivalent. In proximity to the nuclear envelope, the synaptonemal complex is surrounded for a length of 1.5 μ by heterochromatin constituted by coarse fibers. The nucleolus always appears beyond this heterochromatic zone, facing that region which corresponds to the secondary constriction and is situated at the beginning of the euchromatic region of the bivalent.

Characteristic relations exist between the chromosome containing the nucleolar organizer and the nucleolus. The secondary constriction, facing the nucleolus, is approximately 0.25 μ long. The former is composed of approximately 30 Å diameter chromatin fibers. Beyond the secondary constriction the euchromatin is formed by fibers of about 100 Å diameter. The 30 Å diameter fibers extend from the lateral component of the synaptonemal complex up to the nucleolus. The latter constantly presents a fibrillar center oriented toward the secondary constriction. The 30 Å fibers penetrate into the fibrillar center where they are no longer visible (fig. 4 et 5).

The fibrillar center is largely surrounded by highly electron dense fibrils (fig. 5 and 6). As the nucleolus gradually increases in size extension of the fibrils can be observed, forming an irregular band. Beginning at the secondary constriction the following can be seen : 1. the fibrillar center, 2. a narrow dense fibrillar zone, 3. a band composed of less dense fibrils, 4. a mixed fibrillar and granular zone, 5. a granular zone (fig.5). Since it is known that synthesis of fibrils precedes that of the granules, the preceding configuration is highly suggestive of nucleolar synthesis

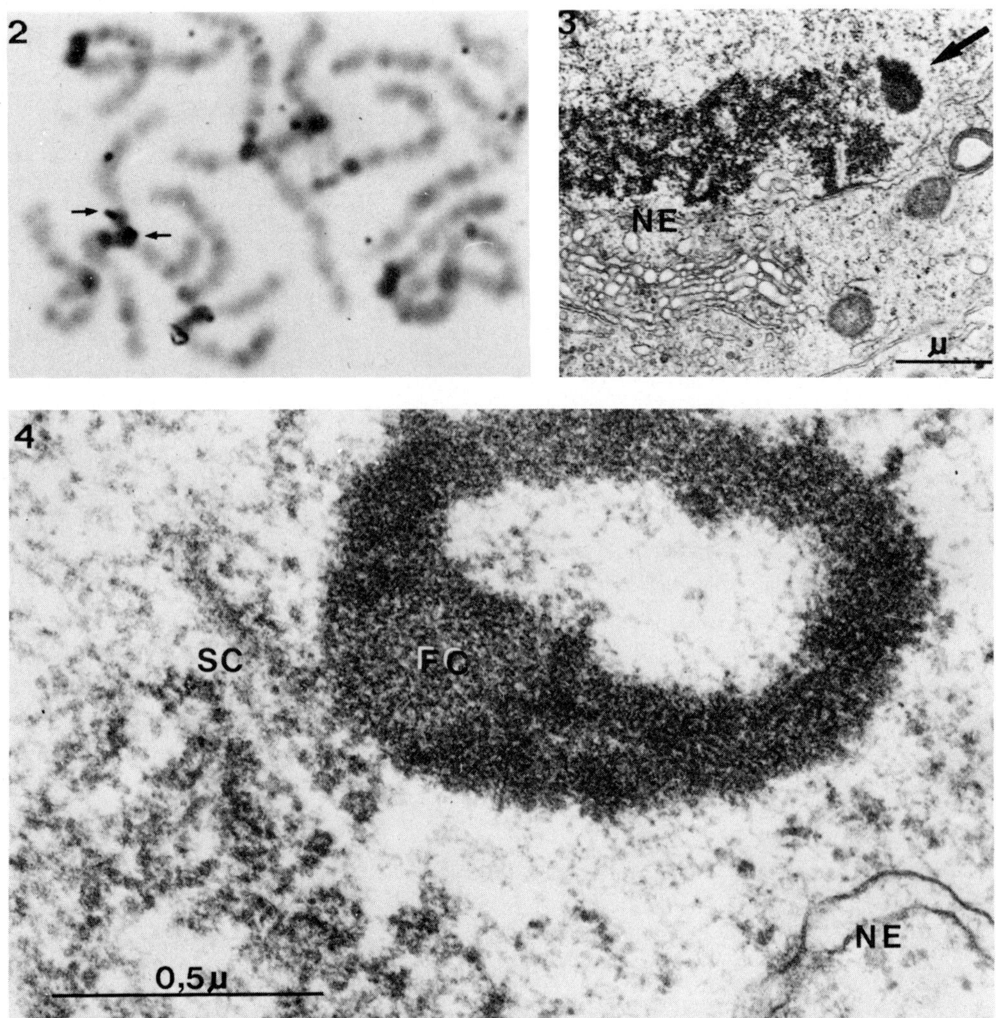

Fig. 2. Pachytene stage of mouse oocyte after in situ hybridization with ^{3}H-rRNA. Arrows indicate 2 clusters of silver grains symmetrically located on each side of the juxta-centromeric region of a bivalent.

Fig. 3. Newly formed nucleolus, constituted by a fibrillar center partially surrounded by dense fibrils, emerging in intimate contact with a bivalent whose heterochromatin is associated with a chromocenter anchored to the nuclear envelope (NE). Several synaptonemal complexes are visible within the chromocenter.

Fig. 4. The fibrillar center (FC) of the growing nucleolus is penetrated by fibers which emanate from the juxta-centromeric region of the bivalent. SC = synaptonemal complex. NE = nuclear envelope.

occuring in close relation with the components of the fibrillar center, i.e. with those fibers which emanate from the secondary constriction and are included in the latter.

Two nucleoli are most often synthesized in a simultaneous manner, symmetrically disposed on either side of the secondary constriction (fig.6). The two originally separate nucleoli often fuse at their granular distal portions.

At the onset of diplotene the chromosomes begin to unravel. The disjoined chromosomes only present a single axial core. The nucleoli are more voluminous and progressively take on a reticular appearance. The relations between the secondary constriction and the fibrillar center are no longer visible. In the one day-old mouse oocyte very large reticular nucleoli can be seen presenting those constituents described by Chouinard[22] : 1. a mixed fibrillar and granular component forming the main portion of the nucleolar reticulum ; 2. several fibrillar centers ; 3. dense fibrils surrounding each fibrillar center (fig. 7).

High resolution autoradiography

While no incorporation of nucleolar RNA precursors is observed at leptotene and zygotene[16] active uptake of ^{3}H-uridine is seen at late pachytene and early diplotene. Following incubation for 30 min., labeling occurs over the dense fibrillar component adjacent to the fibrillar center. No grains are seen overlying the fibrillar center. The granular constituent is unlabeled (Fig. 8).

When incubation is extended to 45 min. labeling also involves the granular component of the nucleolus.

DISCUSSION

At pachytene in situ hybridization reveals that the ribosomal cistrons are located in the secondary constriction region of 3 bivalents. Ultrastructural studies in pachytene nuclei demonstrate that nucleolar synthesis certainly occurs in close relation to the 30 Å diameter chromatin fibers emanating from the secondary constriction. These fibers penetrate into the fibrillar center of the nucleolus.

The relations between the secondary constriction and nucleolar fibrillar center are highly comparable to those previously described in the quail oocyte[12]. In the latter, chromatin fibers originating in a microchromosome containing the nucleolar organizer also penetrate into the fibrillar center of the nucleolus. Comparison of in situ hybridization and ultrastructural data in the quail oocyte clearly shows that the ribosomal cistrons are located in the fibrillar center[13]. It is interesting to note that similar configuratinns can also be seen in the meiocytes of plants wherein the fibrillar center, traversed by the synaptonemal complex, contains the nucleolar organizer[23].

In interphase nuclei of somatic cells, autoradiography following ^{3}H-actinomycin D binding and ^{3}H-uridine incorporation supports the localization of the ribosomal cistrons in the fibrillar center of the nucleolus[24,25].

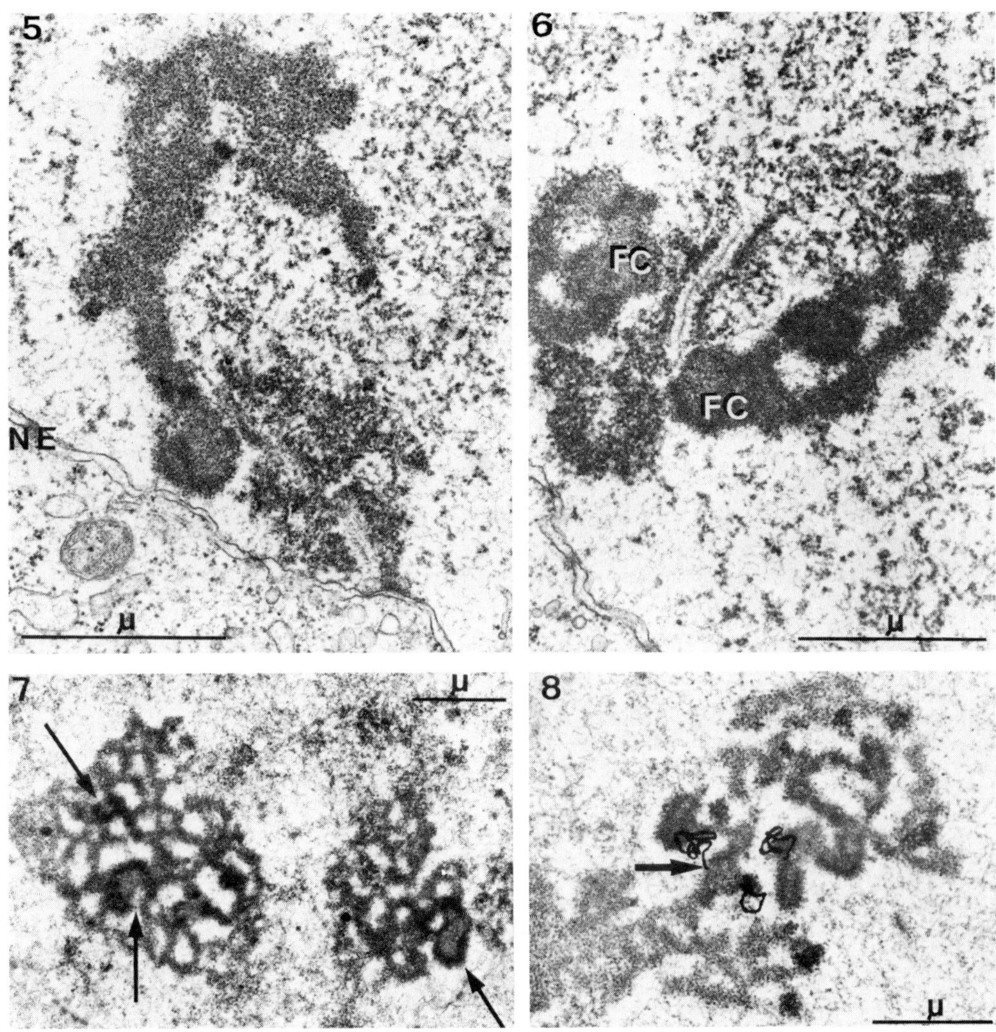

Fig. 5. Nucleolar formation occuring in close connection with the secondary constriction of a bivalent whose paracentromeric heterochromatin is attached to the nuclear envelope (NE). The nucleolus displays from its proximal region to its distal end : the fibrillar center, penetrated by fibers emanating from the chromosome and surrounded by very dense fibrils, a band-shaped fibrillar zone, a fibrillo-granular zone and a granular zone.

Fig. 6. Two nucleoli are symmetrically formed at the secondary constriction of each chromosome whose pairing forms the bivalent. FC : fibrillar center.

Fig. 7. Mouse oocyte at early diplotene. The reticulated nucleoli show several fibrillar centers (arrows) surrounded by dense fibrils.

Fig. 8. Autoradiograph after 30 min. ^{3}H-uridine incorporation. The silver grains are located in the dense fibrils surrounding the fibrillar center (arrow). The latter is unlabeled.

In the mouse oocyte, penetration of chromosomal fibers into the fibrillar center can only be interpreted by accepting that the latter is the region where the rDNA fibers unwind thus allowing rRNA transcription to occur. This interpretation is supported by the observation, in certain oocytes, of two clusters of silver grains lying on either side of the secondary constriction following in situ hybridization. Such a localization exactly corresponds to that of the two fibrillar centers, each of which belongs to a nucleolus undergoing synthesis.

Those fibers which penetrate into the fibrillar center probably form loops. This configuration was previously postulated for human diplotene chromosomes[26]. Kierszenbaum and Tres[27] using a whole mount technique, demonstrated the lampbrush character of meiotic chromosomes in the mouse, with the fibers radiating at both sides of the axis of each homologue forming lateral loops.

The presence of DNA in the nucleolar fibrillar center has been demonstrated in different cell types : Ehrlich's tumor cells[25], mouse fibroblasts[28,29], plant cells[30,31], and quail oocytes[32].

The presence of rDNA transcription in the fibrillar center is presently controversial[33]. Our studies using incorporation of ^{3}H-uridine demonstrate that transcription of the ribosomal cistrons essentially takes place at the periphery of the fibrillar center. Indeed, following brief incorporation labeling is observed over the periphery of the fibrillar center, mainly on the dense fibrils. A hypothetical shift of the preribosomal RNA after its transcription in the central region of the fibrillar center is difficult to accept under our experimental conditions where ^{3}H-uridine incorporation was not followed by a chase. Moreover, since the studies of Miller and Beatty[34] it is known that nascent RNA is not immediately detached from the rDNA thus allowing observation of transcription units. It should thus be accepted that the localization of silver grains corresponds to the site of rDNA transcription. This conclusion leads us to believe that the rDNA loops emanating from the secondary constriction extend through the bulk of the fibrillar center up to the periphery of the latter. It is in this peripheral region that rDNA is transcribed into rRNA. This would explain the constant presence of dense fibrillar strands in direct contact with the fibrillar center Such strands would be composed of both DNP and newly synthesized RNP. Our observations thus lead to the conclusion that the fibrillar center and its surrounding dense fibrils constitute a functional unit.

ACKNOWLEDGEMENTS

This investigation was supported by CNRS (ERA N° 397) and by DGRST.

REFERENCES

1. Heitz,E. (1931) Planta (Berl.) 12, 774-844.
2. Heitz,E. and Bauer,H. (1933) Z. Zellf. mikr. Anat. 17, 67-82.

3. Hsu, T.C., Brinkley, B.R. and Arrighi, F.E. (1967) Chromosoma (Berl.) 23, 137-153.

4. Pardue, M.L. (1974) Cold Spring Harbor Symposia on Quantitative Biology, 38, 475-482.

5. Ford, C.E. (1966) In : Tissue grafting and radiation H.S. Mickle and J.F. Loutit eds, Academic Press, New York, pp. 197-206.

6. Eicher, E.M. (1972) Genetics, 69, 267-271.

7. Dev, V.G., Grewal, M.S., Miller, D.A., Kouri, R.E., Hutton, J.J. and Miller, O.J. (1971) Cytogenetics, 10, 436-451.

8. Henderson, A.S., Eicher, E.M., Yu, M.T. and Atwood, K.C. (1974) Chromosoma (Berl.) 49, 155-160.

9. Elsevier, S.M. and Ruddle, F.H. (1975) Chromosoma (Berl.) 52, 219-228.

10. Henderson, A.S., Eicher, E.M., Yu, M.T. and Atwood, K.C. (1976) Cytogenetics, 17, 307-316.

11. Henderson, A.S., Warburton, D. and Atwood, K.C. (1972) Proc. Nat. Acad. Sci. USA, 69, 3394-3398.

12. Mirre, C. and Stahl, A. (1976) J. Ultrastruct. Res. 56, 186-201.

13. Knibiehler, B., Navarro, A., Mirre, C. and Stahl, A. (1977) Exp. Cell Res. (in press).

14. Vagner-Capodano, A.M., Pinna-Delgrossi, M.H. and Stahl, A. (1977) Biol. Cellul. (in press).

15. Mirre, C. and Stahl, A. (1977) Biol. Cellul. 29, 6a.

16. Mirre, C. and Stahl, A. (1977). J. Cell Sci. (Submitted for publication).

17. Luciani, J.M., Devictor-Vuillet, M., Gagné, R. and Stahl, A. (1974) J. Reprod. Fert. 36, 409-411.

18. Ascione, R. and Arlinghaus, R.B. (1970) Biochim. Biophys. Acta 204, 478-488.

19. Gall, J.G. and Pardue, M.L. (1969) Proc. Nat. Acad. Sci. USA 63, 378-383.

20. Granboulan, P. (1965). In : The use of radioautography in investigations on protein synthesis, Leblond, C.P. ed., Academic Press, New York, 4, 43-63.

21. Salpeter, M.M. and Bachmann, L. (1972). In : Principles and techniques of electron microscopy - Biological applications, Hayat, M.A. ed., Van Nostrand Reinhold Co., New York, 2, 221-278.

22. Chouinard, L.A. (1971) J. Cell Sci. 9, 637-663.

23. Jordan, E.G. and Luck, B.T. (1976) J.Cell Sci. 22, 75-86.

24. Goessens, G. and Lepoint, A. (1974) Exp. Cell Res. 87, 63-72.

25. Goessens, G. (1976) Exp. Cell Res. 100, 88-94.

26. Baker, T.G. and Franchi, L.L. (1967) Chromosoma (Berl.) 22, 358-377.

27. Kierszenbaum, A.L. and Tres, L.L. (1974) J. Cell Biol. 63, 923-935.

28. Pouchelet, M., Gansmüller, A., Anteunis, A. and Robineaux, R. (1975) C.R. Acad. Sci. Paris 280, 2461-2463.

29. Anteunis, A., Pouchelet, M., Gansmüller, A. and Robineaux, R. (1975) C.R. Acad. Sci. Paris, 281, 901-903.

30. Lafontaine, J.G. and Lord, A. (1973) J. Cell Sci. 12, 369-383.

31. Lord, A., Nicole, L. and Lafontaine, J.G. (1977) J. Cell Sci. 23, 25-42.
32. Mirre, C. and Stahl, A. (1977) J. Ultrastruct. Res. (Submitted for publication).
33. Recher, L., Sykes, J.A. and Chan, H. (1976) J. Ultrastruct. Res. 56, 152-163.
34. Miller, O.L. and Beatty (1969) Science 164, 955-957.

Chromosomes Today Volume 6, A. de la Chapelle and M. Sorsa eds.

THE SEQUENCE OF TRANSCRIPTIONAL ACTIVITY OF THE OOCYTE CHROMOSOMES IN A MOTH

WALTHER TRAUT
Abteilung Biologie, Ruhr-Universität, D-4630 Bochum, Fed.Rep.Germany

ABSTRACT

During oogenesis in the adult flour moth, Ephestia kuehniella, transcription of the chromosomes is turned off to zero activity in definite steps: (1) activity along all chromosomes without autoradiographically detectable gaps, the chromosomes are of the lampbrush type, (2) gaps of activity during a short period at the beginning of vitellogenesis, (3) activity restricted to two sites throughout the main period of vitellogenesis, (4) no active site detectable after nurse cell degeneration. - The number and location of the nucleolus organizers in early pachytene correspond with the two active chromosomal sites during vitellogenesis.

INTRODUCTION

The chromosomes of mature and unfertilized animal eggs are transcriptionally inactive. As shown in biochemical studies, transcription of heterogeneous RNA, rRNA and tRNA in embryogenesis is turned on not all at the same time but at different stages of early development[3]. The opposite phenomenon, the turning off of transcriptional activity during oogenesis as a regular sequence of steps, is shown here by use of cytological methods. The ovary of the adult flour moth, Ephestia kuehniella, presents a useful system for such a study since it contains all stages from young previtellogenetic oocytes to mature eggs in the same ovariole, and individual oocyte nuclei can be isolated.

MATERIALS AND METHODS

Ovaries of adult moths of a wildtype strain of Ephestia kuehniella kept in the laboratory were used in this study. Oocyte nuclei were isolated manually with fine needles in a drop of Ringer solution and then quickly dried onto the slide by pulling them outside the drop. This procedure was followed by a passage through 70% and 96% ethanol and subsequent air drying.

For autoradiography the ovaries were kept for 15 min at room temperature in Ringer solution (0.9% NaCl, 0.042% KCl, 0.025% $CaCl_2$,

0.02% $NaHCO_3$, 0.25% glucose) containing 200 μCi/ml of 5-^{3}H-uridine (Amersham - Buchler, specific activity 29 Ci/mmol). They were then immediately transferred into cold Ringer solution (≈4°C) without uridine. Within 15 min after this transfer the whole mount preparation of the nuclei were made on subbed slides. Posttreatment included 5 min extraction in cold 5% TCA, an ethanol passage and air drying, staining in lactic acetic orcein, removal of the dye in 70% ethanol, a passage through 96% ethanol and again air drying. Kodak AR10 stripping film was used with an exposure time of 14 days.

For autoradiography of semithin sections, ovaries were fixed in glutaraldehyde and OsO_4 and embedded in araldite. The 0.5 μm sections were stained with carbol fuchsin prior to autoradiography.

RESULTS AND DISCUSSION

Although the ovary of Ephestia is a meroistic insect ovary in which nurse cells are expected to take over RNA synthesis for the oocytes[1], all three types of cells present, oocytes, nurse cells, and follicle cells, are active in RNA synthesis (Fig. 1). Nurse cell and follicle cell nuclei incorporate ^{3}H-uridine in their whole area. The oocyte nuclei are labelled in regions only where by chance chromosomes are included in the sections.

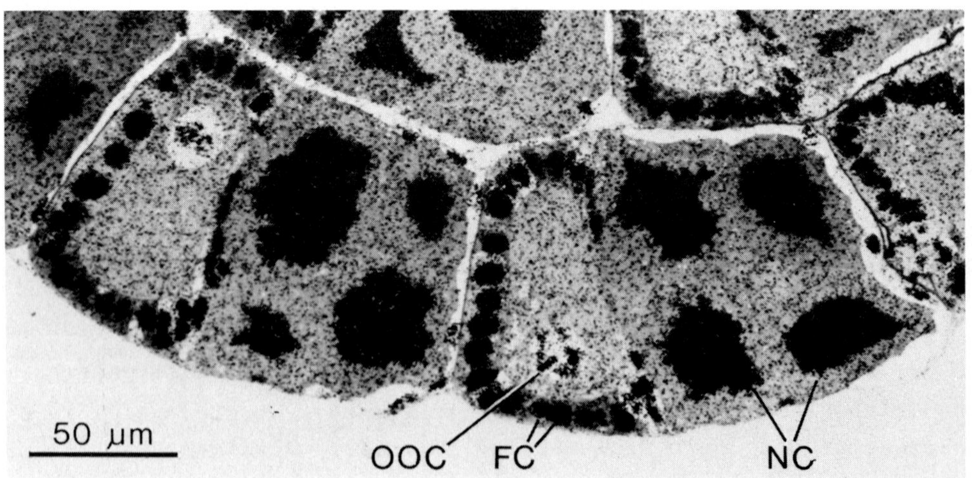

Fig. 1. RNA synthesis in the ovary of Ephestia kuehniella. Microautoradiograph after 15 min of incubation in ^{3}H-uridine; 0.5 μm section prestained with carbol fuchsin. OOC : oocyte nucleus, NC : nurse cell nucleus, FC : follicle cell nucleus.

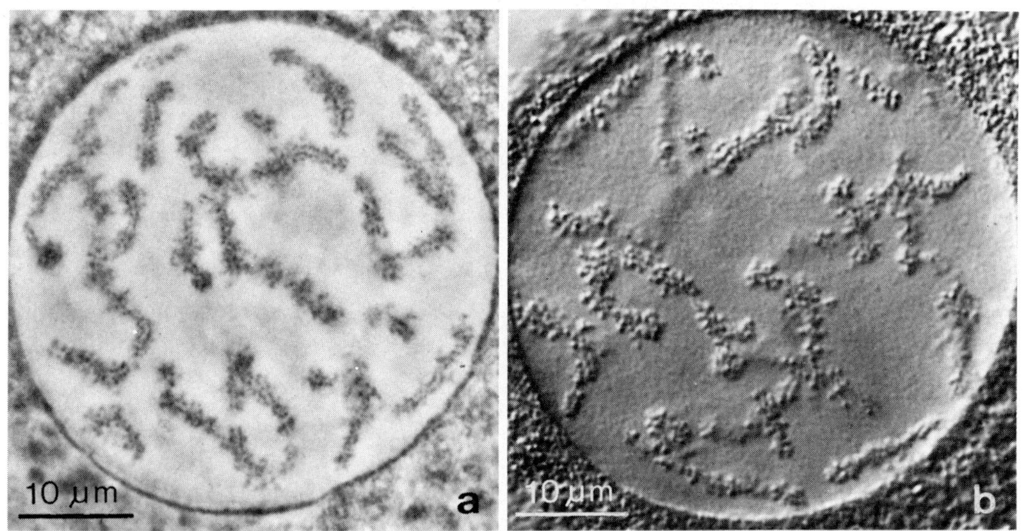

Fig. 2. Nuclei of young previtellogenetic oocytes, whole mount preparations, lactic acetic orcein, a) phase contrast, b) Nomarski interference contrast.

Transcriptional activity of the oocyte nucleus, however, is more easily followed using preparations of whole nuclei. In young previtellogenetic stages of oogenesis the chromosomes are of the lampbrush type (Fig. 2a and b). The loops emanating from the chromosomal axis are short, about 1-2 μm in length. Particular loops are longer, These lampbrush chromosomes are bivalents in the pachytene stage. Female meiosis in Ephestia is achiasmatic[7]. The homologues remain fully paired throughout oogenesis, at the end of which they shorten to be transformed into metaphase I bivalents without an intervening diplotene stage.

In Giemsa-stained preparations the loops are not as obvious but can still be seen in young oocytes as fringes on the bivalents (Fig. 3). In older vitellogenetic oocytes or later stages the fringes have disappeared. Instead, a rather regular chromomere pattern becomes visible (Fig. 4). Irregular chains of bivalents are apparent in this stage, connected by an unreliably stained material. Comparable terminal associations of chromosomes were found by Moses et al.[5] in spreads of synaptonemal complexes and by Dutrillaux[4] in methaphase chromosomes. They were shown to be random associations between different chromosomes. Nucleoli are not visible in light microscopic preparations even in the fluorescence microscope after acridine orange staining.

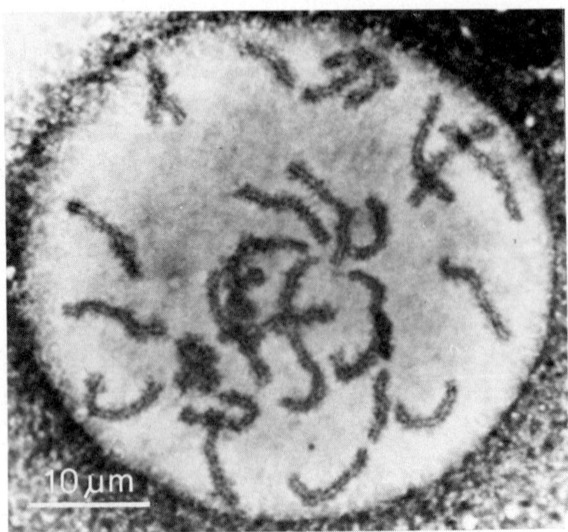

Fig. 3. Oocyte nucleus of a previtellogenetic stage of oogenesis, whole mount preparation, Giemsa.

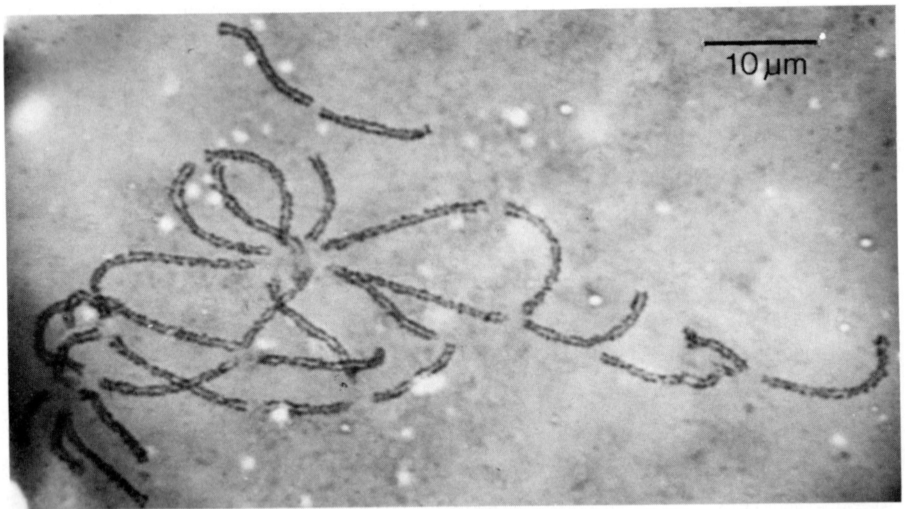

Fig. 4. Oocyte nucleus of a late stage of oogenesis, at the time of nurse cell degeneration, whole mount preparation, Giemsa.

Fig. 5. A living ovariole of Ephestia kuehniella and labelling patterns of oocyte nuclei characteristic for the indicated regions of the ovariole; b - e: microautoradiographs after short term ^{3}H-uridine incubation, whole mount preparations all at the same magnification. S4 - S8: designation of the stages of oogenesis according to Cruickshank[2].

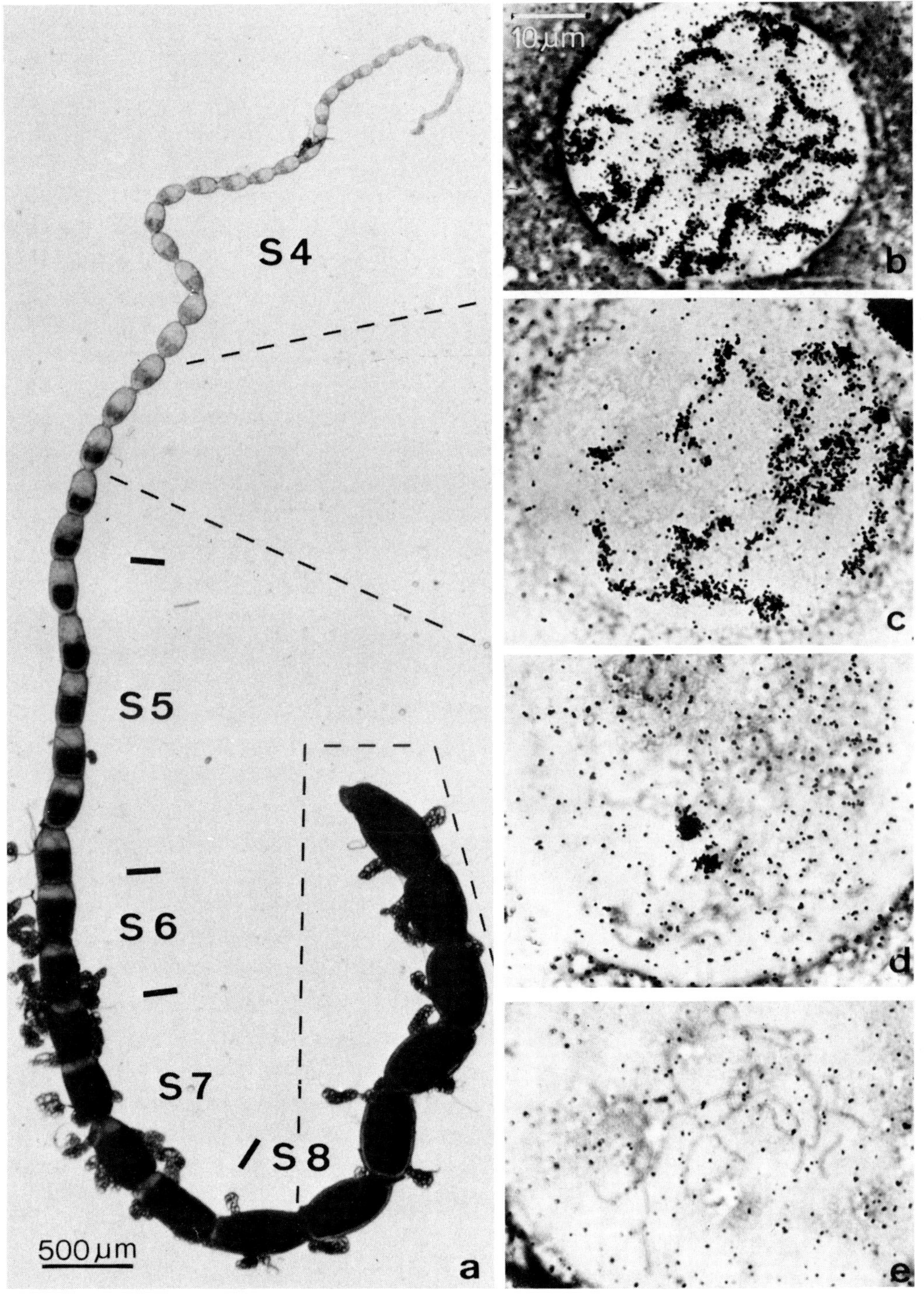
10 μm
b
S4
c
S5
S6
d
S7
S8
500 μm
a
e

The dried nuclei are flat enough to be used in microautoradiography after short term ^{3}H-uridine incubation. Four different patterns of labelling can be distinguished in autoradiographs[6]: (1) labelling of all chromosomes in full length (Fig. 5b), (2) labelling of all chromosomes but with gaps (Fig. 5c), (3) labelling restricted to precisely two sites (Fig. 5d), and (4) no labelling at all (Fig. 5e).

These patterns represent consecutive stages of transcriptional activity of the chromosomes. Every stage can be classed with a different stage of oogenesis (Fig. 5a-e). From the youngest oocytes which can be handled in the way described until the visible beginning of vitellogenesis the chromosomes are fully active. During a short period, at the beginning of vitellogenesis gaps of activity appear in the chromosomes. This may be merely a stage of transition to the next stage in which two sites are active and remain so throughout vitellogenesis till nurse cell degeneration. After nurse cell degeneration no transcriptionally active site is further detectable in the oocyte nucleus. Transcriptional activity in the oocyte nucleus is obviously turned off in steps.

Transcriptional activity in all stages and at all sites is suppressed by incubation in 10 µg/ml actinomycin D 20 min prior to

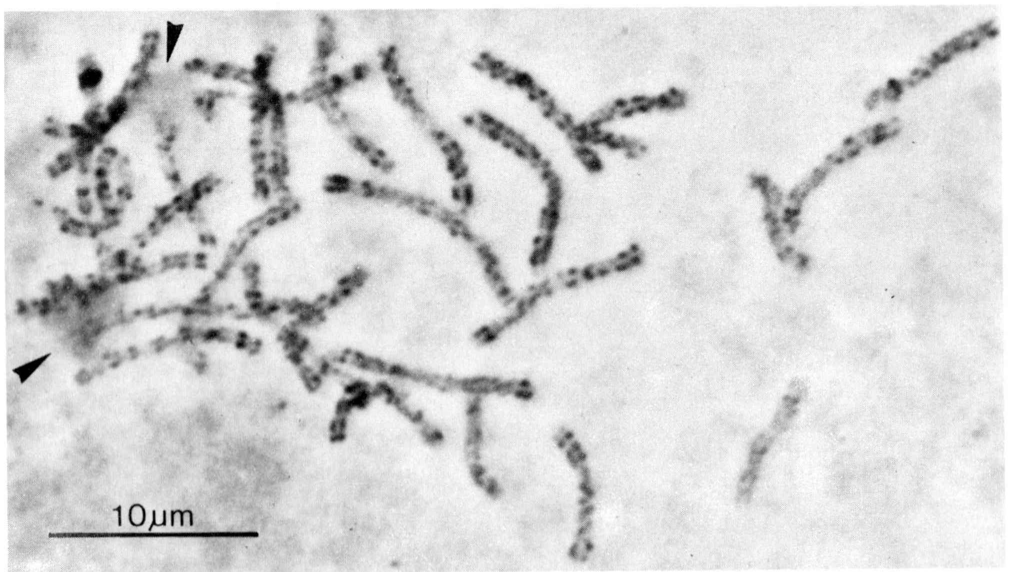

Fig. 6. Early pachytene stage of a larval ovary of Ephestia kuehniella, lactic acetic orcein, phase contrast.

the addition of ^{3}H-uridine. An attempt to characterize the synthesis of RNA at different sites and stages as either α-amanitin-sensitive or α-amanitin-resistant failed. Even high doses of α-amanitin, up to 60 µg/ml, using the same pre-incubation time as with actinomycin D, do not influence transcriptional activity in any of the three types of cells present in the ovary. Probably α-amanitin does not enter the cells.

In another approach it was intended to identify the chromosomes carrying the two sites active during vitellogenesis. Chromosomes cannot be distinguished from one another in mitotic metaphase preparations. As in other Lepidoptera species they are too small. In the adult ovary the chromosomes are long enough but they all look alike. It is impossible to identify one of the 30 bivalents merely by measuring its length. Fortunately in the larval ovary the bivalents while already in the pachytene stage, exhibit sufficient details of the chromomere pattern for mapping. Some of the bivalents can be recognized easily in different preparations. Among them are two nucleolus bearing chromosomes (Fig. 6, arrows), the nucleolus organizers being in a terminal position.

The corresponding number and location of the nucleolus organizing regions and the transcriptionally active sites during vitellogenesis lead to the suspicion that in fact these two may be identical, though no typical nucleolus is present in the oocytes of adult moths. The rRNA genes therefore may be the last ones to be turned off in the sequence of steps that eventually leads to the inactivity of the chromosomes in the mature egg.

ACKNOWLEDGEMENT

This work was supported by a grant of the Deutsche Forschungsgemeinschaft.

REFERENCES

1. Bier, K. (1965) Zool. Jb. Physiol., 71, 371-384.
2. Cruickshank, W.J. (1971) J. Insect Physiol. 17, 217-232.
3. Davidson, E.H. (1976) Gene activity in early development (2nd edit.) Academic Press, New York, pp. 1-452.
4. Dutrillaux, B. (1977) this volume.
5. Moses, M.J., Statton, G.H., Gambling, T.M. and Starmer, C.F. (1977) Chromosoma 60, 345-375.
6. Traut, W. (1975) Cytobiologie 11, 172-180.
7. Traut, W. (1977) Genetica 47, 135-142.

Chromosomes Today Volume 6, A. de la Chapelle and M. Sorsa eds.

GENETIC IMPLICATIONS OF THE FUSION OF HUMAN CELLS WITH PLANT PROTOPLASTS

A. LIMA-DE-FARIA
Institute of Molecular Cytogenetics, University of Lund, Lund (Sweden)

ABSTRACT

The fusion of human cells with plant protoplasts is not a dramatic phenomenon. There are many similarities between humans and higher plants at the biochemical level. Hemoglobin is present in flowering plants and steroids with hormone action occur in *Haplopappus* and other species. Neurotransmitters which in animals carry the information between nerve cells also occur in *Citrus* and other plant genera.

It is suggested that the genes responsible for these biochemical processes have arisen in plants and animals independently of any function that they may have. Their formation has been dictated by the molecular organization of the DNA and of the chromosome.

INTRODUCTION

In a first series of experiments HeLa cells were fused with carrot protoplasts by means of polyethylene glycol and by culturing the fused components in a plant medium at 20°C. Heterokaryons were formed containing human and carrot nuclei which were surrounded by a cell wall. Nuclear contact also occurred and the percentage of heterokaryon formation ranged from 0.3 to 0.6%[1].

A similar experiment was carried out, independently of ours, involving the fusion of human cells with tobacco hybrid protoplasts by means of polyethylene glycol. It produced essentially the same results[2].

A second series of experiments carried out at our laboratory was planned with a view to favouring the human cells. The fusion agent was inactivated Sendai virus and the plant species was *Haplopappus gracilis* which is known to tolerate high temperatures[3]. The culture medium used after fusion was a modified MEM medium employed in culturing human cells and the temperature was 37°C. The fusion of the human cells with the plant protoplasts has led to the formation of heterokaryons containing one human nucleus and one *Haplopappus* nucleus. These appeared with a frequency of circa 12%. In the heterokaryons a cell wall had formed which surrounded the two types of cytoplasms that had mixed[4].

These results may seem surprising but this is not necessarily so. The difference between humans and plants lies largely in the formation of specialized organs. We shall see that structurally and biochemically there are many similarities between the two types of organisms.

There are three fields in which these similarities are most evident: 1) the cell ultrastructure, 2) the occurrence of specific genes and 3) certain biochemical pathways.

1) SIMILARITIES AND DIFFERENCES IN CELL ULTRASTRUCTURE

The cells of plants and animals may seem very different under the light microscope but the differences are not so great when their components are studied with the electron microscope and are analysed biochemically.

Chloroplasts and cell walls are present in plants but not in human cells. Apart from this the similarities between the two types of cells are pronounced.

The molecular organization of the cell membrane in mammals and plants is essentially the same[5]. The endoplasmic reticulum of the cytoplasm of plants and humans does not display basic differences. This is the component of the cytoplasm where many biochemical reactions occur and which conveys metabolites and macromolecules in and out of the cytoplasm. The mitochondria are also essentially alike in plants and humans. They perform similar functions in the cell metabolism and have a similar ultrastructure[6]. The dictyosomes are the characteristic morphological elements of the Golgi apparatus. The dictyosomes of rat and of maize have the same ultrastructural configuration as do those of many other plant and animal species[7]. The nuclear envelope in plants and higher animals is also very similar if not identical and the chromosomes of plants are almost indistinguishable from those of humans. The chromosomes have the same basic organization in the two types of organisms: 1) they have two arms, 2) they have a centromere, 3) they possess telomeres, 4) they go through the same type of mitotic and meiotic cycles and 5) display the same basic phenomena such as crossing-over, chiasma formation and nucleolus organization. In connection with the chromosomes there is an organelle in higher animals not present in flowering plants: the centrioles. Although higher plants do not have centrioles their chromosomes move to the spindle poles as effectively as chromosomes in animal cells.

Hence, apart from centrioles, chloroplasts and cell walls the other components of plant and human cells are essentially alike.

These comprise: the cell membrane, the endoplasmic reticulum, the mitochondria, the ribosomes, the Golgi apparatus, the nuclear envelope and the chromosomes. Human chromosomes in a plant nucleus or in a plant cytoplasm do not find themselves in a very foreign environment.

2) SIMILAR GENES IN PLANTS AND HUMANS

The development of molecular biology and in particular the rapid expansion in the field of molecular cytogenetics has made it possible to isolate biochemically some specific genes in eukaryotes.

Two types of genes have proved to be the easiest to isolate and analyse: the cistrons for 18S and 28S ribosomal RNA and the histone genes.

The ribosomal genes are present in humans, other animals, plants and bacteria (review in[8]). The genes are so similar in the different groups of species that hybridization of RNA with DNA can be carried out by using, for instance, hamster 18S and 28S RNA and DNA of *Haplopappus*. This hybridization is most efficient mainly because of the large number of ribosomal genes found in flowering plants[9]. Sinclair and Brown[10] have also obtained a high degree of hybridization between frog ribosomal RNA and barley and wheat DNA.

The histone genes have recently been analysed biochemically in detail[11,12]. These genes are also present in plants and humans. The histone fractions H3 and H4 have characteristic electrophoretic mobilities even when obtained from organisms so diverse as peas and cows. The amino acid sequences of H3 and H4 in these two species are very similar. H4 shows only two substitutions out of 102 amino acids and H3 four substitutions out of 135 amino acids[13].

Cytochrome C is another protein which is present in prokaryotes and eukaryotes and has changed little throughout millions of years of evolution. Cytochrome C is a protein of vital importance which functions in mitochondrial electron transfer. The amino acid sequence analysis discloses that about 60% of the total number of amino acids at corresponding positions are identical in wheat and human chains[14].

Plants also produce hemoglobin, a well-defined protein. This protein which has been studied in detail occurs mainly in leguminous plants and is called leghemoglobin[15]. The amino acid sequence of the leghemoglobin of soybean has been established and when compared with the gamma chain of human hemoglobin shows many points of similarity. A comparison between the regions surrounding the haem-binding histidines of leghemoglobin and the human gamma chain (globin) has been made based on the genetic code. This showed that the differences in most cases involve only a single base pair giving a mutation frequency of 1·06. This means that mainly point mutations must have taken place in this part of the globin chain during its evolution[16].

The fact that plants have hemoglobin genes but do not have red blood cells is of significance for understanding the organization of the eukaryotic chromosome. In our opinion the eukaryotic chromosome has such a rigid organization, established by the chromosome field, that it cannot evolve except in the direction dictated by its own molecular architecture. A gene, such as the hemoglobin gene, is formed independently of any function that it may have in the plant. The chromosome simply cannot produce anything else. The gene remains there with no obvious function. Later in the course of evolution, as other genes are produced by a process dictated by the molecular architecture, their combination may lead to a meaningful function. This may have happened, for instance, in the vertebrates when the genes for hemoglobin together with other genes led to the formation of red blood cells.

This rigid organization of the eukaryotic chromosome is supported by the finding that the ribosomal genes occupy the same position relative to kinetochores and telomeres in the eukaryotic chromosome, irrespective of whether a given chromosome belongs to the complement of an alga or of a human[17].

DNA sequence organization has been extensively carried out in animals. The genome of *Nicotiana tabacum* (one of the species used in human/plant cell hybridization) was investigated by DNA/DNA reassociation for its spectrum of DNA repetition components and pattern of DNA sequence organization. Zimmerman and Goldberg[18] who made this study came to the conclusion that "the tobacco genome bears remarkable similarity to that found in the genomes of most animal species investigated to date". The animal species mentioned include man.

Not only specific genes but the overall organization of the DNA are similar in certain higher plants and in humans.

3) SIMILAR BIOCHEMICAL PROCESSES IN PLANTS AND HUMANS

Plants, although they do not have such highly evolved organs as animals, possess very similar chemicals and some almost identical biochemical pathways.

To start with, although plants do not have a nervous system they harbour in their cells the neurotransmitters dopamine, noradrenaline (norepinephrine) and serotonin. The nerve cells of humans and other animals communicate with one another by secreting neurotransmitters which traverse the tiny gap between two nerve cells and produce a change in the electrical activity of the receiving cell. These are messenger molecules which are released from one cell, travel a certain distance, and interact with the surface of a second cell. Among these messenger molecules are dopamine, noradrenaline and serotonin which have a basic role in the communication between the nerve cells of animals, humans included[19]. Serotonin is also a potent animal vasoconstrictor found particularly in the brain, intestinal tissue, blood platelets and mast cells. Moreover, it stimulates uterine contraction[20].

In plants dopamine and noradrenaline are found as derivatives of phenylpropan - amino acids. Noradrenaline is abundant in *Citrus* plants such as *Citrus aurantium* and is also present in *Portulaca oleracea*.

Serotonin is a derivative of the amino acid tryptophan. Oxidation of tryptophan leads to the formation of 5-hydroxytryptophan which results in the formation of serotonin. An intermediate compound in this chemical process, 5-hydroxy-tryptamine, is present in large quantities in bananas[21]. Hence, although plants do not have a nervous system they have three of the main neurotransmitters.

Some steroids are known to have androgenic and estrogenic activity in the animal kingdom, i.e. they function as sex hormones which are secreted chiefly by the testis and ovary. Both androgens and estrogens have been isolated from testis and ovarian tissue. Among the sex hormones which influence the female sex cycles are estrone and estriol, two estrogenically active compounds which are found in particularly high concentrations in the urine of pregnant women. A third estrogen, β-estradiol, also present in the urine of pregnant women is the normally secreted ovarian hormone[20].

Both estrone and estriol are found in plants. Three steroids with sexual hormone action have been described in flowering plants: the first is androstanetriol which is present in *Haplopappus heterophyllus* (Compositae). This species happens to belong to the same genus that we used in our human/plant cell hybridizations. The second is estrone which has been extracted from palm fruits and the third is estriol present in willows (*Salix*)[21].

The function of these three steroids with sexual hormone action is not yet understood in plants, but we should not forget that flowering plants also have ovaries and that although these ovaries are not identical with those of mammals they have, however, the same basic function.

Ecdysone is an insect molting hormone which was first isolated from the silkworm, *Bombyx mori*. Six additional molting hormones have been isolated from a variety of insects and crustaceans.

As Hikino and Takemoto[22] point out "until five years ago who would imagine that a plant might contain an insect-molting-hormone? Had there been someone with such a fantastic idea, he would have been able to discover immediately that a number of botanical extracts showed insect-molting-hormone activity". In the last few years β-ecdysone has been isolated from ferns and from flowering plants. The molting hormone is present in plant species which have no close taxonomic relationship and is widely distributed since it has been found in some 80 families.

The function of ecdysone in plants is unknown. Again we have a situation where a plant produces a substance which may not necessarily have any function. The function only develops later in the course of evolution when other genes become available which in combination produce a coordinated effect. A gene or a series of genes arise along the DNA as a consequence of its chemical properties and independently of any possible function in the plant or animal.

A corollary of such a postulate is that humans may possess a series of genes with no obvious function but that in the future, as man evolves, these non-functional genes may become integrated into larger gene constellations with a resulting meaningful function.

The fusion of human cells with plant protoplasts is therefore not as dramatic as appears at first sight. The differences between the two groups of organisms may not be as profound as we have imagined.

Eukaryotic plant-like organisms may have existed approximately one billion years ago in the Protozoic era[23]. Plants and animals must have diverged prior to this time. This seems to imply parallel lines of evolution characterized by the maintainance and acquisition of similar genes and similar biochemical pathways.

REFERENCES

1. Dudits, D., Rasko, I., Hadlaczky, G. and Lima-de-Faria, A. (1976) Hereditas, 82, 121-124.
2. Jones, C.W., Mastrangelo, I.A., Smith, H.H., Liu, H.Z. and Meck, R.A. (1976) Science 193, 401-403.
3. Eriksson, T. (1965) Physiologia Plantarum 18, 976-993.

4. Lima-de-Faria, A., Eriksson, T. and Kjellén, L. (1977) Hereditas 87 (in the press).

5. Robertson, J.D. (1969) in Handbook of Molecular Cytology, Lima-de-Faria, A. ed., North-Holland, Amsterdam, pp. 1403-1443.

6. Munn, E.A. (1969) in Handbook of Molecular Cytology, Lima-de-Faria, A. ed., North-Holland, Amsterdam, pp. 875-913.

7. Favard, P. (1969) in Handbook of Molecular Cytology, Lima-de-Faria, A. ed., North-Holland, Amsterdam, pp. 1130-1155.

8. Birnstiel, M.L., Chipchase, M. and Speirs, J. (1971) in Progress in Nucleic Acid Res. and Mol. Biology vol. 11, Davidson, J.N. and Cohn, W.E. eds., Academic Press, New York, pp. 351-389.

9. Ståhle, U., Lima-de-Faria, A., Ghatnekar, R., Jaworska, H. and Manley, M. (1975) Hereditas, 79, 21-28.

10. Sinclair, J.H. and Brown, D.D. (1971) Biochemistry, 10, 2761-2769.

11. Gross, K., Probst, E., Schaffner, W. and Birnstiel, M. (1976a) Cell, 8, 455-469.

12. Gross, K., Schaffner, W., Telford, J. and Birnstiel, M. (1976b) Cell, 8, 479-484.

13. Elgin, S.C.R. and Weintraub, H. (1975) in Annual Review of Biochemistry, vol. 44, Snell, E.E. ed., Annual Reviews Inc., Palo Alto, pp. 725-774.

14. Dayhoff, M.O. (1971) in Chemical Evolution and the Origin of Life, Buvet, R. and Ponnamperuma, C. eds., North-Holland, Amsterdam, pp. 392-419.

15. Ellfolk, N. and Sievers, G. (1971) Acta Chem. Scand., 25, 3532-3548.

16. Ellfolk, N. (1972) Endeavour, 31, 139-142.

17. Lima-de-Faria, A. (1976) Hereditas, 83, 1-22.

18. Zimmerman, J.L. and Goldberg, R.B. (1977) Chromosoma, 59, 227-252.

19. Nathanson, J.A. and Greengard, P. (1977) Scientific American, 237, 108-119.

20. White, A., Handler, P. and Smith, E.L. (1964) Principles of Biochemistry, McGraw-Hill, New York, pp. 1-1106.

21. Metzner, H. (1973) Biochemie der Pflanzen, Ferdinand Enke Verlag, Stuttgart, pp. 1-376.

22. Hikino, H. and Takemoto, T. (1974) in Invertebrate Endocrinology and Hormonal Heterophylly, Burdette, W.J. ed., Springer-Verlag, Berlin, pp. 185-203.

23. Schopf, J.W. (1968) J. Paleontol., 42, 651-688.

ACKNOWLEDGEMENTS

This work was supported by research grants from the Swedish Natural Science Research Council.

EFFECTS OF MUTAGENS ON CHROMOSOMES

Chromosomes Today Volume 6, A. de la Chapelle and M. Sorsa eds.

INTRODUCTION

A.T. NATARAJAN

Department of Radiation Genetics & Chemical Mutagenesis, State University of Leiden, P.O. Box 722, Leiden, and J.A. Cohen Institute of Radiation Pathology and Radiation Protection (The Netherlands)

A great majority of mutagens produce effects on eukaryotic nucleus which can be detected cytologically in many ways such as, inhibition of cell division, arrest of metaphase, numerical and structural chromosome aberrations, sister chromatid exchanges (SCE's), premature chromosome condensation, etc. In addition, incorporation of halogeneated pyrimidine bases in the DNA enables differential staining of the chromatids containing such bases. The study of the effects of mutagens on chromosomes is important both from theoretical and practical points of view. For example, (a) from the type and frequencies of aberrations produced, one can gain insight into the mode and time of action of the agent used as well as the organization of interphase nucleus and metaphase chromosomes, (b) since most of the mutagenic carcinogens produce chromosome aberrations, cytological tests can be employed to pre-screen mutagens. Much progress has been made in this field in the last few years and this has recently been reviewed[1,2]. The papers which follow this introduction will deal with the current research in this area. I would like to consider here briefly some general problems in this field.

PRIMARY LESIONS AND DNA REPAIR

Evidence has accumulated that the DNA in a G_1 chromosome behave as one long double strand maintaining the polarity through the centromere in biarmed chromosomes[3]. It is clear that DNA is the primary target molecule for the origin of chromosome aberrations as well as SCE's. Most of the lesions produced by the mutagens in the DNA are repaired and some are 'mis repaired' leading to detectable chromosome aberrations and/or SCE's. Among the induced lesions such as, single strand breaks, double strand breaks, base damage, alkylated bases, thymine dimers, cross links (inter and intra- strand as well as DNA-protein) have all been implicated in the production of chromosome aberrations and SCE's. Thus, there seems to be no qualitative differences between the lesions which lead to chromosome aberrations or SCE's. However, some type of lesions (e.g., strand breaks) may preferentially lead to chromosome aberrations, while other (e.g., base damage) may lead to SCE's. The pathways that lead to the production of SCE's seem to be different from those leading to chromosome aberrations. For instance, (a) SCE formation is exclusively associated with the 'S' phase, whereas the chromosome aberrations arise at all stages of the cell cycle, (b) Except for Bloom's syndrome, all other syndromes characterised by elevated frequencies of spontaneous chromosome aberrations (Fanconi's anaemia, ataxia telangiectasia) are not associated with increased frequencies of SCE's, and (c) Caffeine, which is suspected to inhibit post-replication repair in mammalian cells, does influence the yield of induced chromosome aberrations and not the SCE's. All repair processes known in mammalian cells, such as excision repair, photo reactivation, post replication repair have been implicated in the production of chromosome aberrations. Photoreactivation of UV induced dimers reduce the frequencies of induced chromosome aberrations and SCE's.[4,5] UV irradiation of cells derived from xeroderma pigmentosum patients (deficient in excision repair) induce higher frequencies of chromosome aberrations and SCE's than in the normal human cells.[6,7] Fanconis anaemia cells are more susceptable for the production of chromosome aberrations and not SCE's induced by cross linking agents.[8,9] So far it has not been possible to pinpoint any type of DNA lesion or a specific repair process associated exclusively to chromosomal aberrations or SCE's.

LOCALISATION OF CHROMOSOME ABERRATIONS AND SCE's.

Most of the radiation induced chromosome aberrations are randomly distributed along the length of the chromosomes, while the chemically induced ones tend to localise in the constitutive heterochromatic regions. Though base composition in these regions was considered to be an important factor for this localisation, this idea was abandoned and the possible role of repetitive DNA in these regions as a causative factor was postulated.[10] Following the advent of chromosome banding techniques, many authors have reported that the induced aberrations tend to be concentrated in the 'R band'regions or interbands![11] Using sequential banding of the same chromosomes, (Q and R banding) it has been shown that the localisation of breaks is very subjective and the localisation points for one aberration may vary depending on the technique used to band the chromosomes.[12]

Frequencies of spontaneously occuring SCE's have been found to be higher in constitutive heterochromatic regions of Microtus agrestis,[13] facultative heterochromatin of human[14] as well as at the junction of eu-heterochromatic regions in Indian muntjac.[15] Further studies are necessary before a generalisation can be made on this point. However, it appears that the spontaneous frequency of SCE's depends on the DNA content of the genome, and as expected, all mammals seem to have similar frequencies of SCE's. The claim that SCE's do not occur in Drosophila, may be due to very low DNA content of this organism when compared to mammals.

CELL CYCLE STAGES AND THE TYPES OF ABERRATIONS INDUCED

It is accepted that irradiation of G_1 cells and G_2 cells will lead to the production of chromosome and chromatid type of aberrations respectively. Exceptions to this generalization have been found, e.g., (a) irradiation of blood lymphocytes (G_o) of ataxia telangiectasia patients lead to the production of both chromosome and chromatid types of aberrations,[15] (b) Irradiation of mature sperm produces, in addition to chromosome type aberrations, some chromatid aberrations, as studied cytologically in mouse oocytes[17] and genetically in Drosophila.[18] Similarly, alkylating agents induce only chromatid aberrations irrespective of the cell stage treated. Exception to this has also been found in dry plant seeds[19] and early mouse sperm treated with polyfunctional alkylating agents.[20] Further studies of these exceptions, may shed light on the time sequence of chromosomal repair processes and their influence on the type of aberrations produced in these systems.

APPLIED ASPECTS

In recent years the idea that the risk of induced reciprocal translocation in man after radiation exposure can be assessed using the frequencies of dicentrics has gained some currency stimulated by the findings of Brewen and colleagues,[21] that the frequencies of dicentrics in the blood lymphocytes increased linearly with an increase in the effective chromosome arm number of the different species studied in this respect. Thus, in man with 81 arms, the frequencies of dicentrics following a given dose of X-rays will be double the value obtained in mouse with 40 arms. Since dicentrics and reciprocal translocations are produced at equal frequencies following radiation this would then provide a useful method for estimating the frequencies of reciprocal translocations. Though many exceptions were found to this, it was maintained that man is twice as sensitive as mouse. BrdU labelling techniques enables one to distinguish the first division cells from the second division ones and it was found by us that in the fixation time used in their studies, 34% of the mouse cells were in their 2nd division. When mouse cells were fixed at 36 hours and compared to human cells fixed at 48 hours, the frequencies of dicentrics were similar.[22] It is claimed that if one studies, human lymphocytes fixed at 42 hours, a higher value of dicentrics is obtained thus giving a 2:1 ratio for mouse to man.[23] Though there are some differences in yield between individuals and between fixation times, they never reach a two fold increase in man and the mouse.[24] In future studies it will be important to employ BrdU technique to identify and score the first division cells to make valid comparisons.

Though chromosome aberrations in lymphocytes proved to be a good indicator of in vivo exposure to ionizing radiation, this system does not appear to be very sinsitive to monitor human exposure to chemical mutagens.[25] Frequencies of SCE's in PHA stimulated lymphocytes, increase during treatment of patients with potent chemical mutagens but drop to control level a few weeks after stopping the treatment.[26] This is in contrast to findings with ionizing radiations, in which case the aberrations are known to persist for years, which indicates that the chemically induced lesions are repaired effectively during the recovery, unlike X-ray induced lesions which are fixed at the G_1 stage itself in the lymphocytes.

SCREENING OF MUTAGENS AND CARCINOGENS

The fact that most of the mutagenic carcinogens are able to induce chromosome aberrations and SCE's has been exploited to devise in vivo and in vitro cytological techniques to prescreen potential mutagens. The in vitro systems, such as treating mammalian cells in culture are easy and quick;[27] the introduction of rat liver microsomes to this system, has enhanced its value,[28] as one can screen compounds which need metabolic activation to become effective. Complimentary in vivo mammalian techniques (micro nuclei, bone marrow chromosome aberrations, blood lymphocytes etc.,) are essential however, as in vitro techniques do not take into account the naturally operating detoxification mechanisms.

REFERENCES

1. Natarajan, A.T. (1976) Biol. Zbl. 95, 139-156.
2. Kihlman, B.A. (1977) Caffeine and chromosomes, Elsevier, Amsterdam, pp 1-504.
3. Lin, M.S. and R.L. Davidson (1975) Nature, 254, 354-356.
4. Griggs, H.J. and M.A. Bender (197) Science 179, 86-88.
5. Kato, H. (1974) Nature 249, 552-553.
6. Parrington, J.M. et al. (1971) Ann. Human Genet. (Lond.) 35, 149-160.
7. De Weerd-Kastelein, E.A. et al. (1977) Mutation Res. (in press).
8. Sasaki, M.S. and A. Tonomura (1973) Cancer Res. 33, 1829-1836.
9. Latt, S.A. et al. (1975) Proc. Natl. Acad. Sci. U.S.A. 72, 4066-4070.
10. Natarajan, A.T. and G. Ahnstrom (1972) in Constitutive Heterochromatin in Man. R.A. Pfeiffer Ed. F.k. Schauttauer Verlag. Stuttgart. New York, pp 215-223.
11. Buckton, K.E. (1976) Int. J. Radiat. Biol. 29, 475-488.
12. Dubos, C. et al. (1977) Mutation Res. (in press).
13. Natarajan, A.T. and I. Klasterska (1975) Hereditas 79, 150-154.
14. Schnedl , W. et al. (1976) Human Genetics 32, 199-202.
15. Carrano, A.V. and S. Wolff (1975) Chromosoma 53, 361-369.
16. Taylor, A.M.R. et al. (1976) Nature 260, 441-443.
17. Brewen, J.G. and A.T. Natarajan (unpublished results)
18. Maddern, R.H. and B. Leigh (1976) Mutation Res. 41, 255-268.
19. Ramanna, M.S. and A.T. Natarajan (1976) Chromosoma 18, 44-59.
20. Generosa, W.M. et al. (1977) Proc. 2nd Inter. Conf. Environmental Mutagens, Edinburgh. p. 91.
21. Brewen, J.G. et al. (1973) Mutation Res. 17, 245-254.
22. De Boer, P. et al. (1977). Mutation Res. 42, 379-394.
23. Preston, R.J. and J.G. Brewen (1977) in Symposium on Actions of physical and chemical mutagens on the Somatic Chromosomes of Man, Edinburgh. (in press).
24. Natarajan, A.T. and P.P.W. van Buul, (to be published).
25. Schinzel, A. and W. Schmid (1976) Mutation Res. 40, 139-166.
26. Natarajan, A.T. et al. (1977) in Symposium on 'Action of physical and chemical mutagens on the somatic Chromosomes of Man', Edinburgh (in press).
27. Perry, P. and H.J. Evans (1975) Nature 258, 156-158.
28. Natarajan, A.T. et al. (1976) Mutation Res. 37, 83-90.

Chromosomes Today Volume 6, A. de la Chapelle and M. Sorsa eds.

MOLECULAR MECHANISMS IN THE PRODUCTION OF CHROMOSOMAL ABERRATIONS: STUDIES WITH THE 5-BROMODEOXYURIDINE-LABELLING METHOD

B.A. KIHLMAN, H.C. ANDERSSON and A.T. NATARAJAN
Department of Genetics, University of Uppsala, S-750 07 Uppsala 7, Sweden and Department of Radiation Genetics and Chemical Mutagenesis, State University of Leiden, Leiden, The Netherlands

ABSTRACT

Lateral roots of *Vicia faba* were exposed first to 5-bromodeoxyuridine (BrdUrd) and then to thymidine to produce cells in which the chromosomes had one chromatid with unsubstituted DNA (TT-chromatid) and one containing unifilarly BrdUrd-substituted DNA (TB-chromatid). Such cells were exposed in G_2-prophase to long-wave ultraviolet light (which breaks BrdUrd-substituted but not normal DNA) and the resulting chromosomal aberrations were scored in slides prepared as Feulgen squashes or according to the fluorescent-plus-Giemsa (FPG) technique. The great majority of the 78 aberrations analysed in FPG-stained slides were located in chromatid segments of the TB-constitution or associated with sister chromatid exchange.

INTRODUCTION

There is now conclusive evidence that the structural element of the eukaryotic chromosome is the chromatin fibre which consists essentially of DNA and histone. Apparently, each chromatid contains only one, regularly folded chromatin fibre, that is, the chromatid is uninemic. The DNA in the chromatin fibre is likely to consist of one long double helix, which is responsible for the structural integrity of the fibre and, consequently, of the chromatid. From this it follows that DNA is the chromosomal component involved in aberration production. This notion is strongly supported by results obtained in studies on the production of chromosomal aberrations by physical and chemical agents. Observations made during the last few years further indicate that mechanisms for repair of DNA damage are involved in the formation of chromosomal aberrations. There is neither time nor need to go into all this evidence, which has recently been reviewed[1,2]. A few words, however, must be said about the two main types of chromosome-damaging effects produced by physical and chemical agents.

The great majority of the chemical mutagens appear to be capable of inducing chromosome lesions at all stages of the cell cycle. These lesions, however, seem to require DNA and chromosome replication to be expressed as chromosomal aberrations. Thus, the DNA damage produced by these agents at the post-synthesis phase G_2 gives chromosomal aberrations at the second and not at the first mitosis after treatment.

The aberrations are always of the chromatid type, irrespective of the stage of the mitotic cycle at which they are produced. Most important of the agents producing this type of effect are alkylating agents and ultraviolet light (UV) (for references, see 1 and 2).

Evans has coined the terms 'misreplication' and 'misrepair' and suggested that the lesions produced by UV and by alkylating agents develop into chromosomal aberrations as a result of errors during replication, rather than as a result of errors in connection with repair of DNA damage[2]. The molecular mechanisms of misreplication will not be further discussed here.

In contrast to alkylating agents and UV, ionizing radiation produces sub-chromatid aberrations in cells exposed during prophase, chromatid-type aberrations in G_2 and S cells and chromosome-type aberrations in cells exposed during G_1. The effect does not require DNA and chromosome replication to be expressed as chromosomal aberrations (i.e., it is S-independent) and usually appears in the first mitosis after treatment. Although most typically produced by ionizing radiation, a rather similar effect is obtained after treatment with the antibiotics bleomycin, phleomycin and streptonigrin, and also by some methylated oxypurines at temperatures below 30°C (for references, see 1).

Since chromosomal aberrations are produced by these agents independently of DNA synthesis, they are likely to have arisen by misrepair, rather than by misreplication. It is further now generally accepted that the repair process involved is some form of recombinational repair which is likely to have much in common with the molecular mechanisms involved in genetic recombination during meiosis. Several models for genetic recombination have been suggested. Perhaps most favoured of these is the model developed by Holliday[3], according to which the first step is the production of single strand DNA breaks in each of two homologous non-sister chromatids. The breaks are produced in strands of the same polarity. The strands, freed by breakage and unwinding from their original partners, reassociate immediately with the complementary, unbroken strands in the homologous chromatids and the strand interruptions are sealed by polynucleotide ligase. The resulting recombinant complex, frequently referred to as 'half-chromatid chiasma' or 'Holliday structure', is resolved by further breakage-reunion events. Since the position of the 'outside' and 'inner' crossing strands in the recombinant complex can be interchanged by rotation, all four strands in the complex can be regarded as equivalent. Therefore, an endonuclease that recognizes and cleaves crossed strands, produces the parental and recombinant configuration of outside markers with equal frequency[4,5].

The application of this model to the formation of chromosomal aberrations requires certain modifications. Exchange-type aberrations usually occur between sites in non-homologous chromatids or between non-homologous sites in homologous chromatids. The complementary base sequences required for pairing between such sites would be provided by the repetitive DNA sequences that are distributed throughout the genome in eukaryotic organisms, alternating with unique sequences (for, references, see 1). Since 'clean' single strand breaks are rapidly repaired, the lesion of importance for aberration formation is more likely to be a single strand gap or a 'dirty' break which is widened into a gap by exonuclease action.

Sparsely ionizing radiation, such as X-rays, produces single and double strand breaks in DNA in ratios between 10:1 and 20:1. However, perhaps the most common lesions produced by ionizing radiation are endonuclease-sensitive sites that are converted into single strand breaks through the action of endonucleolytic enzymes.

It has been suggested that of the various lesions produced by ionizing radiation, the double strand break is of particular importance for aberration formation[2]. The reason for this would be that single strand breaks and base damage is produced also by chemical substances which, unlike ionizing radiation, have S-dependent effects, that is, the lesions produced by these chemicals require a period of normal DNA synthesis to be expressed as chromosomal aberrations. It has further been suggested that one double strand break should be sufficient to initiate an exchange which means that a chromatid containing a double strand DNA break must be capable of interacting with an undamaged chromatid[2,6]. Resnick[7] has suggested that double strand DNA breaks are normally repaired by a mechanism involving recombination between two homologous chromosomes, one of which is undamaged.

It occured to us that useful information about the mechanisms involved in the formation of chromosomal aberrations might be obtained by a system in which chromosomal aberrations are induced by long-wave UV-light in 5-bromodeoxyuridine (BrdUrd)-substituted chromosomes. When the material is first grown for one cell cycle in the presence of BrdUrd and then for another in the presence of thymidine, the metaphase chromosomes have one chromatid with normal, unsubstituted DNA and one chromatid containing DNA with one BrdUrd-substituted strand. The former chromatid will hereafter be called 'TT' ('T' for thymidine) and the latter 'TB' ('B' for 5-bromodeoxyuridine). Light of wave lengths between 310 and 350 nm induces breaks only in the TB chromatid. Since the two chromatids can be distinguished from each other by differential staining with the fluorescent-plus-Giemsa (FPG) technique[8,9], the system provides a unique opportunity to find out whether an undamaged chromatid can take part in exchange formation with a damaged one. Experiments of this type were initiated by Dr. Natarajan and the senior author in Leiden last spring, using Chinese hamster cells as experimental material. The studies were continued this summer in Uppsala by Christer Andersson and the senior author. In these later experiments, the material was lateral roots of *Vicia faba* rather than Chinese hamster cells.

MATERIAL AND METHODS

The material consisted of lateral roots of the field bean, *Vicia faba*, var. *minor* (Weibulls Åkerböna Primus), grown as described previously[10].

Lateral roots were first exposed to an aqueous (tap water) solution containing 100 μM BrdUrd, 0,1 μM 5-fluorodeoxyuridine and 5 μM uridine and then transferred to a solution containing 100 μM thymidine and 5 μM uridine[9]. Roots were fixed in cold methanol-acetic acid, 3:1, at 24 to 26 hours after being transferred to the thymidine solution. Fixation was usually preceded by a 3-hour exposure to 0,05% colchicine. In most of the experiments seedlings were exposed for 60 minutes at 4 to 6 hours before fixation to long-wave UV-light from 2 Philips 40 watt TL09 Black light fluorescent lamps, which produced light primarily in the 320-380 nm range. During exposure the seedlings were in open Petri dishes containing enough water to keep the roots moist. The dishes were slowly rotated to ensure that all roots received approximately the same dose. The distance between the roots and the light source was 8 cm. Lateral roots in front of and behind the main root were removed before exposure.

After fixation slides were prepared as Feulgen squashes from roots of one or two seedlings. The chromosomes from the roots of the remaining seedlings were differentially stained, essentially according to the modified FPG technique previously described[9]. Some new modifications were made, however. The pectinase treatment used for maceration of the roots was prolonged to 2 hours. After staining with '33258 Hoechst' and mounting in 0,5 x SSC, the preparations were exposed for 15 min. to long-wave UV at a distance of 6 cm from the light source. The slides were then stored for 1 day under ordinary light and temperature conditions in the laboratory and for an additional 3 days at 4°C in the dark. After this treatment, the preparations could be directly stained for 7 min. in a solution containing 3% Giemsa (Gurr's R66) in 0,017 M phosphate buffer, pH 6,8. Usually, however, the slides before staining were either incubated for 60 min. at 55°C in 0,5 x SSC or exposed for 60 min. to long-wave UV.

RESULTS

In the preliminary experiments already referred to, Natarajan and the senior author made the following observations; (1) after growing cultures of Chinese hamster (CHO) in the presence of BrdUrd, long-wave UV-light produced a mixture of chromosome- and chromatid-type aberrations in G_1 cells, chromatid aberrations in S and G_2 cells and sub-chromatid aberrations in G_2-prophase; (2) cells with chromosomes having both chromatids of the TB constitution were several times more sensitive to long-wave UV-light than cells with chromosomes having one TT and one TB chromatid. The first observation is at variance with the findings of Bender et al.[11]. These authors observed only chromatid-type aberrations after exposure of BrdUrd-substituted hamster chromosomes to long-wave UV. The fact that chromatid aberrations were produced in G_1 cells suggests that part of the lesions induced were not converted into chromosomal aberrations until the cells had entered the S-phase. The more than doubled sensitivity of cells containing chromosomes with both chromatids of the TB-constitution in comparison with cells containing chromosomes with one TB and one TT chromatid suggests that the majority of the aberrations observed are the result of an interaction between lesions.

Table 1 shows that in root-tip cells of *Vicia faba* that were exposed to long-wave UV at 26 hours before fixation, that is at a time when they should have been in G_1 or early S, only chromatid-type aberrations were found. At this period the root-tip cells were considerably more sensitive to long-wave UV than when exposed during G_2-prophase. The reason why we found no chromosome-type aberrations could be that the unsynchronized and heterogeneous population of cells in the root meristem is not a very suitable system for studies on effects in G_1 cells. On the other hand, because of their relatively long G_2 period and prophase, the bean root-tip cells provide an excellent material for studies on effects in these stages of the cell cycle. Moreover, the material is very favourable for studies on the inter-

TABLE 1

Types and frequencies of chromosomal aberrations induced by long-wave ultraviolet ('black') light in chromosomes of *Vicia faba* having one TB-chromatid (containing DNA with one 5-bromodeoxyuridine-substituted strand) and one TT-chromatid (containing normal, unsubstituted DNA). After each treatment at least 100 cells were scored for chromosomal aberrations and 25 (300 chromosomes) for sister chromatid exchanges (SCE).

Treatment	Time between light exposure and fixation hours	Abnormal metaphases %	Aberrations per 100 cells: chromatid breaks	iso-chromatid breaks	sub-chromatid exchanges	chromatid exchanges: complete	chromatid exchanges: incomplete	SCE per cell
BrdUrd, 1h light	4	12	1	0	13	1	0	22.5
	5	18	3	2	10	3	1	25.3
	5	26	3	6	15	11	3	21.5
	6	21	7	4	5	7	2	22.3
		Sum	14	12	43	22	6	$\bar{x}$ 22.9
BrdUrd, no light	–	0	0	0	0	0	0	21.6
	–	0	0	0	0	0	0	25.0
	–	0	0	0	0	0	0	21.4
	–	0	0	0	0	0	0	21.0
								$\bar{x}$ 22.3
no BrdUrd, 1h light	–	0	0	0	0	0	0	–
BrdUrd, ½h light	26	25	4	17	0	10	0	64.9
BrdUrd, no light	–	0	0	0	0	0	0	19.5

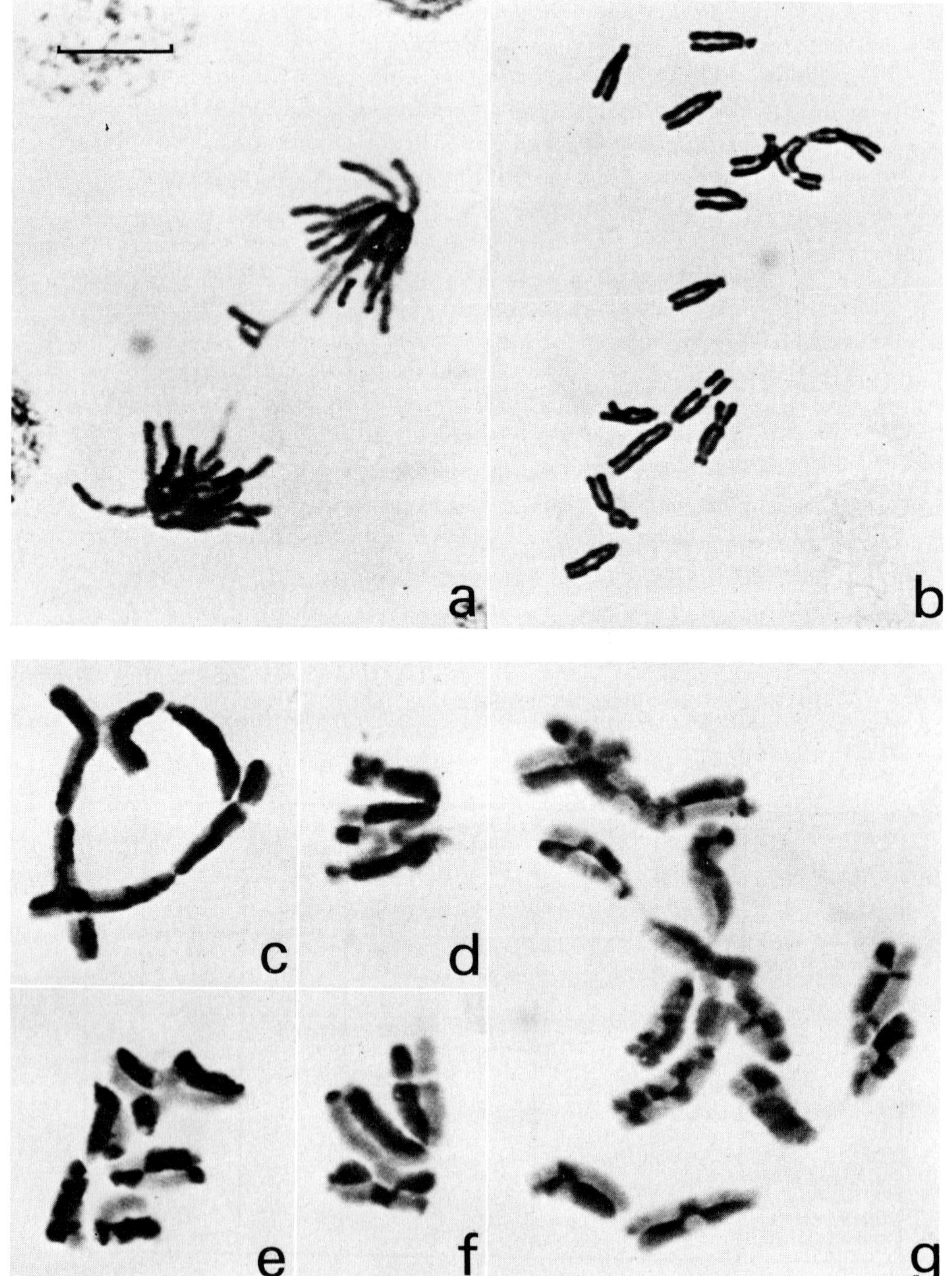
a
b
c
d
e
f
g

action between TT and TB chromatids, since the proportion of exchange-type aberrations is high and a differential staining of these two types of chromatid is easily obtained. The system, however, should be used for qualitative studies only, since the poor penetration power of the long-wave UV in combination with variations in the position of the roots in relation to the light source and in the thickness of the roots from different beans make quantitative studies very unreliable.

Table 1 shows that of the aberrations obtained 4-6 hours after exposure to UV, sub-chromatid exchanges occured with the highest frequency, followed by chromatid exchanges. As with the production of chromatid aberrations in Chinese hamster cells, the fact that sub-chromatid exchanges were observed in cells fixed as late as six hours after exposure suggests that not all of the lesions induced were immediately converted to aberrations. The occurence of isolocus breaks (Fig 1b) should be noted, because unless the points of breakage are located in TB segments close to a sister chromatid exchange, this fact would seem to indicate an interaction between TT and TB chromatids. The last column in the table shows the frequencies of sister chromatid exchange (SCE) observed. In cells fixed 4-6 hours after exposure, the frequencies of SCE are not significantly higher than in unexposed root-tip cells, whereas in cells fixed at 26 hours, there is a three-fold increase of SCE. This confirms the findings of other authors that SCEs are formed during the S-phase (for references, see 12).

Table 2 shows the types of chromatid segments involved in aberration formation. The isochromatid breaks observed in this analysis are not included in the table, since in most cases the distance between the centric and acentric fragments was too large to make it possible to decide whether or not the points of breakage were associated with an SCE. However, in the two cases in which such an analysis was possible, the isochromatid break was clearly associated with an SCE.

It can be seen in Table 2 that 12 of 13 chromatid breaks were in TB segments and 1 associated with an SCE (indicated by the symbol 'TT/TB' in the table). Of the 26 sub-chromatid exchanges observed, 23 were between TB segments (Fig 1c),

---

FIGURE 1

Aberrations induced by long-wave UV-light in *Vicia* chromosomes having one unsubstituted (TT) and one unifilarly BrdUrd-substituted (TB) chromatid. a and b were prepared as Feulgen squashes, c-g according to the FPG technique (TT darkly, TB lightly stained). a, anaphase with side-arm bridge caused by sub-chromatid exchange; b, metaphase containing 2 isolocus breaks and 1 interchange; c and d, sub-chromatid (c) and chromatid exchange (d) involving TB chromatids; e and f, chromatid exchanges in which one point of breakage is located in a TB chromatid and the other associated with an SCE, e is complete, f incomplete; g, chromatid exchange formed by interaction between a TB and a TT chromatid.
The bar in the photomicrograph represents 10 μm.

TABLE 2

Frequencies of chromosomal aberrations involving various types of chromatid segments. TT = segment containing unsubstituted DNA; TB = segment in which one strand of the DNA double helix contains 5-bromodeoxyuridine; TT/TB = segment in which sister chromatid exchange has taken place.

Type of chromatid segment involved in aberration	Frequency of aberrations			
	chromatid break	sub-chromatid exchange	chromatid exchange	sum
TT	0	-	-	0
TB	12	-	-	12
TT/TB	1	-	-	1
TT and TT	-	0	0	0
TT and TB	-	0	4	4
TB and TB	-	23	21	44
TT and TT/TB	-	0	0	0
TB and TT/TB	-	1	10	11
TT/TB and TT/TB	-	2	4	6
Sum	13	26	39	78

1 had one point of breakage in a TB segment and the other associated with SCE, whereas the remaining 2 had both points of breakage associated with SCE. Of the 39 chromatid exchanges analyzed, again the majority (21) had both points of breakage located in TB segments (Fig 1d), 10 had one point of breakage in a TB segment and the other associated with an SCE (Fig 1e and f), whereas 4 had both points of breakage associated with SCE. Of particular interest is that 4 cases were found in which one point of breakage was in a TB segment and the other in a TT segment (Fig 1g).

CONCLUSIONS

We have shown that exposure of BrdUrd-substituted chromosomes of both mammalian and plant cells to long-wave UV-light results in the same types of aberration as are produced by ionizing radiation. Thus, chromosome-type aberrations are produced in G_1 cells, chromatid-type aberrations in S and G_2 and sub-chromatid aberrations in prophase. However, a difference between BrdUrd-long-wave UV and ionizing radiation is that a proportion of the lesions induced by the former system seem to require a longer period to be transformed into aberrations. This results in the production of chromatid aberrations in G_1 and sub-chromatid aberrations in G_2. That this is not just a question of delayed cells is demonstrated by the fact that the chromatid aberrations induced in G_1 may occur in the same cell as chromosome-type aberrations and the sub-chromatid aberrations induced in G_2 in the same cell as chromatid aberrations.

When the chromosomes were differentially stained with a modified FPG technique to make it possible to distinguish between TT (normal, unsubstituted) and TB (unifilarly BrdUrd-substituted) chromatids, it was found that the majority of the aberrations induced were located in TB segments or close to an SCE. Four of the 65 exchanges analyzed (6,2 %) were formed by interaction between a TB and a TT segment.

The major lesion produced by long-wave UV in BrdUrd-substituted DNA is a single strand break. The initial event is the photodissociation of the bromine atom. The uracilylradical thus produced extracts a hydrogen atom from the nearest deoxyribose to form uracil. The resulting sugar radical then undergoes further reactions leading to the single strand break[13]. The repair of the damage thus induced requires the replacement of at least two nucleotides: the one having uracil for a base and the one with a damaged deoxyribose.

In TB-DNA, that is DNA containing one unsubstituted and one BrdUrd-strand, practically all breaks are found in the BrdUrd-strand. However, in a number of different materials, a few breaks have been reported to occur also in the unsubstituted strand[13]. Whatever the cause of these breaks, it is quite clear that double strand breaks are extreamly rare after exposure of TB-DNA to long-wave UV-light and, therefore, are not like to be the lesion responsible for the formation of chromosomal aberrations. Thus, the effect of long-wave UV-light on BrdUrd-substituted chromosomes belongs to the same category as the effects of bleomycin, streptonigrin and methylated oxypurines, which produce an effect very similar to that of ionizing radiation without being capable of producing double strand breaks.

In our opinion, these facts make the idea of double strand DNA breaks as the main lesion leading to chromosomal aberration less attractive also in the case of ionizing radiation. On the other hand, it is true that the interaction between damaged and undamaged chromatids predicted by the double strand break hypothesis has been observed by us, although it occurs at a relatively low frequency. It would, however, be premature to conclude on the basis of this observations that exchanges may be formed between damaged and undamaged chromatids. Other possible explanations are (1) as a result of a minute double SCE, undetectable in the light microscope, the apparently unsubstituted chromatid segment contains a small sector of TB-DNA in which a break has been induced; (2) although long-wave UV-light does not produce any single strand breaks in unsubstituted DNA, it may cause some more subtle change which, although harmless by itself, can interact to give chromatid exchanges with the single strand breaks produced in chromatids containing TB-DNA.

ACKNOWLEDGEMENTS

These studies were supported by grants from the Swedish Natural Science Research Council, O.E. Nycanders forskningsfond and the Dutch Organization for Pure Scientific Research.

REFERENCES

1. Kihlman, B.A. (1977) Caffeine and Chromosomes, Elsevier Scientific Pub. Co., Amsterdam, pp. 1-504.
2. Evans, H.J. (1977) Proc. 2nd Int. Conf. Environmental Mutagens, Edinburgh, 1977, Elsevier/North-Holland, in press.
3. Holliday, R. (1964) Genet. Res., 5, 282-304.
4. Sigal, N. and Alberts, B. (1972) J. Mol. Biol., 71, 789-793.
5. Meselson, M.S. and Radding, C.M. (1975) Proc. Nat. Acad. Sci. USA, 72, 358-361.
6. Leenhouts, H.P. and Chadwick, K.H. (1974) Theor. Appl. Genet., 44, 167-172.
7. Resnick, M.A. (1976) J. theor. Biol., 59, 97-106.
8. Perry, P. and Wolff, S. (1974) Nature, 251, 156-158.
9. Kihlman, B.A. and Kronborg, D. (1975) Chromosoma, 51, 1-10.
10. Kihlman, B.A. (1975) Mutat. Res., 31, 401-412.
11. Bender, M.A., Bedford, J.S. and Mitchell, J.B. (1973) Mutat. Res., 20, 403-416.
12. Kato, H. (1977) Int. Rev. Cytol., 49, 55-97.
13. Hutchinson, F. (1973) Quart. Rev. Biophys., 6, 201-246.

Chromosomes Today Volume 6, A. de la Chapelle and M. Sorsa eds.

NON-RANDOM LOCALIZATION OF CHROMOSOME DAMAGE IN HUMAN CELLS AND TARGETS FOR CLASTOGENIC ACTION.

ANTON BRØGGER
Genetics Laboratory, Norsk Hydro's Institute for Cancer Research, Montebello, Oslo 3, Norway

ABSTRACT

The patterns of chromosome damage in human cells induced by X-rays and chemicals or occurring spontaneously are compared. Bands hit a number of times having a probability of at most 0.01 are denoted goal targets. Examination of coincident goal targets raises the question of agent specificity. Clearly specific patterns are found for X-rays and mitomycin C. The many coincident mercaptoethanol/psoralen-UVA targets may indicate that chromatin/nuclear matrix or nuclear membrane attachments are sites of clastogenic action.

The preferential occurrence of sister chromatid exchange in junctions between heterochromatin and euchromatin may be comparable to the preferential location of chromosome damage in G/R band junctions. Most regions with repetitive DNA (such as centromeric and telomeric chromatin) seem vulnerable to chromosome damage, the Y chromatin being an exception.

INTRODUCTION

The phenomenon that aberrations do not occur at random in the chromosomes has been known for a long time and has been studied in several organisms (references to the first literature may be found in 17).

Non-random distribution of X-ray-induced breaks was observed by Sax[40] in Tradescantia and by Kaufmann[24] in Drosophila. Ford[15] found that nitrogen mustard had a pronounced selective action on particular chromosome regions in Vicia, and his studies were extended to several chemicals by Revell[37].

Virus-associated chromosome breakage also showed a non-random pattern[34]. Quite striking is for example the consistency with which adenovirus 12 induces gaps and breaks in region 17q21 in human cells[29].

Banding techniques have increased the accuracy of the localization of chromosome damage points. During the last few years a number of detailed observations of localization of chromosome damage in human chromosomes have been published. In this paper we will therefore be mainly concerned with results of human chromosome damage obtained by banding methods, using data from the literature and from own studies. The results with different clastogens will be compared and used in a discussion of the nature of the targets for the clastogenic action.

IDENTIFICATION OF GOAL TARGETS

In banded chromosomes there are four types of target for the damage: the G band, the interband (or R band), the G band/R band junction and the centromere:

	No. in human genome	
centromere	24	
G band	144	321
interband (R band)	177	
G/R junction	273	
total targets:	618	

The amount of genetic material in each of these targets is very different. For simplification of the statistical analysis of aberration localization we have considered each target to have an equal probability of being hit. The probabilities of hitting one specific target 0, 1, 2, ... times have been calculated using the formula

$$P_n = \frac{N!}{(N - n)! \; n!} p^n q^{N - n}$$

where N - the number of aberrations, n - the number of hits in one target, p = 1/321 or 1/618, and p + q = 1. If a target has been hit a number of times having a probability of at most 0.01, the target is denoted a goal target. The limit 0.01 is arbitrarily chosen as the level of significance. These considerations and calculations give us a number of goal targets for each clastogen. The localization of these may be compared between the different materials in order to reveal any coincident goal targets. We have done this by indicating the goal targets for each clastogen on transparent sheets and then looking for any merging markings.

SIZE OF THE GOAL TARGET

We localize a number of aberrations to the same site (goal target) in the metaphase chromosome. On the molecular level these sites may be far apart. Each band comprises an average amount of DNA of $5.8\ pg/321 \cdot 2 = 9.034 \cdot 10^{-3} pg$ or $\sim 8.15 \cdot 10^{6}$ base pairs, i.e. a stretch of DNA $\sim$2.8 mm long. To be accurate all we can say is that the localization of damage to certain sites in the metaphase chromosome indicates a clustering of events in the basic chromosome fibre.

TABLE 1

COINCIDENT GOAL TARGETS

Ref.	Agent	Break points	Goal targets	FANC	$SPONT_{lym.}$	$SPONT_{pat.}$	X-ray+CPH	TEPA+ECHH	PUVA	MC Ex	MC Br	MC Ex	MC Br	QM	CHA	MET	X-ray	X-ray	X-ray	X-ray	X-ray
11	X-ray	380	15	0	0	0	0	0	0	0	0	0	0	0	0	0	1	2	2	0	0
20	X-ray	415	8	4	1	0	1	0	2	0	0	0	0	0	0	2	1	0	2	2	
26	X-ray	197	11	3	1	0	1	0	2	1	0	0	1	0	2	2	0	1	2		
38	X-ray	393	27	3	3	1	1	2	2	3	5	4	1	1	4	0	9	2			
8	X-ray	351	15	2	2	3	0	1	0	3	1	1	0	1	1	0	2				
9	X-ray	755	27	(7)	2	3	4	5	4	4	2	5	0	1	2	2					
7	MET	171	26	4	0	0	2	0	(13)	0	0	0	0	1	0						
36	CHA	214	8	4	1	0	0	2	0	1	1	1	0	2							
39	QM	200	9	3	1	0	1	2	1	2	2	1	0								
31	MC Br.	157	2	1	0	1	0	0	0	2	0	0									
	MC Ex.	295	14	2	2	2	3	1	0	6	0										
5	MC Br.	340	6	2	1	0	0	2	1	3											
	MC Ex.	315	9	2	1	2	2	2	0												
43	PUVA	324	30	3	0	2	2	5													
26	TEPA+ ECHH	254	17	(8)	1	3	2														
30	X-ray+ CPH	221	11	4	1	1															
32, 35	SPON pat.	289	9	2	0																
1	SPON lymph.	359	20	4																	
25	FANC	423	31																		

Abbreviations: MET-mercaptoethanol, CHA-chlorambucil, QM-quinacrine mustard, MC-mitomycin C, PUVA-psoralen/longwave UV, TEPA-tris (1-aziridinyl)-phosphine oxide, ECHH-epichlorohydrin, CPH-cyclophosphamid, SPONT-spontaneous aberrations, FANC-Fanconi's anemia.

COINCIDENT GOAL TARGETS

Table 1 shows the result of the comparison of the goal targets with different clastogens using data from the literature. It appears that very few coincident targets are found, with some exceptions: PUVA/MET, X-ray/Fanconi's anemia and TEPA+ECHH/Fanconi's anemia. This general lack of coincidence might indicate an agent specificity in the clastogenic action. However, with one exception there are few coincident targets also between the different X-ray data. This points rather to heterogeneity in the materials, in the experimental conditions and scoring of aberrations as the source of the various apparently unique patterns. Between mitomycin C and X-ray there are a number of coincident goal targets. These are particularly located in the centromeres of chromosomes 1,5,6,7,9,14,15 and 16. Mitomycin C is the only chemical which affects centromeres, whereas X-rays preferentially seem to damage centromeric and telomeric chromatin[8,11,38].

Generally there seems to be more coincident goal targets between the chemicals than between X-rays and the chemicals. Altogether 70 targets are involved. Of these 45 are affected by two agents, 15 by three, and 10 by four or more; the maximum being 8 agents affecting 9q12. The targets 1q12 and 9q12 are known to contain late replicating chromatin with repetitive DNA. The other eight are R bands (Table 2).

TABLE 2
BANDS WITH ≥4 DIFFERENT GOAL TARGETS

Band	Agents
1q12	XRAY MC SPONT FANC
3p21	XRAY MET SPONT FANC XRAY+CPH TEPA+ECHH
5q31	XRAY MET PUVA XRAY+CPH FANC
6p21	XRAY XRAY+CPH TEPA+ECHH FANC
6q21	XRAY CHA XRAY+CPH TEPA+ECHH FANC
7q32	XRAY MET PUVA XRAY+CPH
9q12	XRAY MET CHA MC QM TEPA+ECHH SPONT FANC
10q22	XRAY PUVA TEPA+ECHH SPONT FANC
17q21	XRAY PUVA TEPA+ECHH SPONT
19q13	XRAY QM TEPA+ECHH SPONT FANC

With the spontaneous aberrations there are no clues as to whether they are chemically or radiation- induced, but it might be of significance that six of the ten targets most vulnerable to damage

(Table 2) also are affected spontaneously.

Considering that Fanconi's anemia involves a defect in the repair of ionizing radiation induced DNA damage it is interesting to note the high number of coincident targets with X-ray damage in vivo.

Most striking are the coincidence between the gaps induced by MET and the gaps and breaks induced by PUVA (Table 1). MET is not known to affect the DNA; there is no caffeine effect on MET-induced gaps, and MET does not increase the SCE frequency (Brøgger and Waksvik, unpublished). But MET denatures proteins by breaking S-S bonds. PUVA on the other hand leads to psoralen monoadducts and crosslinks in the DNA, whereas the eventual effect upon proteins, if any, is less clear. What may then be the common target?

There is evidence that chromatin is firmly bound to the inner membrane of the nuclear envelope[16]. This is the peripheral condensed chromatin, and we know that pericentromeric and telomeric chromatin may be found here. Franke[16] has raised the question whether the nuclear membrane contains compounds, probably proteins, that specificially complex with certain chromosome regions. That chromosome condensation begins at the nuclear periphery may reflect this interaction. It remains to be seen whether there exist specific structures, which might be called nuclear membrane/chromatin attachment organelles. We have speculated if these organelles could be the common PUVA/MET targets. But then it should not be expected that the targets occur so irregularly and are not found in all chromosomes.

Other attachments might also be considered. Comings and Okada[13] have suggested that the chromosomes may be attached to all components of the nuclear matrix: the nuclear pore-lamina complex, the intranuclear as well as the nucleolar matrix.

We have mentioned that the apparent agent specificity found in our comparison might be due to heterogeneity of the materials. It is also possible that we are led astray by the small number of aberrations analysed in each investigation. There seems, however, to be at least one example of a true agent specificity. Mitomycin C-induced exchanges are preferentially formed between the regions 1q12, 9q12 and 16q11. Neither the monofunctionally alkylating agent methyl methane sulphonate nor the crosslinking effect of psoralen/UVA seem to induce these types of exchange (Table 3). The lateral extension of the chromatid[6] also seems to be unique to mitomycin C.

TABLE 3

COMPARISON BETWEEN MMS-, PUVA- AND MC-INDUCED CHROMATID EXCHANGE[6,43]

Exchange between secondary constrictions in	MC	MMS		PUVA	
	obs.	obs.	exp.o)	obs.	exp.o)
1 - 1	14	3	12.0	0	2.3
1 - 9	7	0	6.0	0	1.2
1 - 16	1	2	0.9	0	0.2
9 - 9	23	0	19.7	0	3.8
9 - 16	2	0	1.7	0	0.3
16 - 16	2	0	1.7	0	0.3
Other exchanges	35	67	30.0	14	5.8
Total exchanges	84	72	72.0	14	13.9

o) Expected number if same frequency as with MC

ARE ALL ABERRATIONS OCCURRING IN R BANDS?

It was first observed by Caspersson et al.[11] and later confirmed by others (see 39) that most aberrations occur in the G interbands (R bands) or at the R/G junctions. Table 4 shows that the same pattern is found with aberrations induced by psoralen/UVA[43].

TABLE 4

THE LOCALIZATION OF PUVA-INDUCED CHROMOSOME DAMAGE IN GIEMSA BANDS, INTERBANDS AND BAND/INTERBAND JUNCTIONS[43]

Aberration type	Localization in			Sum
	G band	interband (R band)	G/R junction	
Constrictions/gaps PUVA	4	40	16	60
Constrictions/gaps PUVA/CAFF	14	104	78	196
Breaks PUVA/CAFF	5	31	32	68
Total	23	175	126	324
Per cent	7.10	54.01	38.89	100.0

Two questions may be raised:

(1) Is it possible that the G band aberrations in fact are localized in interbands? In preparations of late prophase chromosomes it is possible to identify several lightly stained bands in almost each

of the G bands recognized in metaphase[45].
(2) Is the G/R junction really a different target, or is this localization artefactual owing to our inability to identify very small amounts of chromatin? Thus it might be that the apparent G/R junction aberrations in fact are localized to R bands.

These considerations indicate the possibility that all aberrations occur in R band chromatin. On the other hand, Buckton[8] has pointed out that there seems to be a bias of observation in favour of localizing break points to lightly stained bands. In her study of X-ray-induced aberrations she found more aberrations in lightly stained bands both with G and R banding technique. But analysis with a sequential G and R banding of the same chromosome showed no preference for the R bands. It would be very interesting to see if this proves to be the case also for chemically induced aberrations.

RELATION TO SISTER CHROMATID EXCHANGE

There is in some way a close connection between chromosome damage and SCE, even if at least part of the enzymatic machinery is different[44]. Kato[23] has presented evidence in favour of the hypothesis that some chromosome deletions are the result of failure to complete SCE.

SCE is also occurring non-randomly, even if the data are in some respects not in agreement. A lower incidence of SCE is found in heterochromatin in the kangaroo rat[4], Chinese hamster and Microtus montanus[22] and in the Indian muntjac[10], whereas a higher SCE incidence in heterochromatin is found in Microtus agrestis[33]. In human chromosomes SCE incidence seems to be low at centromeres and telomeres and high in mid-regions[41] and in G/R junctions[14]. An increased frequency in G bands has been found by Smyth and Evans[41], contrary to the observations of Latt[27].

It is difficult to generalize from conflicting evidence. Some data point to an inverse occurrence of SCEs and chromosome damage, the SCEs occurring preferentially in constitutive heterochromatic regions, whereas chromosome damage is more frequent in some centromere and telomere regions and in human R bands, which may represent euchromatin. An interesting point is the preferential occurrence of SCE in junctions between heterochromatin and euchromatin[4,10], which perhaps is comparable to the preferential location of chromosome damage in G/R band junctions.

RELATION TO DISTRIBUTION OF DNA TYPES AND DNA REPAIR

Chromosome damage may be considered to be the result of unfinished repair or misrepair of DNA. The non-random pattern of damage may then reflect (1) a non-random damage of DNA or (2) a non-random repair of DNA, or a combination of both mechanisms. Harris et al.[18] have found a non-random distribution of DNA excision repair in human diploid fibroblasts after exposure to N-methyl-N-nitrosourea, 7-bromomethyl-benz(a)anthracene, N-acetoxy-2-acetyl-aminofluorene or ultraviolet light. More DNA repair was taking place in the cell's inner euchromatin than in the peripherally located heterochromatin.

Studies by Bodell and Banerjee[3] indicate that this may be due to the distribution of satellite DNA in the two types of chromatin. They have found a reduced DNA repair activity in mouse satellite DNA after treatment with MMS and N-methyl-N-nitrosourea. This was not due to a difference in MMS alkylation. Further studies by Bodell[2] indicate more damage and/or repair in the internucleosomal DNA.

It is not clear how we can relate events at the nucleosomal level, involving some 200 base pairs with chromosome bands involving up to 10 millions. But the distribution of heterochromatin and satellite DNA is clearly relevant to our problem.

The banding pattern is not due to different amounts of DNA along the metaphase chromosome[12,42]. But the bands may reflect different states of chromatin in the interphase nucleus. Since the G bands appear to contain late replicating DNA[28], they may be of a heterochromatic nature. The C bands and T bands reflect centromeric and telomeric chromatin, which probably is part of the peripherally located heterochromatin and contains repetitive DNA. Hsu[21] has suggested that this zone of condensed chromatin under the nuclear membrane has a bodyguard function in protecting the euchromatin from damage by absorbing damage, particularly from chemical clastogens. As we have seen, certain areas which contain repetitive DNA sequences are more frequently involved in chromosome damage. But the Y chromatin seems to be an exception.

Hatch and Carrano[19] have proposed a mechanism which particularly accounts for the localization of damage to euchromatin/heterochromatin junctions. The rapid repair of DNA in euchromatin results in few aberrations and SCEs. The very slow or blocked repair in heterochromatin leads to incomplete aberrations, whereas the delayed repair in junctions allows formation of SCEs and complete aberrations.

ACKNOWLEDGEMENT

Thanks are due to Dr.J.Stene, Copenhagen for statistical advice.

REFERENCES

1. Ayme, S., Mattei, J.F., Mattei, F.G. Aurran, Y. and Giraud, F. (1976) Hum.Genet. 31, 161-175.

2. Bodell, W.J. (1977) Nucl.Acids Res. (in press).

3. Bodell, W.J. and Banerjee, M.R. (1976) Nucl.Acids Res. 3, 1689-1702.

4. Bostock, C.J. and Christie, S. (1976) Chromosoma 56, 275-287.

5. Bourgeois, C.A. (1974) Chromosoma 48, 203-211.

6. Brøgger, A. (1974) Hereditas 77, 205-208.

7. Brøgger, A. (1975) Hereditas 80, 131-136.

8. Buckton, K.E. (1976) Int.J.Radiat.Biol. 29, 475-488.

9. Buckton, K.E. (1977) in Symp. on Actions of Physical and Chemical Mutagens on the Somatic Chromosomes of Man, Evans, H.J. and Dolphin, G.W. eds., Edinburgh Univ.Press (in press).

10. Carrano, A.V. and Wolff, S. (1975) Chromosoma 53, 361-369.

11. Caspersson, T., Haglund, U., Lindell, B., and Zech, L. (1972) Exptl.Cell Res. 75, 541-543.

12. Caspersson, T., Farber, S., Foley, G.E., Kudynowski, J., Modest,E.J., Simonsson, E., Wagh, U., and Zech, L. (1968) Exptl.Cell Res. 49, 219-222.

13. Comings, D.E. and Okada, T.A. (1976) Exptl.Cell Res., 103, 341-360.

14. Crossen, P.E., Drets, M.E., Arrighi, F.E. and Johnston, D.A. (1977) Hum.Genet. 35, 345-352.

15. Ford, C.E., (1949) Proc. 8th Int.Congr.Genet. Suppl. to Hereditas pp. 570-571.

16. Franke, W.W. (1974) Int.Rev.Cytol.Suppl. 4, pp. 71-236.

17. Funes-Cravioto, F., Yakovienko, K.N., Kuleshov, N.P. and Zhurkov,V.S. (1974) Mutation Res. 23, 87-105.

18. Harris, C.C., Connor, R.J., Jackson, F.E. and Liebermann, M.W. (1974) Cancer Res. 34, 3461-3468.

19. Hatch, F.T. and Carrano, A.V. (1977) in Abstr.2.Int.Conf.Environ. Mutagens, Mutation Res. (in press).

20. Holmberg, M. and Jonasson, J. (1973) Hereditas 74, 57-67.

21. Hsu, T.C. (1975) Genetics 79, 137-150.

22. Hsu, T.C. and Pathak, S. (1976) Chromosoma 58, 269-273.

23. Kato, H. (1977) Chromosoma 59, 179-191.

24. Kaufmann, B.P. (1946) J.Exp.Zool. 102, 293-320.

25. Koskull, H. von, and Aula, P. (1973) Cytogenetics and Cell Genet. 12, 423-434.

26. Kucerova, M. and Polivkova, Z. (1976) Mutation Res. 34, 279-290.

27. Latt, S.A. (1974) Science 185, 74-76.

28. Latt, S. (1975) Somatic cell genetics, 1, 293-321.

29. McDougall, J.K. (1974) J.Gen.Virol., 12, 43-51.

30.Morad, M., El Zawahri, M. (1977) Mutation Res., 42, 125-139.

31.Morad, M., Jonasson, J., and Lindsten, J. (1973) Hereditas 74, 273-282.

32.Nakagome, Y. and Chiyo, H. (1976) Amer.J.Hum.Genet., 28, 31-41.

33.Natarajan, A.T. and Klasterska, I. (1975) Hereditas, 79, 150-153.

34.Nichols, W.W., Levan, A., Hall, B., Östergren, B. and Kihlman, B. (1963) in Symp. on Cytogenetics of Cells in Culture, Pasadena, U.S.A.

35.Nielsen, J. and Rasmussen, K. (1976) Hereditas, 82, 73-78.

36.Reeves, B.R. and Margoles, C. (1974) Mutation Res., 26, 205-208.

37.Revell, S.H. (1953) in Symp. on Chromosome Breakage, Suppl. to Heredity, 6, 107-124.

38.Sagi, M., Cohen, M.M. and Schaap, T. (1977) in Symp. on Actions of Physical and Chemical Mutagens on the Somatic Chromosomes of Man, Evans, H.J. and Dolphin, G.W. eds., Edinburgh Univ. Press. (in press).

39.Savage, J.R.K., Bigger, T.R.L. and Watson, G.E. (1976) in Chromosomes Today, Pearson, P.L. and Lewis, K.R. eds., John Wiley & Sons, New York, pp. 281-291.

40.Sax, K. (1938) Genetics, 23, 494-516.

41.Smyth, D.R. and Evans, H.J. (1976) Mutation Res., 35, 139-154.

42.Sumner, A.T., Evans, H.J. and Buckland, R.A. (1973) Exptl.Cell Res., 81, 214-222.

43.Waksvik, H., Brøgger, A. and Stene, J. (1977) Hum.Genet. (in press).

44.Wolff, S., Bodycote, J., Thomas, G.H. and Cleaver, J.E. (1975) Genetics, 81, 349-355.

45.Yunis, J.J. (1976) Science, 191, 1268-1270.

Chromosomes Today Volume 6, A. de la Chapelle and M. Sorsa eds.

INTERESPECIES VARIATION IN THE FREQUENCY OF SCE

N.O.BIANCHI,MARTHA S.BIANCHI,ESTHER A.LEZANA,JUAN E. ZABALA SUAREZ
Instituto Multidisciplinario de Biología Celular (IMBICE).,C.C.403,
La Plata,1900, Argentina

ABSTRACT

The BrdU-Giemsa method was employed to analyse the basal frequency of SCEs in human, pig and rabbit lymphocytes. The results obtained showed a significant variability in the rate of exchanges in these species. Rabbit and man were the species with the lowest and highest frequency of SCEs. Pigs showed intermediate values. The interspecific differences found may reflect a variability in the efficiency of the mechanisms of DNA repair or a variability in the sensitivity to BrdU.

INTRODUCTION

The DNA damaged by physical or chemical agents can be repaired and restituted _in toto_ or can be misrepaired. The second alternative usually gives rise to chromosome aberrations and/or to dramatic increases in the level of sister chromatid exchanges (SCEs).

Some human diseases (Fanconi anemia[1],ataxia telangiectasia[2],xeroderma pigmentosa[3]) and some clastogenic agents (X rays,bleomycin[4]) produce high levels of chromosome aberrations and none or low increases in the rate of SCEs. On the other hand, some other diseases (Bloom syndrome[5]) or clastogenic agents (UV radiation[6],most carcinogenic agents[4,7]) give rise to remarkable increases in the frequency of SCEs with low or moderate increases in the rate of chromosome aberrations. Owing to this last phenomenon, SCEs have been employed recently as a test system to detect induced chromosome damage. Human lymphocytes in culture have been usually selected as target cells for the study of SCE induction. However, cells from some other species such as rabbit[8] and hamster[9,10] have also been used for the foregoing purpose. Up to the present there is no available information regarding the basal rates of SCEs in different mammal species. Accordingly, this report is aimed at studying the basal frequency of exchanges in lymphocytes from normal human beings, pigs and rabbits.

MATERIAL AND METHODS

Lymphocyte cultures were started with blood from five adult normal human beings and five adult pigs. Half ml. of total blood was added to 4.5 ml of culture medium. For human lymphocytes the culture medium was TC 199 plus 10% fetal calf serum; in the case of pig lymphocytes the medium had 30% of fetal calf serum and was enriched with 20 mg% of glutamine and 13 mg% of arginine. All cultures received 0.2 ml of PHG, 50 U/ml of penicillin, 50 µg/ml of streptomycin and 5 µg/ml of BrdU. Incubation was performed at 37°C in complete darkness. Cells were harvested at 72 hs. previous treatment with 0.1 µg/ml of colchicine for 3 hs.

Several culture methods were tried for rabbit lymphocytes. However, over a total of 16 animals analysed only one gave a sufficient number of metaphases to analyse the rate of SCE. The culture medium employed for blood culture in this rabbit was similar to that described for pigs. Lymphocyte cultures from the remaining four animals were performed with spleen cells according to the following technique. Rabbits were killed and the spleen removed and placed in a sterile Petri dish. The organ was then punctured through the capsule with a hypodermic needle and injected with 5 ml of complete culture medium. This procedure usually produces the breakage of the spleen capsule in several regions and the culture medium together with most of the spleen cells in suspension can be recovered from the Petri dish. The cell suspension is then centrifuged at 700 rpm for 7'. The bottom of the cell pellet contains red cells, while the upper part contains white cells. The supernatant is discharged and the white cells (mainly lymphocytes) are transferred to another centrifuge tube, washed twice with complete culture medium and resuspended in 3ml of medium. Half ml aliquots of this cell suspension are then seeded into culture bottles containing 4.5 ml of a culture medium similar to that employed for pig blood cultures. PHG, antibiotics and BrdU were employed in the concentrations detailed above. Culture time was 72 hs. and cell harvesting was performed after 3 hours of colchicine treatment.

Chromosome spreads were performed by the same method in the three species. Cells were fixed with 3 to 1 ethanol: acetic acid and slides were prepared by the air drying method. The BrdU-Giemsa method of Korenberg and Freedlander[11] was employed to differentiate the uni- and bifilarly substituted chromatids. The frequency of SCEs was determined in a total of 25 cells per individual. The analysis of variance and the Student t test were used to analyse the interindividual and interspecific differences in the rate of SCEs.

RESULTS

The average frequency of SCEs per cell in human lymphocytes varied from 9.4 ±0.6 in the individual with the lowest rate up to 17.44 ±1.28 in the case with the highest frequency. The mean for the whole group of individuals was 12.75 ±0.53 (Table I).

TABLE I

AVERAGE OF SCEs PER CELL IN NORMAL HUMAN BEINGS

Case	Mean $\overline{X}$	S.E.	Case	Mean $\overline{X}$	S.E
A	17.44	1.277	D	10.72	0.738
B	14.36	0.952	E	9.4	0.6
C	11.60	1.417	Total for the group	12.75	0.527

The analysis of variance showed that the interindividual differences in the rates of SCEs were significant ($p<0.01$). Moreover, the t test showed that differences between the pairs of individuals A-B, B-C, C-D, C-E and D-E were not significant, while differences between all the other possible pair combinations were significant (Table II).

TABLE II

DIFFERENCES IN THE FREQUENCY OF SCEs BETWEEN PAIRS OF INDIVIDUALS

Pairs	A-B	A-C	A-D	A-E	B-C
Means ($\overline{X}i-\overline{X}j$)	3.08	5.84	6.72	8.04	2.76
t	1.934	3.062	4.556	4.556	1.573
p	>0.05	<0.005	<0.001	<0.001	>0.10

Pairs	B-D	B-E	C-D	C-E	D-E
Means ($\overline{X}i-\overline{X}j$)	3.64	4.96	0.88	2.20	1.32
t	3.021	4.408	0.551	1.430	1.388
p	<0.005	<0.001	>0.50	>0.10	>0.10

The samples in pigs and rabbits were much more homogeneous than in human beings. In pigs the rates of SCEs in the individuals with the lowest and highest frequencies were 5.56 ±0.06 and 7.88 ±0.73 respectively, with a mean of 7.27 ±0.307 (Table III):moreover, the analysis of variance showed that the interindividual variability was not significant ($p>0.05$).

TABLE III
AVERAGE OF SCEs PER CELL IN PIGS

Case	Mean $\overline{X}$	S.E.	Case	Mean $\overline{X}$	S.E
F	7.88	0.733	I	7.44	0.513
G	7.80	0.783	J	5.56	0.589
H	7.68	0.713	Total for the group	7.272	0.307

Interindividual differences in rabbits were not significant either. The individual with the lowest rate had 3.64 ±0.365 while the highest frequency observed was 5 ±0.258 SCEs. The mean for the group was 4.464 ±0.158 (Table IV).

TABLE IV
AVERAGE OF SCEs PER CELL IN RABBITS

Case	Mean $\overline{X}$	S.E.	Case	Mean $\overline{X}$	S.E.
K	5	0.258	N	4.36	0.336
L (x)	4.92	0.479	0	3.64	0.365
M	4.40	0.238	Total for the group	4.464	0.158

(x) Blood lymphocyte culture, the remaining individuals are spleen lymphocyte cultures.

Our results point out that man is the species with the highest rate of SCEs, rabbit the species with the lowest frequency, and pigs have intermediate values of SCEs. When the t test is employed to analyse this variation, independently of the pair of species compared the differences are highly significant (Table V).

TABLE V
DIFFERENCES IN THE FREQUENCY OF SCEs BETWEEN PAIRS OF SPECIES

Pairs	Human-Pig	Human-Rabbit	Pig-Rabbit
Means ($\bar{X}i-\bar{X}j$)	5.478	8.286	2.808
t	100.42	168.38	90.33
p	<0.001	<0.001	<0.001

It is at least theoretically possible that the interspecies variability in the frequency of SCEs may be somehow related with the amount of DNA per nucleus. Accordingly, the average of SCEs per cell was divided by the amount of nuclear DNA in the species and the differences compared using the t test. The amount of nuclear DNA in the leukocytes of humans pigs and rabbits were taken from the Cell Biology Data Book.[12] The figures given by this publication are:human lymphocytes 5.72 pg/nucleus (average from three figures), index SCE/DNA: 2.229; pig leukocytes 6.85 pg/nucleus, index SCE/DNA: 1.061; rabbit lymphocytes 5.75 pg/nucleus, index SCE/DNA:0.776. Once more,the t test indicates that the interspecies differences in the indices are significant (Table VI).

TABLE VI
DIFFERENCES IN THE INDICES SCE/DNA BETWEEN PAIRS OF SPECIES

Pairs	Human-Pigs	Human-Rabbit	Pig-Rabbit
Means ($\bar{X}i-\bar{X}j$)	1.168	1.453	0.285
t	21.41	29.53	9.23
p	<0.001	<0.001	<0.001

DISCUSSION

The molecular mechanisms involved in the production of SCEs are not fully understood. However, it has been suggested that the origin of SCEs is closely related to the mechanisms of post-replicational repair or with errors in the excision-type repair system. Recently, Wolff et al[3] were unable to demostrate increases in the frequencies of SCEs in various cell lines defective in the excision repair or post-replication repair mechanism. Accordingly, it has been concluded that these types of deficiency are inadequate to explain the production of SCEs. On the other hand, Latt et al[13] have reported that the treatment of Fanconi's anemia cells with a bifunctional alkylating agent produces less than half the increase in SCEs found in identically treated normal lymphocytes and a higher yield of chromatid breaks. These results are interpreted by the authors as arising from a defect in some type of DNA repair that would produce an increase in open and a deficiency in repaired DNA breaks. In the light of all the above information it seems reasonable to assume that SCEs are probably related to some form of DNA recombinational repair process different from those so far known. This being the case, the observed variability in the frequencies of SCEs may perhaps reflect interindividual (in the case of human beings) or interspecies differences in the efficiency of the recombinational repair system involved in the production of exchanges.

Brewen and Peacock[14] by studying the configuration of ring chromosomes were able to demonstrate that SCEs are a spontaneous event. However, these authors, as well as others[15,16], have also shown that 3HTdR and probably BrdU may increase the spontaneous rate of exchanges. Accordingly, another factor to be taken into account in considering the causes influencing the yield of SCEs, is BrdU itself. Within certain limits, there is a correlation between the dose of this compound in blood cultures and the frequency of SCEs observed.[17,18] Thus, it is possible to speculate that interspecies differences in the cellular sensitivity or in the uptake of the drug, may produce variable rates of SCEs even in cases in which the BrdU concentration in the cultures is maintained constant.

Whether the interspecies differences in the rate of SCEs reflect a variability in the mechanisms of DNA repair or in the sensitivity to BrdU is an important matter to be settled. The first possibility has direct implications to understand the response of different mammal species to clastogenic agents. The second has only the limited value of demonstrating that BrdU may follow variable metabolic pathways in different mammals. At the present time it is impossible to decide which of these two possibilities is valid. However, there are some facts which seem to support the former hypothesis.

The radiosensitivity of lymphocytes, measured as yield of dicentrics per dose of X rays, varies in different mammal species and human beings.[19] Thus, for instance, lymphocytes from trisomic patients have higher radiosensitivity than lymphocytes from normal human beings.[20] On the other hand, lymphocytes from pig, rabbit and other species exhibit lower radiosensitivity than lymphocytes from man.[21,22] When the data on radiation is confronted with the existing information on the basal rates of SCEs the presence of a positive correlation can be easily established. Down syndromes,[23] normal human beings, pigs and rabbits exhibit decreasing frequencies of SCEs; moreover, the same ordering is also found when the radiosensitivity of lymphocytes from the same individuals and species is compared.[20,21,22]

It is now known that the cellular mechanisms involved in the production of SCEs and radiation-induced chromosome aberrations are essentially different.[3,4] Despite of this, the positive correlation between SCEs and chromosomal radiosensitivity deserves further investigation. The molecular mechanisms and the enzymes involved in the processes of DNA repair are far from being fully understood. Therefore, if some enzymes are common to several or all the cellular processes of DNA repair certain positive correlation between these processes is to be expected. Accordingly, at the light of the foregoing information, the interspecies differences in radiosensitivity and frequency of SCEs could perhaps be ascribed to interspecies variations in the type or efficiency of the mechanisms of DNA repair.

SUMMARY

The BrdU-Giemsa method was used to analyse the frequency of SCEs in five normal human beings, five pigs and five rabbits. A total of 25 second mitosis were scored for SCEs in each individual. Man was the species with the highest frequency of SCEs (average 12.75 SCEs per cell), and also showed significant differences in the interindividual rates of SCEs. Rabbit was the species with the lowest rate of SCEs (average 4.64 SCEs per cell). Pig had an intermediate frequency of exchanges (7.27 SCEs/per cell). No significant interindividual differences in the rate of SCEs were detected in pigs or rabbits.

The interspecies variability in the frequency of exchanges may reflect interspecies differences in the mechanisms of DNA repair or in the sensitivity to BrdU.

ACKNOWLEDGEMENTS

This work was supported by grants from the International Atomic Energy Agency, the Organization of American States, the "Consejo Nacional de Investigaciones Científicas y Técnicas" and the "Comisión de Investigaciones Científicas de la Provincia de Buenos Aires".

We wish to acknowledge the technical assistance of Mr. J. Serraino.

REFERENCES

1. Hayashi,K. and Schmid, W. (1975),Humangenetik, 29,201-206.
2. Galloway, S.M. and Evans, H.J. (1975), Cytogenet Cell Genet.,15, 17-29.
3. Wolff, S., Bodycote, J., Thomas, G.H. and Cleaver, J.E. (1975), Genetics, 81,349-355.
4. Perry, P. and Evans, H.J. (1975), Nature, 258,121-125.
5. Chaganti, R.S.K., Schonberg, S. and German, J. (1974), Proc. Natl. Ac. Sci. 71,4508-4512.
6. Kato, H. (1973), Exptl.Cell Res. 82,383-390.
7. Solomon, E. and Bobrow, M. (1975), Mutation Res., 30,273-278.
8. Stetka, D.G. and Wolff, S. (1976), Mutation Res., 41,333-342.
9. Stetka, D.G. and Wolff, S. (1976), Mutation Res., 41,343-350.
10. Natarajan, A.T., Tates, A.D., Van Buul, P.P.W., Meijers, M. and De Vogel, N. (1976), Mutation Res. 37,83-90.
11. Korenberg, J.R. and Freedlander, E.F. (1974), Chromosoma 48,355-360.
12. Altman, P.L. and Katz, D.D. (1976), Cell Biology I, Fed.Amer.Soc. Expl.Biol., Bethesda, Maryland, p.p. 367-378.
13. Latt, S.A. (1973) Proc. Natl.Ac.Sci., 70,3395-3398.
14. Brewen, J.G. and Peacock, W.J. (1969), Mutation Res., 7,433-440.
15. Wolff, S. (1964), Mutation Res., 1,337-348.
16. Miller, R.C., Aaronson, M.M. and Nichols, W.W. (1976), Chromosoma, 55,1-11.
17. Kato, H. (1974), Nature, 251,70-72.
18. Wolff, S. and Perry, P. (1974), Chromosoma, 48,341-353.
19. Sankaranarayanan, K. (1976), Mutation Res., 35,371-386.
20. Sasaki, M.S., Tonomura, A. and Matsubara, S. (1970) Mutation Res., 10,617-633.
21. Brewen, J.G., Preston, R.J., Jones, K.P. and Gosslee, D.G. (1973) Mutation Res., 17,245-254.
22. Scott, D. and Bigger, T.R.L. (1974) Chromosoma 49,185-203.
23. Lezana, E.A., Bianchi, N.O., Bianchi, M.S. and Zabala-Suarez, J.E. (1977) Mutation Res. in press.

Chromosomes Today Volume 6, A. de la Chapelle and M. Sorsa eds.
© 1977 Elsevier/North-Holland Biomedical Press, Amsterdam, The Netherlands

WHAT ARE SISTER CHROMATID EXCHANGES?

H.J. EVANS
MRC Clinical and Population Cytogenetics Unit, Western General Hospital, Crewe Road, Edinburgh EH4 2XU, UK.

ABSTRACT
The development of new staining techniques to visualise sister chromatid exchange (SCE) in cells exposed to mutagens has led to a better understanding of the mechanisms involved in the formation of such exchanges. SCE are induced by a wide variety of different physical and chemical agents and their incidence provides a sensitive indicator of DNA damage in proliferating mammalian cells. It is shown that lesions which affect one or both strands of the DNA can result in the development of SCE, but only when damaged DNA undergoes replication. The nature of the lesions; the frequency and distribution of SCE in mammalian cells; the sensitivity of the cells to their induction by mutagens, are discussed and possible mechanisms involved in the formation of SCE during replication considered.

Since Dr. Kihlman was going to talk about possible molecular mechanisms involved in the production of chromosome aberrations, and since I had previously discussed this topic at another recent symposium[1], I thought it would be more profitable if I spent most of my time talking about sister chromatid exchanges and gave only a brief mention to induced structural chromosomal aberrations. I am not, therefore, going to talk generally about the "actions of carcinogens/mutagens on mammalian chromosomes" as in my original title, but I shall specifically concern myself with the actions of carcinogens/mutagens in producing sister chromatid exchange (SCE).

1. The visualisation of SCE and confirmation of Taylor's original studies.

The notion that exchanges might take place between identical loci on sister chromatids goes back to the very early discussions on mechanisms involved in crossing over, when such exchanges were considered by some to be associated with the processes involved in meiotic recombination. Evidence in support of such exchange events was provided from studies on maize by McClintock[2] in her work on the instability of ring chromosomes at mitosis, and by Schwartz[3] in his studies on the behaviour of ring chromosomes at meiosis. However, the first really decisive demonstration of the occurrence of equal symmetrical sister chromatid exchanges (SCE) that do not result in an alteration in overall chromosome morphology, and their first utilisation in revealing facets of chromosome sub-structure, was made by Taylor[4] in his autoradiographic studies of DNA replication and segregation in plant mitotic chromosomes.

Taylor observed SCE's as switches in radioactive label between sister chromatids at the second (M2) mitosis following ^{3}H-thymidine labelling during the S-phase of the M1 cycle. Later studies by others showed that many of the SCE's observed in autoradiographic experiments were actually induced by the endogenous radiation from the incorporated tritium[5]; that the frequency of SCE was increased in cells exposed to X-rays[6], ultraviolet light[7,8] and certain chemical mutagens[9]; and that a proportion of gross chromatid aberrations (including deletions) seemed to involve a sister chromatid exchange event at the site of aberration[10]. An inherent difficulty in all these studies, however, was the technical complexity of the experiments and the relatively low resolution of the autoradiographic techniques.

Over the last few years, techniques have been developed which enable us to distinguish between sister chromatids without resorting to the use of radioisotopes and autoradiography. These involve exposing cells to 5-bromodeoxyuridine (BUdR) for two rounds of replication so that M2 chromosomes each possess one chromatid unifilarly substituted with BUdR and its sister bifilarly substituted. Such chromatids fluoresce, or stain, differentially with various fluorochromes and with Giemsa[11,12], the bifilarly substituted chromatid staining pale and binding less dye than its sister (Fig. 1). The mechanism of differential staining is still not clearly understood, although it is well known that BUdR-substituted chromatin has altered protein binding properties which are probably largely responsible for the reduced dye-binding efficiency[13].

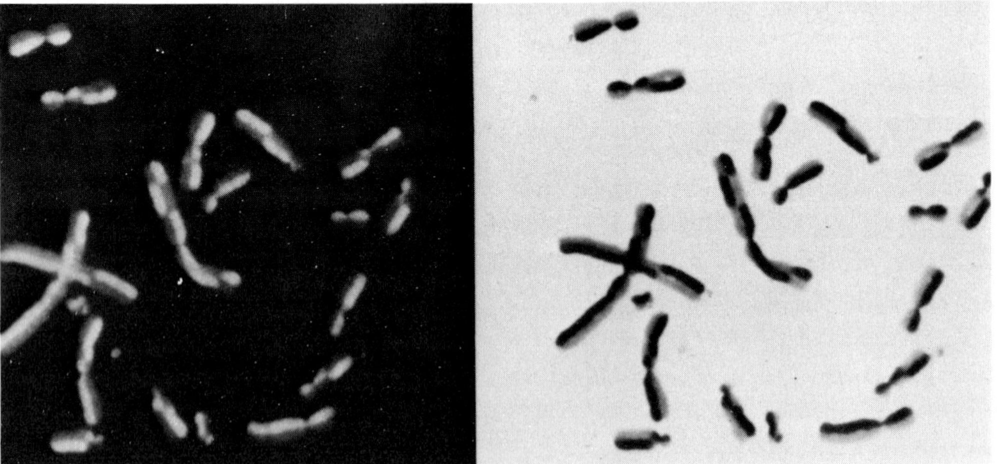

Fig. 1. Chinese hamster cell line (CHO) metaphase cells stained first with Hoechst 33258 (left hand picture) and then with Giemsa (FPG technique) after two rounds of replication in BUdR. Note the differential fluorescence/staining of sister chromatids and the points of exchange (SCE).

The new staining techniques to give "harlequin chromosomes" clearly reveal SCE and provide a powerful means for their study. It was quickly demonstrated by Wolff and Perry[14,15] in our laboratory that the SCE's observed by autoradiography are the same as those that are found by the new staining techniques, but, because of the greater resolution of the latter methods, more exchanges can be discerned with the staining procedures. At least a proportion of the SCE's observed in BUdR treated human or Chinese hamster cells are induced by the analogue itself, as indeed are many of the SCE events observed in cells exposed to radio-isotopes, since the frequency of SCE events depends upon the concentration of BUdR used and is influenced by exposure of the cells to light.

The simplicity and power of the technique to detect SCE's using staining methods, quickly led to confirmation and extension of Taylor's original findings on the replication and segregation of labelled DNA in chromatids. SCE's that occur in the first cell cycle after labelling are duplicated to give "twin" exchanges that affect both sister chromosomes in tetraploid M2 cells, whereas exchanges that occur in the second cell cycle following label are "single" events affecting only one chromosome. Taylor's studies on single and twin exchanges led him to two important conclusions: (i) that the chromatid contained two sub-units which had polarity differences that restricted their rejoining, and (ii) that both sub-units of a chromatid are exchanged when an SCE event occurs. The first conclusion follows from the finding of the ratio of two singles to one twin exchange in tetraploid M2 cells, whereas a ratio of 10:1 is expected on random rejoining of the two sub-units. The second follows directly from the appearance of twins at the second mitosis. These conclusions were not generally accepted due to some contradictory autoradiographic data but have now been amply confirmed using the new staining techniques. Moreover, it is clear that if exchanges involving only single sub-units (single-strand exchange) did occur, then these would result in chromosomes showing partially harlequinised, or heterostaining, properties at the first mitosis post-BUdR labelling, and chromosome segments showing three levels of staining at the second mitosis. These have been searched for extensively and not found, re-enforcing the important conclusion that SCE involves an exchange of both strands of the chromatid.

2. The frequency and distribution of SCE

The evidence that SCE events occur spontaneously still rests largely on the early work on ring chromosomes in plants, and the more recent, but equivocal, studies on ring chromosomes in human cells in culture[16]. Although a proportion of SCE observed using harlequin staining techniques must be induced by the BUdR, it is nevertheless pertinent to consider the distribution of these events.

Studies on a variety of plants and animals, including man, have shown that the incidence of SCE is proportional to DNA content[17] and that the distribution between chromosomes conforms to a Poisson. Their incidence is, therefore, a function of chromosome length, but detailed studies on man reveal that the smaller chromosomes consistently show a lower than expected frequency[18] (Fig. 2). Our own studies have also shown that although there may be as much as a three-fold difference in the mean frequency of SCE's between identically handled samples taken from different people, their incidence is both sex and age independent.

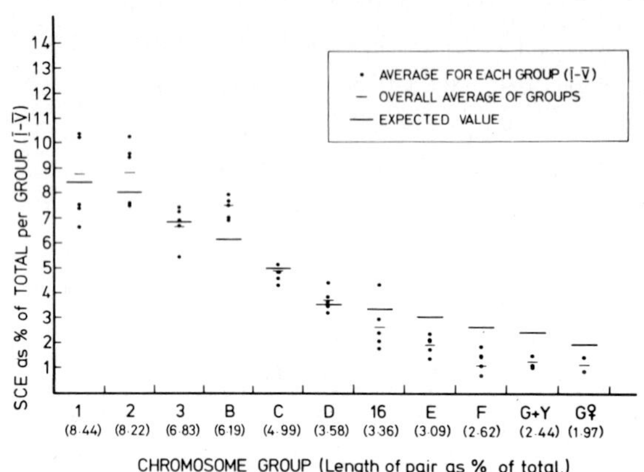

Fig. 2. Distribution of SCE between chromosomes in the human complement expressed as a percentage of the total SCE scored in cells from each of five different groups of people[18]. Total mean frequencies are indicated by the thin horizontal bars and the thick bars indicate frequencies expected if SCE incidence is a simple function of chromosome length.

A major exception to the relative uniformity of SCE frequencies in man was originally reported by Chaganti *et al.*[19] who reported an enormously high incidence of SCE in cells originating from patients with Bloom's syndrome - a syndrome characterised by a high spontaneous aberration frequency. Other human "chromosome breakage" syndromes, such as Fanconi's anaemia and Ataxia telangiectasia, do not share the high "spontaneous" SCE incidence[18] and its occurrence in Bloom's syndrome may be associated with an abnormality in DNA synthesis in these cells[20,21].

Although the distribution of SCE between chromosomes approximates to random, the distribution within certainly does not. In man there is an excess of SCE in the mid and some terminal regions of the chromosomes and a deficiency at the centromere and in the constitutive heterochromatin (Fig. 3). Depending upon the species studied, there are conflicting reports as to whether SCE's are located on the pale G-banded regions, or at the interfaces between pale and dark G-bands, but clear evidence of an excess of SCE at C-band interfaces in the muntjac[22] and the kangaroo rat[23]. The distribution of SCE's in many ways parallels the distribution of "spontaneous" gross chromosomal aberrations, but the significance of this non-random distribution in the generation of SCE events is not immediately obvious.

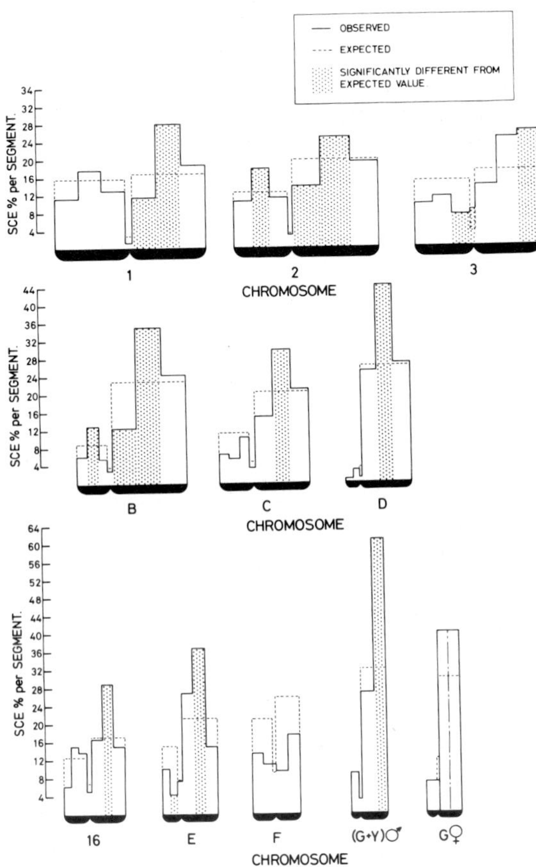

Fig. 3. Distribution of sites of SCE within chromosomes in the human complement[18].

3. <u>The induction of SCE by UV light, X-rays and chemical mutagens.</u>

Perhaps the most important development of the use of the harlequin staining technique, has been to study the response of chromosomes to the induction of SCE following exposure to a variety of different mutagenic/carcinogenic agents.

The original studies on the effect of ultra violet light on inducing SCE's were extended by studies on Chinese hamster chromosomes by Kato[7] and Wolff and colleagues[24] who exposed cells to UV light at different times in the cell cycle. Cells exposed in G1 and S showed increases in SCE at the first mitosis post-irradiation, but cells exposed in G2 did not. However, if the G2 cells were allowed to progress through a second cell cycle then an increased number of SCE's was evident at the second metaphase after irradiation. These results implied the presence of a long-lived lesion which remained in the DNA, but which could only be amplified to form an SCE as a consequence of the chromosomes going through a replication phase.

Paul Perry and I[25] immediately followed these UV studies by carrying out similar experiments on Chinese hamster cells exposed to X-rays. The cells were synchronised by mitotic shake-off and the results of exposure in G2, S and G1 at the interphase prior to the second mitosis following BUdR labelling are shown in Fig. 4.

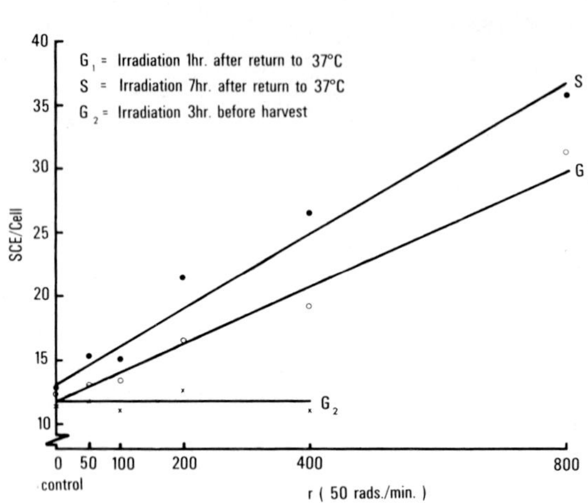

Fig. 4. Incidence of SCE in synchronised Chinese hamster (CHO) cells exposed to X-rays in the G1, S or G2 phases of the cell cycle[25].

Cells exposed in G2 showed lots of chromatid aberrations but no SCE's, whereas cells in S showed chromatid aberrations with some SCE's and cells in G1 showed chromosome-type aberrations and a lower frequency of SCE's. These results contrast with those obtained with ultra-violet light in that they showed that X-rays are very inefficient in inducing SCE's and rarely form the type of long-lived lesion in the DNA that results in sister chromatid exchange. On the other hand they show once again that SCE's cannot be observed at the mitosis immediately following mutagen exposure when the exposure occurs after the completion of DNA synthesis.

In contrast to our X-ray results, when we looked at the responses to a variety of chemical mutagens that were applied to cultured Chinese hamster cells at the beginning of a two cycle treatment with BUdR, we found that all chemical agents known to be mutagenic or carcinogenic under the conditions of our experiment, all resulted in very dramatic increases in sister chromatid exchange and often at extremely low concentrations. The results are summarised in Fig. 5. The initial range of compounds studied can be seen to include monofunctional and bifunctional alkylating agents, agents that produce single stranded breaks in the DNA, and others which intercalate in the DNA helix. Again we found that with the chemical agents SCE's were developed during the S phase, so that lesions present in

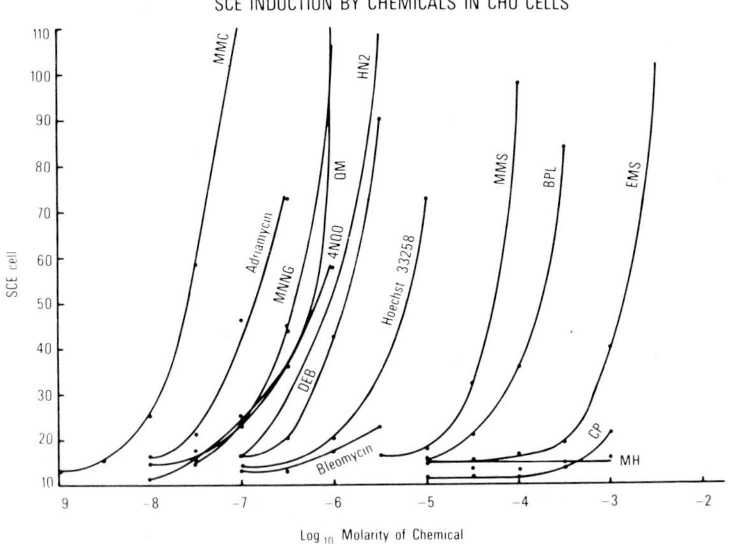

Fig. 5. Incidence of SCE in CHO cells exposed over two cell cycles to a range of mutagens or suspected mutagens. For details see Perry and Evans[25].

G2 did not result in formation of an SCE until the cell proceeded through a further replication cycle. In other words, the lesions that produced the SCE's were long-lived and required to be present during replication to give an SCE.

One important feature of our experiments with chemical mutagens was the demonstration that exposure of cells to a low concentration of a given mutagen could induce vast numbers of SCE, but produce little or nothing in the way of an increase in aberration frequency: this pattern was exactly the reverse of our findings with X-rays, which were very potent in producing aberrations but very inefficient in producing SCE's. These findings have led to a considerable interest in the use of SCE frequencies as a very sensitive general measure of exposure of mammalian cells to mutagens and the simplicity of the SCE detection techniques, and their high frequencies at low levels of exposure are important components in what is proving to be a powerful *in vitro* system for assessing the potency of potential mutagens/carcinogens.

4. The nature of the lesions and the mechanisms involved in the formation of SCE's.

When we turn to consider what kinds of lesions result in SCE, the fact that they may be induced by a very wide variety of different mutagens suggests that the exchanges are the result of a response on the part of the cells replicating machinery to a variety of different lesions in DNA. The substitution of thymine in the DNA by BUdR itself results in SCE, and exposure of cells with such substituted DNA to visible light, gives a very large increase in SCE frequency[15]. Exposure of BUdR-substituted DNA to long wave length UV or visible light results in single strand breaks in the DNA, these may not be long-lived lesions, but it seems very probable that their presence during replication is responsible for most of the SCE's observed following BUdR plus visible light treatment.

In the case of SCE's induced following UV exposure, there are some contradictory data regarding the nature of the lesions involved. Kato[26] has shown that when a cell line from the marsupial rat kangaroo (*Potorous*) was exposed to various UV doses a typical response curve showing an increase in SCE (detected by autoradiography) with increasing dose was obtained. The cell line used was known to possess a photoreactivating enzyme which monomerises thymine dimers on exposure of the cells to visible light, giving a typical photoreactivation or photo-repair response. When cells were exposed to various UV doses followed immediately by a twelve hour exposure to visible light, the yield of SCE was markedly decreased (Fig. 6). Thus, most of the SCE induced by UV light in *Potorous* cells are photoreactiveable by visible light, so that the major lesion responsible for SCE in these cells would appear to be the thymine dimer.

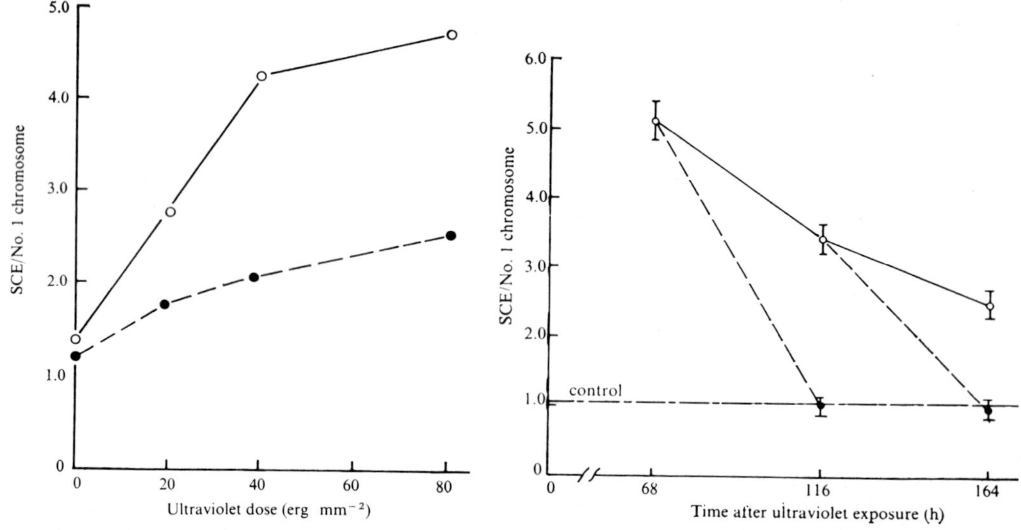

Fig. 6. UV dose response curve for SCE induction in chromosome 1 of a female rat kangaroo cell line. -O-, UV light only; --●--, UV light followed by a 12 hr exposure to visible light. The cells had a doubling time of approx. 38 hrs; were labelled with ^{3}H-thymidine prior to UV exposure and sampled 68 hrs later. From Kato[26].

Fig. 7. Details of cells as for Fig. 6, but they were exposed to a single dose of 60 erg/mm^2 and either maintained continuously in the dark (o) or exposed to visible light for 12 hrs at various times prior to observation (●). From Kato[26].

Potorous cells also undergo repair replication and excise dimers, and other lesions, in the absence of light. When Kato allowed cells to be maintained in the dark after a UV exposure the yield of SCE decreased with time over a three to seven day period, presumably due to excision of UV damage by dark repair processes. This decrease was, however, accelerated if the cells were exposed to visible light for the twelve hour period immediately prior to observation (Fig. 7), indicating that a significant proportion of dimers remain unexcised by the excision repair system; that these presumably continue to generate SCE over a number of cell cycles; and that may be removed or repaired at any time by photoreactivation.

The studies on photorepair of UV lesions resulting in SCE, clearly point to the thymine dimer as the lesion mainly responsible for SCE induction. However, other studies on Chinese hamster[24] and human[27] cells with different inherent capacities for excision repair, show either no correlation, or a less convincing correlation between diminished repair capacity and increased SCE yield. In the studies on cell lines from humans with xeroderma pigmentosum (XP), an Xp variant cell line with normal excision repair and defective in post-replication repair, showed an expected normal SCE response. However, although some lines from complementation groups defective in excision repair showed a good inverse correlation between levels of UV-induced unscheduled DNA synthesis and SCE induction, cells from other complementation groups which were deficient in excision repair (as measured by unscheduled DNA synthesis) showed no such correlation. More work is clearly warranted here.

In our own studies using repair deficient human cells[28], we have examined the response of blood lymphocyte chromosomes from a xeroderma pigmentosum patient to the induction of SCE by UV light and various chemical mutagens. With UV light we get an enhanced response and find a similar increase in SCE incidence relative to controls when XP cells were exposed to a variety of chemical mutagens - a finding that has also been made by Wolff *et al.*[29].

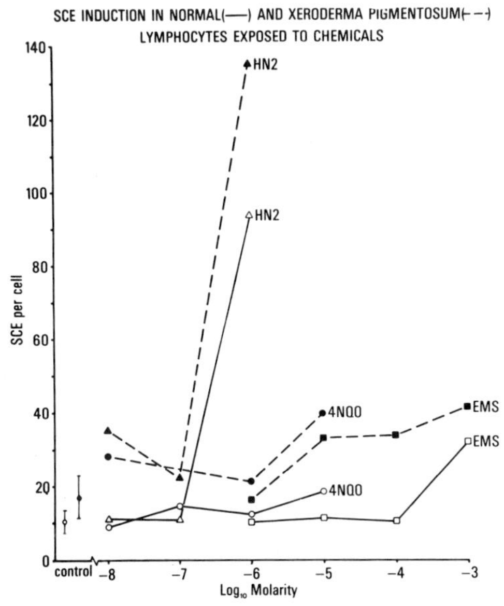

Fig. 8. Incidence of SCE in normal (-) and xeroderma pigmentosum (---) blood lymphocytes after *in vitro* exposure to nitrogen mustard (HN2), 4-nitroquinoline oxide (4NQO) and ethyl methane sulfonate (EMS).

Our experiments (Fig. 8) show that cells from our XP patient are more sensitive to the induction of SCE's by both monofunctional (ethylmethane sulfonate) and bifunctional (nitrogen mustard) alkylating agents, and to the "UV-like" agent 4-nitroquinoline oxide. Each of these compounds produce lesions in the DNA which are subject to removal in normal cells, but which would appear to be removed less efficiently in XP cells. It is also of interest to note here that in all our experiments, and in some of the data published by others in the literature, the levels of SCE in BUdR-treated XP cells not exposed to light, or to other known mutagens, is slightly, but consistently, higher than in controls. This indicates that XP cells are also less efficient in handling BUdR-induced anomalies in the DNA than normal cells.

I think that from all these studies it is difficult to avoid the general conclusion that there must be indeed a variety of different lesions in the DNA that can result in the formation of SCE. How do these lesions produce an exchange?

To answer this question I think we need to consider four of the most important conclusions that we can draw from the data I have presented. These are: (i) many different lesions can result in SCE, they may be long-lived and can be present in the DNA over a number of cell cycles. (ii) Many of the lesions clearly affect only one strand of the DNA and may not themselves be single strand breaks, but their presence presumably results in a localised distortion of the helix. (iii) SCE is only produced at the time of replication of the DNA, so that SCE's are generated as a consequence of the presence of a lesion during replication. (iv) SCE does not involve single strand exchange between sisters, but does necessitate an exchange of both duplexes of DNA.

These conclusions imply that SCE's originate as a response of the replication machinery to by-pass an abnormality in the DNA that would interfere with the normal replication. It would seem probable that the exchanges are initiated at a replication fork, and must involve an exchange between a newly synthesised strand on one chromatid and the parental strand of its sister, and thus a controlled breakage of one (if one is undamaged) or both parental strands. A model depicting such events is shown in Fig. 9

5. Concluding comments

In conclusion I think I should emphasise that the very high frequencies of SCE's induced at low, sub-lethal, mutagen concentrations would imply that the exchange process itself parallels the exchange process occurring at meiosis, since it must rarely result in errors that are expressed as mutations or chromosome aberrations. It would seem probable, however, that the breakage and exchange of duplexes that is involved in SCE formation may sometimes be imperfect, but we have little evidence on this point. We should note, however, that some features of SCE show parallels with gross structural aberrations; such as their distribution

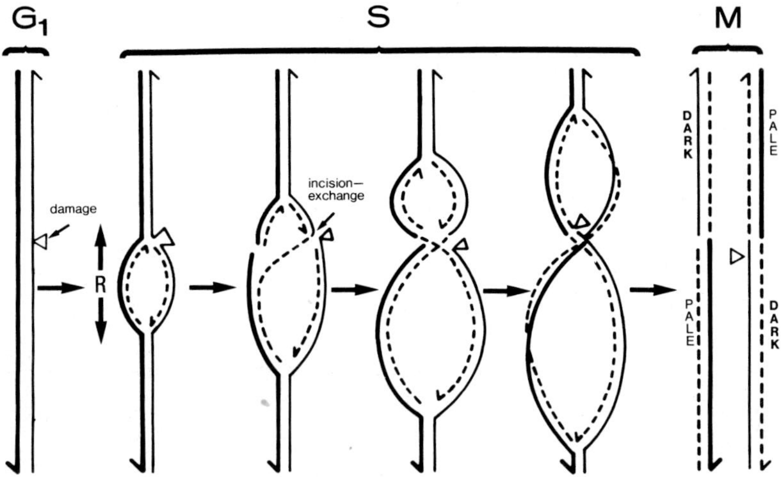

Fig. 9. Schematic representation of the formation of an SCE in a replicating unit.

within the chromosome complement; their occasional presence at the sites of X-ray-induced aberrations; and the fact that some of the lesions that can result in the formation of aberrations can also lead to SCE. On the other hand, and as I have already indicated, there are certainly differences between some of the mechanisms involved in producing these two classes of change, as is clearly evident from our X-ray results. Further information on the similarities and differences between these two categories of chromosome change is desirable and will, I am sure, be forthcoming as a result of the very considerable interest in the use of chromosome aberrations and particularly SCE's in mutagen testing procedures.

REFERENCES

1. Evans, H.J. (1977) in Environmental Mutagens, Scott, D., Sobels, F.H. and Bridges, B.A. eds., North-Holland, Amsterdam (in press).
2. McClintock, B. (1938) Genetics, 23, 315-376.
3. Schwartz, D. (1953) Genetics, 38, 251-260.
4. Taylor, J.H. (1958) Genetics, 43, 515-529.
5. Gibson, D.A. and Prescott, D.M. (1972) Exptl. Cell Res., 74, 397-402.
6. Gatti, M. and Olivieri, G. (1974) Mutation Res., 17, 101-112.
7. Kato, H. (1973) Exptl. Cell Res., 82, 383-390.
8. Rommelaere, J., Susskind, M. and Errera, M. (1973) Chromosoma (Berl.), 41, 243-247.

9. Kato, H. (1974) Exptl. Cell Res., 85, 239-247.

10. Heddle, J.A. and Bodycote, D.J. (1970) Mutation Res., 9, 117-126.

11. Latt, S.A. (1973) Proc. Nat. Acad. Sci. (US), 70, 3395-3399.

12. Perry, P. and Wolff, S. (1974) Nature, 251, 156-158.

13. Evans, H.J. (1977) in Advances in Human Genetics, Vol. 8, Harris, H. and Hirschhorn, K. eds., Plenum Press, New York, pp. 347-438.

14. Wolff, S. and Perry, P. (1974) Chromosoma (Berl.), 48, 341-353.

15. Wolff, S. and Perry, P. (1975) Exptl. Cell Res., 93, 23-30.

16. Brewen, J.G. and Peacock, W.J. (1969) Mutation Res., 7, 433-440.

17. Kato, H. (1977) Chromosoma (Berl.), 59, 179-191.

18. Galloway, S. and Evans, H.J. (1975) Cytogenet. Cell Genet., 15, 17-29.

19. Chaganti, R.S.K., Schonberg, S. and German, J. (1974) Proc. Nat. Acad. Sci. (US), 71, 4508-4512.

20. Hand, R. and German, J. (1975) Proc. Nat. Acad. Sci. (US), 72, 758-762.

21. Gianelli, F., Benson, P.F., Pawsey, S.A. and Polani, P.E. (1977) Nature, 265, 466-469.

22. Carrano, A.V. and Wolff, S. (1975) Chromosoma (Berl.), 53, 361-369.

23. Bostock, C. and Christie, S. (1976) Chromosoma (Berl.), 56, 275-287.

24. Wolff, S., Bodycote, J. and Painter, R.B. (1974) Mutation Res., 25, 73-81.

25. Perry, P. and Evans, H.J. (1975) Nature, 258, 121-125.

26. Kato, H. (1974) Nature, 249, 552-553.

27. de Weerd-Kastelein, E.A., Keijzer, W., Rainaldi, G. and Bootsma, D. (1977) Mutation Res. (in press).

28. Perry, P., Jager, M. and Evans, H.J. (1977) unpublished.

29. Wolff, S., Rodin, B. and Cleaver, J.E. (1977) Nature, 265, 347-349.

Chromosomes Today Volume 6, A. de la Chapelle and M. Sorsa eds.

REMARKS AND DATA ON SOME METHODS TO MONITOR THE IN VIVO INDUCTION OF CHROMOSOME ABERRATIONS IN MAMMALS

WERNER SCHMID

Division of Medical Genetics, Department of Pediatrics, University of Zurich, Zurich (Switzerland)

ABSTRACT

The limitations of aberration scoring in cultured lymphocytes from persons exposed to known or potentially mutagenic agents are discussed. Chromosomal non-disjunction in somatic cells, due to impairment of the spindle apparatus by spindle poisons can routinely be scored by the micronucleus method in bone marrow cells of small mammals. Non-disjunction in germ-line cells, due to a variety of ill-known causes require much more tedious studies; some relevant data are presented from karyotype studies on Chinese hamster preimplantation embryos.

Besides of presenting experimental data, I should like to say a few things intended for the ears of the general medical geneticist who is not specialized in mutation research and who, understandably, at times must be bewildered by what he is learning from the mutation scene.

First, some words about chromosome studies in lymphocyte cultures of persons who have been exposed to known or potential mutagens; later I will devote my time mostly to the problem of induced non-disjunction in mammals.

Lymphocyte cultures from exposed persons are of value in cases of massive whole-body exposure to ionizing radiation. Under these circumstances the method - which explicitely consists in scoring dicentric chromosomes - can serve as a biological dosimeter. The results are, however, difficult or impossible to interprete in cases of partial body exposure to ionizing radiation and this, of course, is the prevalent situation in persons who come for genetic counselling. The difficulties arise from the fact that 1.) only about one percent of a person's lymphocyte population is circulating in the peripheral blood at a given time and 2.) that it is quite difficult to estimate how much of the lymphocyte forming tissue actually was exposed to radiation. Obviously, in such cases the number of dicentrics scored in blood cultures has little or no correlation with the radiation dose received by the gonads. Conclusions in respect to possible genetic damage in germ cells must be sought from calculations based on reconstruction of the situation at the time of irradiation. If there is a probability that the gonads received a considerable dose I offer to my patients prenatal diagnoses, whether or not any dicentrics are found in peripheral lymphocyte cultures.

The problems of partial body irradiation were reviewed by Lloyd (1977)[1] at a recent Symposium in Edinburgh, based on ten years experience at Harwell.

Even more difficult to interprete are findings in lymphocyte cultures of persons which were exposed to either proven or suspected chemical mutagens. Anyone who has followed the pertinent literature of the last ten years is aware of the many conflicting results that were reported about persons exposed to LSD, heavy metals, tranquillizers, spray adhesives, vinyl chloride etc.

Several groups have tried to bring more light into that problem by studying lymphocyte cultures of persons exposed therapeutically to various doses of well-known clastogens. One such study was published by Schinzel and myself (1976)[2] and involved 67 patients in Zurich. Some criticism of that study centered around the fact that 3-day cultures were scored instead of 2-day cultures which are known to give a higher yield of aberrations in radiation-exposed persons. We had chosen the 3-day cultivation period 1.) because most of the studies in the field had used the same cultivation time and 2.) because the yield of metaphases in patients under cytostatic treatment is extremely low at 2 days.

In the meantime Lawler (1977)[3] reported the results of a similar study she carried out in London and in which she studied the metaphases in 2-day cultures. Her results are quite comparable to ours in respect to incidence and types of aberrations. I have not seen any new data that would induce me to change any of the conclusions we drew from our own study (Schinzel and Schmid, 1976)[2]:

Observations

1. Incidence of chromosome type aberrations increased in only 2/3 of the patients treated with high doses of the strong clastogenic drugs (alkylating drugs, antibiotics).
2. Chromosome type aberrations generally not increased in patients treated with antimetabolites and spindle poisons.
3. Incidence of chromatid breaks and exchanges generally normal in all groups.
4. In exceptional cases with high breakage rates no correlation with incidence of chromosome type aberrations.

Conclusions

1. Normal results do not exclude exposition to massive doses of clastogens.
2. Elevated incidence of chromatid-type aberrations are not a characteristic finding in persons exposed to a wide variety of strong chemical mutagens.
3. This system is inadequate for monitoring weak or questionable mutagens in exposed populations.

I certainly would not go as far and say that the in vivo lymphocyte method is unsuitable under any circumstances. Perhaps if one and the same investigator would spend years on a single problem, do all the analyses by himself, use extremely constant laboratory conditions, then some useful results might turn up.

In practice, however, in safety evaluation of chemical compounds, the method simply does not work for 3 main reasons:
1.) there are severe limitations in the method itself. Numerous clastogens are well known to act almost exclusively on dividing cells, usually during S-phase. This has been studied sufficiently on bone marrow cells and on cells in vitro. It is a wild speculation that this situation should be entirely different if lymphocytes are involved. It is practically unknown whether and if, to what extent, these agents actually have long-lasting clastogenic effects after contact with cells in the G^{o} phase. Studies on SCE induction have not clarified that problem. If higher exchange rates were observed, these disappeared about one week after cessation of therapy and may have been due to the presence of traces of the drugs. 2.) The background rate of aberrations, mostly those of the chromatid type, but of the chromosome type as well, vary too much from one laboratory to the next, from time to time and from one investigator to the other. 3.) In a last point I seem to be in agreement even with advocates of the method. I can quote e.g. Purchase (1977)[4] who in Edinburgh said:"A major constraint in applying chromosomal analysis techniques to exposed populations is the expense and the resources required. Simpler effective alternatives will have to be developped and evaluated if this type of monitoring is to become widely used."

One alternative in which I do not believe is the suggestion made by Heddle (1977)[5] who proposed micronucleus scoring in human blood cultures. In a situation where the expected yield of aberrations is at best at minimal levels such an attempt would, in all probability, create but new confusion. By these words I do by no means want to cast doubt upon the results by Countryman and Heddle (1976)[6] which were obtained on blood cultures irradiated in vitro. I wish to add, however, that several years ago when we were working out the micronucleus test on bone marrow cells of small mammals we tried hard to apply the method to germ cells in testicular material as well as to fibroblast cultures. Strong clastogens were used as test substances. We gave up both attempts due to too low yields in spermatogonia and to too many artifacts in cultured fibroblasts, which, in controls, often exhibit segmented and fragmented nuclei.

What are the alternatives if the lymphocyte method is unsuitable? Experimental methods on bone marrow cells of small mammals on the one hand and SCE scoring in lymphocyte cultures, perfected by the addition of microsome preparations on the other hand. Another solution would be to follow the advice of some of the leading speakers in environmental mutagenesis and stop monitoring for chromosome breaking effects at all. The rationale behind that is, as voiced by De Serres (1976)[7] the observation that all chromosome damaging agents at the same time produce gene mutations which can be picked up by the well-known microbiological assays. Only in a second step, when it comes to the characterization of a mutagenic compound would it be necessary to apply cytogenetic techniques as well. Probably this opinion is correct as far as structural chromosome aberrations are concerned. The problem of non-disjunction will, however, bring us back again to the experimental bone marrow techniques.

The problem of non-disjunction is a very serious one in human pathology. While this is clear to the medical geneticists, it is often underrated by biologists. The proportion of abnormal human phenotypes due to numerical aneuploidies is about ten times higher than those due to chromosome breakage. This does not mean that I want to play down the importance of mutagens affecting chromosome structure. We do not know the role of structural and non-disjunctional events in the process of carcinogenesis. All I want to point out is the importance of non-disjunction in the problem of human congenital defects and the need for studies on that process.

Scoring for induced non-disjunction is, indeed, a great unsolved problem. De Serres (1976)[7] in the already mentioned paper on the application of short-term tests for mutagenicity makes the following statement:"At present there are no assays for non-disjunction which are useful for mass screening programs. This is a serious deficiency since non-disjunction is produced by chemicals which interfere with the spindle apparatus and agents which would produce this type of damage might not necessarily produce gene mutations and chromosome aberrations. A good example of such specificity is that shown by colchicine".

As far as spindle poisons are concerned De Serres is wrong. Chromosome loss due to the effects of spindle poisons can easily and effectively be scored in vivo by the micronucleus test on bone marrow cells. But De Serres may still be right: Non-disjunction in the origin of human malformation syndromes may not be related at all to spindle poisons. The best known factor involved, maternal age, speaks against such a simple assumption. If, however, factors are involved acting indirectly by influencing processes involved in the ageing process, then, indeed, can we not hope to solve the problem by studying the effects of chemicals on the spindle in somatic cell divisions. The same holds for factors involving e.g. synapsis or delayed fertilization of the ovum. If we are faced with all that, and educated guesses go in that direction, then we do not have a simple screening method for the relevant kind of non-disjunction. Then we really have to go into the tedious work of studying meiotic cells and early embryos.

I mentionned that the micronucleus test on bone marrow of mice or other small rodents is a simple and rapid method for screening chromosome loss due to spindle poisons (Schmid, 1976)[8]. If the investigator adheres to our recommended application scheme he can even suspect an action on the spindle apparatus by looking thoroughly at his preparations. Treatment of the animal is begun 30 hours prior to fixation. If all the action of a mutagen occurs during the S-phase of the cell cycle there will be no micronuclei in mature, red erythrocytes. The development from the last S-phase to a red mature erythrocyte takes more than 30 hours. If the test substance, like e.g. Vincristine, acts later, during mitosis, a small proportion of erythrocytes can complete maturation and will show micronuclei. It is quite clear that such a hint obtained in this screening test must be confirmed. An appropriate method is anaphase scoring in cells grown on cover slips.

Figure 1 shows the results from micronucleus tests on Vincristine and Colcemid. (Maier and Schmid, 1976)[9]

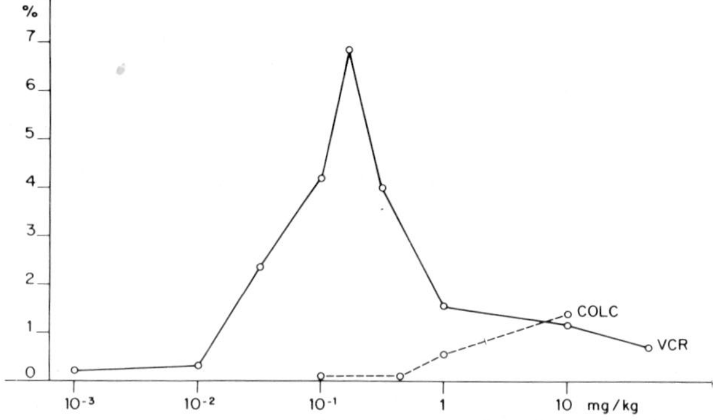

Fig. 1. Dose-effect curves for Vincristine (VCR) and Colcemid (COLC) obtained in micronucleus test (from Maier and Schmid, 1976)[9]

Both substances are suitable for metaphase arrest in cytogenetics but the two substances differ fundamentally in their clinical usefulness and I think these results nicely demonstrate why this is so. Vincristine is a useful drug in cancer therapy, Colchicine or Colcemid was never a success in that respect. Why this difference? Under the influence of low doses of Vincristine single chromosomes are lost and give rise to micronucleated cells which, of course, are doomed. This is the therapeutically useful dose range. Higher doses completely block the spindle apparatus and induce tetraploid nuclei which, in erythroblasts, are expelled without formation of micronuclei. In Colcemid the range with the partial effect on the spindle is largely missing and in higher doses the drug becomes toxic to the animal.

As I said before, I doubt whether spindle poisons are of great relevance to the problem of non-disjunction in the germ lines, but this is not more than a guess. We also do not know whether numerically aneuploid somatic cells play a causative role in cancer. This is a problem that urgently needs elucidation.

I now wish to present some data which are relevant to the problem of non-disjunction in the germ line. For many years Mr. Binkert in my laboratory has performed karyotype studies in preimplantation embryos of the Chinese hamster. In these experiments we always used natural breeding conditions, i.e. without inducing superovulation in the females. The embryos developed in their mothers up to the 4 - 8 cell stage. Then they were rinsed from the oviducts and prepared immediately using Tarkowski's air-drying method. From every embryo and every mitosis that was evaluated cut-out karyotypes were made. Three series of studies were carried out so far:

1.) A control (Binkert and Schmid, 1977 a)[10]

2.) A study on the progeny of males who's germ cells had been treated by a single large dose of an alkylating agent during the sensitive postmeiotic stage. (Binkert and Schmid, 1977 b)[11]

3.) A study in which the males were mated twice a week, up to 68 days after a single treatment with the mutagen. In this way we obtained results from 20 separate mating periods covering from treated mature sperm up to treated spermatogonia. (Binkert and Schmid, in prep.)[12]

Among 226 control embryos there were 5.3 percent abnormal karyotypes. More than half of these were triploids and haploids. Trisomies and monosomies were exceedingly rare. Of course we wonder why this is so different from the supposed situation in man. From studies on human abortions it is assumed e.g. by the Boué's, (1973)[13] that one in every two conceptions in man carries a chromosome aberration, the vast majority of these being trisomies and monosomies. It is not clear whether the optimal timing of copulation in respect to oestrus and ovulation in these rodents plays a decisive role for the difference in comparison to man. In any case, we have observed that spontaneous copulation in our hamsters takes place exactly 5 - 7 hours before ovulation.

What makes me hesitate to believe that this is a major factor are the results from fluorescent marker studies on chromosome 21 in patients with Down's syndrome and their parents which indicate that at least a sizable proportion is due to non-disjunction at the first meiotic division and cannot be explained by delayed fertilization. (Mikkelsen et al. 1976)[14]

The low spontaneous aberration rate in the Chinese hamster is not an exception. It is similarly low in all mammals which have been properly investigated in that respect (Binkert and Schmid, 1977a)[10]. For the study of many problems this low aberration rate in the Chinese hamster is a most valuable asset. An impossible number of animals would have to be studied if we were dealing with a spontaneous rate of 50 percent.

Table 1 presents the results obtained after treatment of the fathers with a single high dose of the alkylating agent Trenimon. The male germ cells had been subjected to the mutagen in postmeiotic stages, 3 - 23 days prior to mating. The result was a high increase in structural aberrations but no increase in genome mutations.

TABLE 1

Chinese hamster preimplantation embryos. The male germ cells were exposed to a single sublethal dose of the alkylating agent Trenimon in postmeiotic stages. (From Binkert and Schmid, 1977b)

	n	%
Total number karyotyped	221	100.0
Normal karyotypes	167	75.6
Abnormal karyotypes	54	24.4
Triploidy	2	0.9
Haploidy	0	0
Trisomy	2	0.9
Monosomy	0	0
Structural aberrations	51	23.1

The structural aberrations were almost exclusively of the chromosome type as is shown in detail in table 2.

TABLE 2

Types of structural aberrations observed in experiment 2 (relates to table 1).

	n	%
Total number of aberrant centric elements in 50 embryos	125	100.0
Presumed simple deletions	68	54.4
Dicentrics	20	16.0
Rings	9	7.2
Others (mostly presumed translocations)	28	22.4

The third experiment, the study over 68 days is not yet entirely completed and the data are preliminary. The results were obtained from 20 separate mating periods. From each of these periods it is intended to collect data on 30 embryos. Until now about half of the preparations have been evaluated.

In table 3 the results were pooled to cover biologically important periods: mature sperm, spermatids, prophase I and spermatogonia. All had the same treatment, a single sublethal dose of Trenimon.

Of interest are the underlined figures since they reveal something that is missed in the dominant lethal tests, because there, these aberrations disappear in the high background rates. There is a definite increase in genome mutations, trisomies and monosomies. And if we look at the structural aberrations, we can see that embryos with a single deletion or with a reciprocal translocation make up roughly two thirds of the total structural aberrations. This is qualitatively quite different from the

TABLE 3

Types of abnormal karyotypes in experiment 3.

	n	Total %	Ploidy %	Genome %	Total struct. aberrat. %	1 single deletion or 1 ring %	1 reciprocal translocation or inversion %
Control	226	5.3	3.1	0.9	1.8	0	0
Sperm (1 - 5 d)	55	7.3	0	0	7.3	1.8	0
Spermatids (8 - 23 d)	95	45.3	5.3	2.0	38.0	14.7	3.2
Prophase I (25 - 44 d)	101	18.0	3.0	4.9	9.9	1.0	4.9
Spermatogonia (46 - 68 d)	144	10.4	0	3.5	6.9	4.2	1.4

results in the sensitive spermatid stage. There, most of the embryos have multiple structural aberrations, about which, in terms of causation of malformation, we are much less concerned.

TABLE 4

Details on genome mutations

	n	Total %	Monosomies %	Trisomies %	Tetrasomies %
Control	226	0.9	0.4	0.4	0
Treated					
1 - 23 d	371	1.1	0.3	0.8	0
25 - 68 d	245	4.1	1.2	2.0	0.8

Table 4 considers genome mutations only. The data from the second and third experiments were pooled. The treated gametes were divided in two groups only with treatment up to 23 days before mating on the one side and treatment 25 - 68 days before mating. Thus, the dividing line is roughly at the end of prophase I. Again it can be seen that genome mutations are not increased in the exposed postmeiotic stages but they seem to be elevated in treated prophase I and in treated spermatogonia. If this trend continues in the rest of our material it would indicate that meiosis after all is not such a good filter protecting from chromosomal aberrations as we had thought before on the basis of dominant lethal tests.

Studies on this material are very tedious indeed. But I think they are important as long as we have no firm correlations with data from simpler testing procedures. The problem of non-disjunction is a relevant one in human pathology. It is no consolation that over 95 percent of foetuses with numerical aberrations are aborted. If the number of live-born numerical aneuploids is ten times higher than of those with an unbalanced structural aberration we must conclude that a ten percent increase in non-disjunction equals the effect of doubling the rate of structural aberrations. In practical terms there seem to be two alternatives: either to devote more efforts to elucidate the causes of non-disjunction in man or circumvent that issue by spreading prenatal chromosome analysis to all pregnancies.

REFERENCES

1. Lloyd, D.C. (1977) A review of the problems in interpreting aberration yields induced by in vivo irradiation of lymphocytes. Proceedings, Symposium on actions of physical and chemical mutagens on the somatic chromosomes of man, Edinburgh July 7, 1977, H.J. Evans ed., in press.

2. Schinzel,A. and Schmid,W. (1976) Lymphocyte chromosome studies in humans exposed to chemical mutagens. The validity of the method in 67 patients under cytostatic therapy. Mutation Res. 40, 139 - 166.

3. Lawler, S.D. (1977) Monitoring for mutagenic hazard by chromosome studies, same Ref. as under 1.

4. Purchase, I.F.H. (1977) Chromosome analysis in exposed populations: review of industrial problems, same Ref. as under 1.

5. Heddle, J.A. (1977) Micronuclei: an assay for chromosomal damage in cultured lymphocytes, same Ref. as under 1.

6. Countryman, P.I. and Heddle, J.A. (1976) The production of micronuclei from chromosome aberrations in irradiated cultures of human lymphocytes. Mutation Res. 41, 321 - 332.

7. De Serres, F.J. (1976) The application of short-term tests for mutagenicity in the toxicological evaluation of environmental chemicals. In: In vitro metabolic activation in mutagenesis testing. F.J. De Serres et al. eds, Elsevier/North-Holland Biomedical Press, Amsterdam.

8. Schmid, W. (1976) The micronucleus test for cytogenetic analysis. Chapter in "Chemical Mutagens" Vol. 4, Plenum Publishing Co., A. Hollaender, ed., 31 - 53.

9. Maier, P. and Schmid, W. (1976) Ten model mutagens evaluated by the micronucleus test. Mutation Res. 40, 325 - 338.

10. Binkert, F. and Schmid, W. (1977a) Preimplantation embryos of Chinese hamster. I. Incidence of karyotype anomalies in 226 control embryos. Mutation Res. 46, 63 - 76.

11. Binkert, F. and Schmid, W. (1977b) Preimplantation embryos of Chinese hamster. II. Incidence and type of karyotype anomalies after treatment of the paternal postmeiotic germ cells with an alkylating mutagen. Mutation Res. 46, 77 - 86.

12. Binkert, F. and Schmid, W. Preimplantation embryos of Chinese hamster. III. Karyotype anomalies observed over a period of 1 - 68 days after single treatment of the

sires with an alkylating mutagen, in prep.

13. Boué, J. and Boué, A. (1973) Anomalies chromosomiques dans les avortements, spontanés, in A. Boué and Ch. Thibault (eds.), Les Accidents Chromosomiques de la Reproduction, Institut National de la Santé et de la Recherche Médicale, Paris, 29 - 55.

14. Mikkelsen, M., Halberg, A. and Poulsen, H. (1976) Maternal and paternal origin of extra chromosome in trisomy 21. Human Genet., 32, 17 - 21

CHROMOSOMES IN MALIGNANCY

Chromosomes Today Volume 6, A. de la Chapelle and M. Sorsa eds.

INTRODUCTION

ALBERT LEVAN
Institute of Genetics, University of Lund, S-223 62 LUND, Sweden

ABSTRACT

A brief review is given of the development of chromosome research in cancer, from the early pathologists of the 1890's to present time. Also, some new data are presented relating to the obscure phenomenon of double minutes, found so far only in connection with malignancy.

CONTRIBUTIONS OF CHROMOSOME RESEARCH TO THE ELUCIDATION OF THE CANCER PROBLEM

The pathologists of the 1890's were well familiar with the fact that malignant growth is characterized by manifold disturbances of nuclear and cellular divisions. Especially von Hansemann[1] in a number of papers ranging from 1890 to 1920 regarded the asymmetric nuclear divisions, frequent in cancer cells, as a key factor in the development of malignancy. Such observations stimulated Boveri to his somatic mutation theory for the origin of cancer[2]. Thus the cancer problem was early recognized as a concern of cell genetics and chromosome research.

Cytogeneticists, with very few exceptions, as Belling[3], Winge[4], did not pay much attention to chromosomes of malignant cells for a long period. This was due entirely to the tardy development of mammalian chromosome methodology. When in the early 1950's cells of malignant effusions - first ascites, later on pleural exudates - became available for chromosome study, this was actually the first mammalian material, in which chromosomes had a technical quality that had long been routine in many other materials, as higher plants, insects etc. Thus it happened that chromosome work with malignant cells opened up the entire field of mammalian and human cytogenetics.

Even the first chromosome investigations in serially grown ascites tumors in rats[5] and mice[6, 7] revealed basic cytogenetic properties that distinguished malignant cells from normal. In malignant cells a high degree of chromosomal variability is often combined with a surprising stability of the stemline karyotype. The tumor stemlines undergo evolutionary changes from the normal karyotype of the original host cell to the usually aberrant karyotypes of advanced cancer stages. On the cellular level, these chromosome mechanisms are true counterparts to those acting on the organism (and species) level during the natural evolution. At all levels, the karyotypic development takes place in response to selective pressures from the environment.

By the chromosomal exploration of human effusion tumors it was ascertained that the karyotypic evolution is nonrandom: certain chromosome types tend to increase in number, others to decrease[8, 9]. Before this, however, Nowell and Hungerford[10] established the significance of the individual chromosome in cancer development by their discovery in 1960 of the correlation between the Philadelphia chromosome and chronic myelogenous leukemia. It would take 11 years before the second case of a similar correlation between a specific structural change of a chromosome and a malignant disease was revealed, viz. that between the $14q^{+}$ marker and Burkitt's lymphoma[11].

In the meantime work with experimental tumors yielded lots of convincing evidence that chromosome variation in cancer is indeed nonrandom and predetermined. Furthermore, in each type of cancer the majority of the chromosomes are largely unaffected, the significant changes being restricted to a few chromosome types. These changes often form a characteristic sequential pattern. Next, experimental work with Chinese hamsters and especially rats led to the finding of a suggestive principle: The chromosome pattern during tumor development is strictly dependent on the inducing factor[12]. Identical chromosome pattern is found in dimethylbenzanthracene-induced rat leukemias[13], sarcomas[14] and carcinomas[15], whereas a different pattern is found in Rous virus-induced rat sarcomas[16]. This principle is important, because it implies that the inducing agent directly interferes with the hereditary material of the hostcell. It is tempting to try and extend this principle to human neoplasms; several attempts have been made, but our ignorance of the inducing agents in most human tumors undoubtedly is a serious obstacle. Still the perfect accordance in most chromosomal features between experimental and human tumors, makes it likely that the inducing agent acts through similar mechanisms in both. Therefore, it may be reasonable to expect that it will eventually be possible to draw conclusions from the chromosome picture to the etiology of human tumors.

DOUBLE MINUTES - A PROBLEM RESTRICTED TO CANCER

To the student of chromosomes, malignant cells have a special allure. Chromosomal deviations that are seldom seen in normal material abound in malignant cells. Still, it has been slightly disappointing - although perhaps expected - that chromosomal abnormalities in tumors are all of the same kind as described earlier in normal cells.

There is, however, one chromosomal deviation that to my knowledge has not been found outside of cancer. This is the still completely unexplained phenomenon of the so-called double minutes (DMs). These fascinating small structures have been reported by many authors from altogether some 60 tumors. The first case - a metastatic exudate from a human pulmonary carcinoma - was found in 1962[17]. The first case of DMs in experimental tumors was 6 Rous virus-induced mouse sarcomas reported 5 years later[18]. On the basis of current work in our laboratory, dealing with a mouse ascites tumor called SEWA[19-22], I shall briefly describe some features of the DMs. The SEWA tumor has been known for several years to be a carrier of DMs and has turned out especially favorable for the study of several aspects of the DM problem.

In the SEWA tumor the DMs vary in number from 0 to over 1,000 per cell, usually they occur in numbers from 1 to 50 per cell. They vary in size and rather resemble small acentric rings. No centromeric structure is seen. They are negatively C-heterochromatic and stain weakly in G-staining. They are Feulgen-positive and their DNA replicates simultaneously and early during S. While they are maintained fairly constant during long-term in vivo passage, they soon become eliminated in vitro.

Last March, one SEWA line in vivo had lost most of its DMs, and at the same time the stemline chromosome number had gone up from 43 to 51. This increase in chromosome number was due to the appearance in most cells of 1 to 12 copies of a new chromosome type, quite different from the ordinary tumor chromosomes (Fig. 1). This new chromosome is a medium-sized metacentric which has a number of properties in common

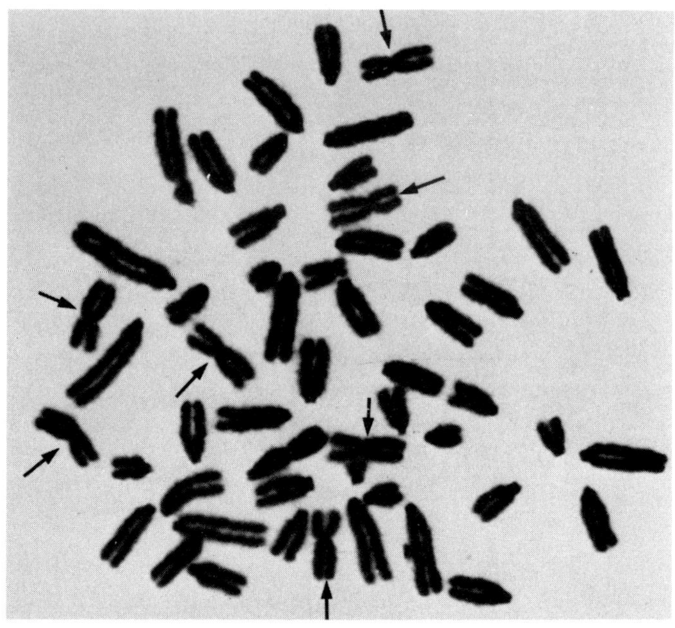

Fig. 1. Metaphase from the SEWA mouse tumor, with seven copies of chromosome originated de novo (arrows).

with the DMs: It has no C-heterochromatin, although in contrast to the DMs it clearly has a centromere. In G-banding it shows faint staining with no distinct bands. The DNA of the new chromosome replicates simultaneously and early during S, except for the centromeric region which is somewhat later. When the SEWA line is explanted in vitro, the new chromosome is rapidly lost and the chromosome number 43 is recovered.

It is evident that the new chromosome is not a derivative of ordinary mouse chromosomes. Instead it shares so many morphologic and functional properties with the DMs that it is difficult not to associate its origin with the DMs. Our tentative hypothesis[22] is that the new chromosome has originated from the DMs on a specific occasion by an unknown mechanism, and that both the DMs and the new chromosome represent amplification of hereditary material stimulating tumor growth.

CONCLUSION

Among the different areas of mammalian cytogenetics that were initiated in consequence of the methodologic progress after 1950, cancer cytogenetics for a long period showed rather slow and hesitant progress. The discovery of the Philadelphia chromosome long remained the sole striking and clear-cut result. Observations in most other tumors at that time were controversial and difficult to interpret, and it took large amounts of hard work to establish understandable chromosome patterns in them.

Perhaps no area of mammalian cytogenetics has gained more than cancer cytogenetics from the different chromosome banding techniques, and the present rapidly accelerating expansion of this area may be seen as a direct consequence of the application of this tool. Large amounts of new data have accumulated and, as far as they can be surveyed at present, they support strongly the notion that the chromosomal evolution in cancer follows meaningful, predetermined patterns. The interaction between the inducing agents and the hereditary matter of the host cell, demonstrated in a few experimental tumors undoubtedly constitutes an important principle in oncogenesis. In general, the karyotypic evolution of tumor stemlines, including the appearance of double minutes, may be regarded as a means for adjusting the genic equilibrium towards increasing malignant capacity.

ACKNOWLEDGEMENTS

Work from our laboratory, reviewed in the present article, has been supported by grants from the Swedish Cancer Society and from the John and Augusta Persson Foundation for Medical Research. I also wish to thank my colleagues in the Cancer Chromosome Laboratory for discussions, criticism and use of unpublished data.

REFERENCES

1. Hansemann, D. von (1890) Über asymmetrische Zellteilung in Epithelkrebsen und deren biologische Bedeutung. Virchows Arch. Pathol. Anat., 119, 299-326.

2. Boveri, T. (1914) Zur Frage der Entstehung maligner Tumoren. Gustav Fischer, Jena, 64 pp.

3. Belling, J. (1927) The number of chromosomes in the cells of cancerous and other human tumors. J. Am. Med. Ass., 88, 396.

4. Winge, Ö. (1930) Zytologische Untersuchungen über die Natur maligner Tumoren II. Teerkarzinome bei Mäusen. Z. Zellforsch. mikr. Anat., 10, 683-735.

5. Makino, S. (1951) Some observations on the chromosomes in the Yoshida sarcoma cells based on the homoplastic and heteroplastic transplantations. A preliminary report. Gann, 42, 87-90.

6. Hauschka, T.S. and Levan, A. (1951) Characterization of five ascites tumors with respect to chromosome ploidy. Anat. Rec., 111, 467.

7. Bayreuther, K. (1952) Der Chromosomenbestand des Ehrlich-Ascites-Tumors der Maus. Z. Naturforsch., 7b, 554-557.

8. Steenis, H. van (1966) Chromosomes and cancer. Nature, 209, 819-821.

9. Levan, A. (1966) Non-random representation of chromosome types in human tumor stemlines. Hereditas, 55, 28-38.

10. Nowell, P.C. and Hungerford, D.A. (1960). A minute chromosome in human CML. Science, 132, 1497.

11. Manolov, G. and Manolova, Y. (1971) A marker band in one chromosome No. 14 in Burkitt lymphoma. Hereditas, 69, 300.

12. Mitelman, F., Mark, J., Levan, G. and Levan, A. (1972) Tumor etiology and chromosome pattern. Science, 176, 1340-1341.

13. Kurita, Y., Sugiyama, T. and Nishizuka, Y. (1968) Cytogenetic studies on rat leukemia induced by pulse doses of 7,12-dimethylbenz(α)anthracene. Cancer Res., 28, 1738-1752.

14. Levan, G., Ahlström, U. and Mitelman, F. (1974) The specificity of chromosome A2 involvement in DMBA-induced rat sarcomas. Hereditas, 77, 263-280.

15. Ahlström, U. (1974) Chromosomes of primary carcinomas induced by 7,12-dimethylbenz(α)anthracene in the rat. Hereditas, 78, 235-244.

16. Mitelman, F. (1971) The chromosomes of fifty primary rat sarcomas. Hereditas, 69, 155-186.

17. Spriggs, A.I., Boddington, M.M. and Clarke, C.M. (1962) Chromosomes in human cancer cells. Brit. Med. J., (2), 1431-1435.

18. Mark, J. (1967) Double-minutes - a chromosomal aberration in Rous sarcomas in mice. Hereditas, 57, 1-22.

19. Levan, G., Mandahl, N., Bregula, U., Klein, G. and Levan, A. (1976) Double minute chromosomes are not centromeric regions of the host chromosomes. Hereditas, 83, 83-90.

20. Levan, A., Levan, G. and Mitelman, F. (1977) Chromosomes and cancer. Hereditas, 86, 15-30.

21. Levan, G., Mandahl, N., Bengtsson, B.O. and Levan, A. (1977) Experimental elimination and recovery of double minute chromosomes in malignant cell populations. Hereditas, 86, 75-90.

22. Levan, A., Levan, G. and Mandahl, N. (1977) A new chromosome type replacing the double minutes in a mouse tumor. Cytogenet. Cell Genet., in press.

Chromosomes Today Volume 6, A. de la Chapelle and M. Sorsa eds.

A POSSIBLE ROLE FOR NONRANDOM CHROMOSOMAL CHANGES IN HUMAN HEMATOLOGIC MALIGNANCIES

JANET D. ROWLEY
Department of Medicine and The Franklin McLean Memorial Research Institute, The University of Chicago, 950 East 59th Street, Chicago, Illinois 60637

ABSTRACT

The fact that nonrandom chromosome changes are found in malignant cells suggests that these changes provide the cells with a proliferative advantage over those with a normal karyotype. The nature of the chromosome changes has been most clearly defined for the myelogenous leukemias, both acute and chronic, and for meningioma; supporting data are accumulating, however, for the lymphoproliferative disorders and for other solid tumors as well. Since the nonrandom changes occur in different chromosomes, it seems reasonable to assume that several gene loci are involved, and that alteration of any one of these loci is sufficient to provide the mutant cells with a proliferative advantage. These cells are not necessarily malignant; furthermore, they may not be the predominant clone. Other evidence supports the notion that the mutant cells arise from a single cell and therefore have a clonal origin.

The most common karyotypic abnormalities in leukemia are translocations, which are presumably reciprocal. In chronic myelogenous leukemia (CML) in the chronic phase, a translocation between Nos. 9 and 22, t(9;22), is typical; a t(8;21) is seen in acute myelogenous leukemia, and a t(15;17) in acute promyelocytic leukemia. In the acute phase of CML, common abnormalities include an isochromosome for the long arm of No. 17 [i(17q)] and other rearrangements, usually involving various chromosomes. Gains of a No. 8 or of a No. 19, or a double Ph^1, are other frequent changes. In ANLL, a gain of a No. 8 or loss of a No. 7 and gain or loss of No. 21 often occur.

We have no direct evidence regarding the genes involved in these aberrations, but many of the affected chromosomes carry genes related to nucleic acid biosynthesis. Many of the aberrations change the position of or alter (usually increase) the number of copies of these genes. The critical effect of this change in gene action within the mutant cell may be to keep it in the mitotic cycle and to prevent the cell from entering the resting stage. This would explain how a single cell with the essential karyotypic change could give rise to the predominant malignant clone.

INTRODUCTION

The role of chromosomal changes in malignant cells has been discussed for more than 60 years.[1] Whereas we still have no answer to this critical question, we are gradually gaining more insight, which may provide clues to the answer within the next few years. The existence of nonrandom chromosomal changes in chronic myelogenous leukemia, some types of acute leukemia, Burkitt lymphoma, and in meningiomas is beyond dispute. The challenging questions, at present, are (1) *how* and (2) *why* nonrandom changes, particularly consistent translocations, occur.

In this paper, I will present a summary of the nonrandom changes in hematologic malignancies and will then discuss some of the possible answers to the questions, *how* and *why*. Since most of our information on chromosomal changes comes from a study of patients with leukemia, it provides much of the evidence for nonrandom patterns. Since a detailed review of both acute leukemia[2] and chronic myelogenous leukemia[3] has been completed recently, only a brief summary will be presented here.

CHRONIC MYELOGENOUS LEUKEMIA (CML)

Chronic phase

Approximately 85% of patients with CML have the Philadelphia (Ph^1) chromosome in their bone marrow cells.[4] This abnormality, first described by Nowell and Hungerford in 1960,[5] consists of a "deletion" of part of the long arm of a G group chromosome. In 1970, the Ph^1 was identified by banding as a 22q- chromosome,[6,7] and it was shown, in 1973, to result from a previously unidentified translocation involving No. 9.[8] Cells from 569 patients with Ph^1+ CML have been examined with banding; 529 (93%) had the usual 9;22 translocation [t(9;22)(q34;q11)].[3] Of the remaining 40 patients 38 had a variant translocation, and two appeared to lack the distal part of No. 22 and thus to have no translocation.[9,10] Of the patients with variant rearrangements, in 17 22q was translocated to some chromosome other than 9 (e.g., 2, 3, 5, 6, 10, 11, 13, 17, 19, and 21), whereas 21 had a complex, three-way translocation.[3] Except for one patient with a 21;22;22 translocation,[11] all the complex rearrangements involved Nos. 9 and 22 and a third chromosome, with the break points in Nos. 9 and 22 occurring in the usual bands. There appears to be no difference in survival between patients with the usual and those with a variant translocation.[12]

Acute phase

Approximately 135 patients in the acute phase of Ph^1+ CML have had a karyotypic analysis, with banding, of their leukemic cells.[3] Thirty-five showed no change in karyotype; their cells had 46 chromosomes including the Ph^1 chromosome, as in the chronic phase. The other 100 patients had additional changes superimposed on the Ph^1+ cell line. Although the majority of the latter patients had 46-48 chromosomes, the modal chromosome number ranged from 44 to over 60. Since evolution of the karyotype occurs quite rapidly, it may be difficult or impossible to identify the initial change in many of these patients. The most common single changes were an isochromosome for the long arm of No. 17, i(17q), and various structural rearrangements, usually reciprocal translocations; each occurred in 12 patients. A second Ph^1 chromosome was noted as the only change in 10 patients, and an extra No. 8 was noted in 5. Most of the changes occur in combination; thus, many patients had a double Ph^1, as well as +8, i(17q), and/or +19. No patient had an additional No. 19 as a single initial change. The most frequent aberrations seen in all 93 patients are summarized in Table 1.

TABLE 1

SUMMARY OF THE MOST COMMON CHROMOSOME CHANGES IN 93 PATIENTS WITH CHRONIC MYELOGENOUS LEUKEMIA IN THE ACUTE PHASE[a]

	Chromosome number				
Chromosome aberration	8	9	17	19	Ph^1
Gain	35	9	6	18	38
Rearrangement	3	2	30[b]		4[c]

[a]Abstracted from Table 2, ref. 3.
[b]Includes 24 patients with an i(17q).
[c]Each was a dicentric Ph^1 chromosome.

ACUTE NONLYMPHOCYTIC LEUKEMIA (ANLL)

Very little information is available regarding the chromosomal pattern in acute lymphocytic leukemia; therefore this section includes the data available for ANLL only. Cells from approximately 210 patients with ANLL have been analyzed with banding[2]; 108 patients (51%) had a chromosomal abnormality which was identified precisely in 105. The chromosomal gains, losses, and rearrangements are summarized in Figure 1. There is evidence that some portion of the apparent chromosomal variability is related to evolution of the karyotype in ANLL. In an attempt to distinguish primary from secondary events, the aberrations noted in 85 patients who had minimal changes, i.e., modal chromosome numbers of 45-47, are indicated in the shaded area of the figure. Although a gain of No. 8 and a loss of No. 7 are the most frequent changes in either case, other aberrations, such as a gain of No. 1 or No. 19, are seen only in patients with higher modal numbers.

In some patients, it is possible to follow the development of other chromosomal changes in the course of serial analyses of bone marrow samples. In our first series of 50 patients with ANLL,[13] 8 showed a change in their karyotype as the disease progressed. In six patients this involved the gain of a chromosome, which was a No. 8 in 5 cases. Thus, an additional No. 8 is a common occurrence both in the evolution of ANLL and in CML in the acute phase.

Two structural rearrangements are sufficiently important to merit special mention. The first occurs in acute myeloblastic leukemia (AML) and is seen in about 10-15% of all patients with aneuploidy. Prior to banding, it was described as -C,+G,+E,-G[14]; Rowley showed that this was a translocation, presumably reciprocal, involving Nos. 8 and 21, t(8;21)(q22;q22).[15] This translocation is unique in that

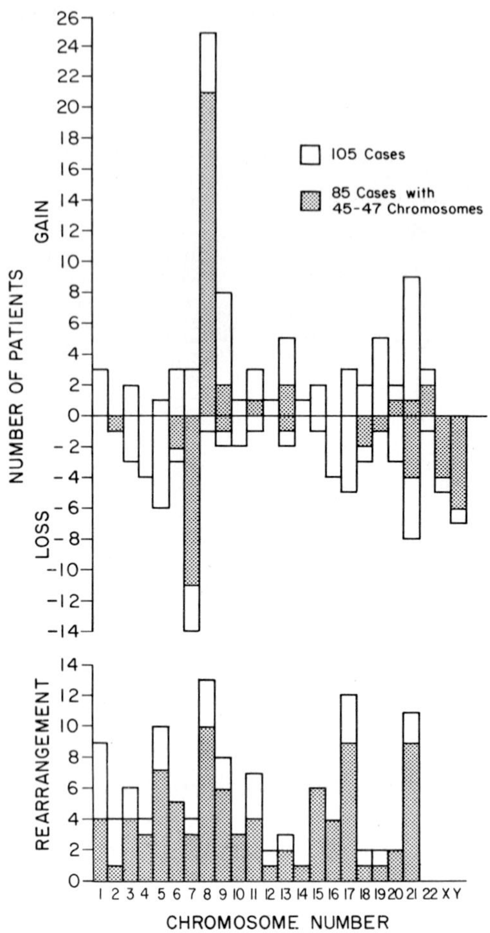

Fig. 1. Diagram of chromosomal changes seen in 105 patients with ANLL. The changes in 85 patients with modal chromosomal numbers of 45-47 are indicated in the dark portion.

its presence is frequently associated with the loss of a sex chromosome, an X in females (33%) and the Y in males (59%), which is otherwise a rare occurrence. This translocation is also of interest because two patients with variant translocations have been identified. In each patient, the translocation involved three chromosomes; Nos. 8 and 21 with breaks in the usual band were two of them.[16] Thus, variant 8;21 translocations follow the same pattern as the three-way variant 9;22 translocations in CML. This provides further evidence that these translocations are under quite precise, although presently unknown controls.

The other consistent translocation has only recently been identified as a 15;17 translocation, t(15;17)(q22;q21), in acute promyelocytic leukemia (APL).[17] Our first two patients with APL were noted to have a deleted 17q; metaphase chromosomes from a third patient had clearer bands, and a structural rearrangement involving No. 15 as well as No. 17 was noted. Giemsa banding in cells from a fourth patient, whose marrow morphology was intermediate between those for AML and APL, showed that the rearrangement was a reciprocal translocation. We have studied one other patient with APL after Hodgkin's disease, who did not have the disseminated intravascular coagulation that is an almost constant feature of APL. The 15;17 translocation was not present, although she had other karyotypic aberrations. Two patients with APL reported by other investigators[18,19] had the same translocation, supporting our suggestion that this is a consistent abnormality associated with a specific, rare type of acute leukemia.

OTHER MALIGNANCIES

Chromosomal abnormalities in other malignancies will be presented in detail by other participants. Two points should be made here because they are pertinent to my discussion: (1) A nonrandom pattern is quite evident in solid tumors with a minimal karyotypic aberration.[20] Thus, in meningioma, the cells from the majority of tumors have 45 chromosomes, lacking one No. 22.[21,22] (2) The majority of Burkitt lymphomas (BL), both African and American, show a consistent translocation, t(8;14)(q24;q32). This translocation has been observed in each of 8 scorable primary African Burkitt lymphomas reported by Zech et al.,[23] 2 others were 14q+, but the quality of the chromosomes was too poor to permit scoring for the 8q, and in each of 5 cell lines (2 established from North American and 3 from African Burkitt lymphomas) reported by McCaw et al.,[24] as well as in a few other lymphoid malignancies.

Thus, in summary, there are nonrandom changes involving gains and losses of whole chromosomes, and also of specific regions of chromosomes. In addition, up to the present time, we have identified four relatively consistent translocations, the 9;22 in CML, the 8;21 in AML, the 15;17 in APL, and the 8;14 in BL.

ABNORMALITIES OF CHROMOSOME NO. 1

Attention has been focused thus far on chromosomal aberrations in specific diseases. It is also profitable to analyze the types of abnormalities that involve a particular chromosome in a variety of hematologic disorders. Chromosome No. 1 is especially prone to rearrangements and therefore provides an excellent example for this type of analysis. Among the patients with acute leukemia, polycythemia vera, and myelosclerosis with myeloid metaplasia whose cells were studied in my laboratory, 10 had an abnormality that led to a gain (9 patients) or loss (1 patient) of part or all of one No. 1.[25] I know of reports of 26 additional patients with similar hematologic disorders who had a gain (25 patients) or a loss (1 patient) of part or all of one No. 1. The clinical and karyotypic data, with authors, are tabulated in ref. 25.

The two patients with deletions were missing 1q32 to qter and 1p32 to pter, respectively. By analysis of the data from the 34 patients who were trisomic for all or part of No. 1, it was determined whether (1) there were any trisomic regions that were common to all patients, (2) there were fragile sites on the chromosome, and whether (3) these fragile sites were similar to those seen in myeloid cells from patients with hematologic disorders who had other structural rearrangements of No. 1. Five of the patients were trisomic for all of No. 1, whereas 29 patients were trisomic for part or all of the long arm; six of these were trisomic for a portion of the short arm as well. Despite the fact that these 34 patients had various hematologic disorders, *every one was trisomic for the region 1q25 to 1q32.*[25] In addition, trisomy for 1q32 to qter or for 1q25 to 1q21 was present in 28 and 30 patients, respectively.

The break points leading to the rearrangements observed in the 29 patients with partial trisomy were clustered in the long arm, particularly 1q21, 1q32, and 1q12, the centromere, and in the short arm 1p22. This clustering does not merely reflect particularly fragile sites, as can be seen from analysis of break points in 13 patients with balanced reciprocal translocations involving No. 1. One break occurred in 1q12 and one at the centromere, but all of the other 11 break sites were located in the short arm, and 6 of these were in 1p36, which was never affected in the patients with partial trisomy. Thus it is possible to show that there is a specific region of chromosome No. 1 that is invariably present in the trisomic state in patients who have an excess of any part of No. 1. Furthermore, the break points involved in rearrangements leading to trisomy are quite different from those leading to balanced reciprocal translocations.

HOW CONSISTENT TRANSLOCATIONS MAY BE PRODUCED

The mechanism for the production of specific, consistent reciprocal translocations is unknown. Possibly, specific translocations are the result of cell selection. This is the simplest hypothesis, and it may prove to be correct. In such a model, chromosome breaks and rearrangements occur continuously at a low frequency. Many of these rearrangements lead to no change in cell metabolism, and the cells therefore do not proliferate preferentially; other rearrangements may be lethal to the cells. Still others provide the cell with a proliferative advantage, and cells with these changes not only persist, but eventually become the predominant cell type. In such a model, the chromosome change is the fundamental, initial event that leads to the neoplastic nature of the cell.

Other possible explanations depend on either (1) chromosomal proximity since translocations may occur more frequently when two chromosomes are close together, or (2) regions of homologous DNA that might pair preferentially and then be involved in rearrangements.

The fact that many of the affected chromosomes, e.g., Nos. 2, 14, 15, 21, and 22, are involved in nucleolar organization supports these proposals. The centromere region of No. 1 also has an association with the nucleolus. All partial trisomies which result from a break in the centromere of No. 1 involve translocations of 1q to the nucleolar organizing region of other chromosomes, specifically Nos. 9, 13, 15, and 22.[25]

On the other hand, proximity or homologous DNA sequences should lead to an increased frequency of these rearrangements in patients with constitutional abnormalities, but this has not been observed. It is possible that either or both of these mechanisms are subject to selection; a translocation might occur because the chromosomes are close together, but only certain specific rearrangements might have a proliferative advantage which results in leukemia and thus allows them to be detected.

One other possible mechanism which I find quite intriguing concerns transposable genetic elements that can cause large-scale rearrangements of adjacent DNA sequences. These consist of controlling elements which have been found in Maize[26,27] and in Drosophila,[28] and of insertion sequences in bacteria.[29] Not only do these elements exert control over adjacent sequences, but the type of control, that is, an increase or a decrease in gene product, is related to their position and orientation in the gene locus. Whereas they can cause nonrandom chromosomal deletions adjacent to themselves, these controlling elements can also move to another chromosomal location, and they may transpose some of the adjacent chromosomal material with them.

There is no direct evidence for the existence of controlling elements in mammalian cells, although the data of Hozumi and Tonegawa[30] showing a different pattern of DNA hybridization in the genomes of mouse embryo cells and a mouse plasmacytoma have been interpreted by Nevers and Saedler[31] as compatible with the presence of such elements. The properties of controlling elements are (1) change in location within the DNA, (2) the transferring of adjacent DNA in this change, and (3) the alteration of the normal mechanism for genetic regulation, depending on site and orientation of the inserted sequences. These properties, plus a selective system to remove changes that do not have a proliferative advantage in hematologic cells, are just those required to explain consistent translocations occurring as somatic mutations.

WHY NONRANDOM CHANGES

Having revealed our complete ignorance as to how consistent translocations occur, we now find the same perplexing situation regarding the reasons for the occurrence of nonrandom changes.

There is good cytological[32] and biochemical[33] evidence that, in an individual patient with chronic myelogenous leukemia or Burkitt lymphoma, the tumor cells have a clonal origin. For example, in the case of CML, initially a single cell has the 9;22 translocation, and when the patient comes to the physician, frequently all cells in division contain the Ph^1 chromosome. Thus the cell with the 9;22 translocation has a proliferative advantage over chromosomally normal cells. The question to be examined now relates to the kinds of gene loci that can provide this proliferative advantage.

First, there are two points that should be emphasized; one concerns the genetic heterogeneity of the human population, and the second, the variety of cells involved in malignancy. There is convincing evidence from animal experiments that the genetic constitution of an inbred strain of rats or mice plays a critical role in the frequency and type of malignancies that develop. Thus, whether 50-day-old female rats given 7,8,12 trimethylbenz(a)anthracene develop breast cancer or leukemia depends in large measure on whether they are of the Sprague-Dawley[34] or Long-Evans[35] strain. Some of the factors controlling the susceptibility of mice to leukemia not only have been identified, but also have been mapped to particular chromosomes, and their behavior as typical Mendelian genes has been demonstrated.[36]

We are much more aware than we used to be of certain genetic traits in man that predispose to cancer, such as Bloom syndrome, Fanconi anemia, and ataxiatelangiectasia.[37] We recognize the existence of cancer-prone families[38]; of inherited genetic susceptibility to specific types of malignancy, such as retinoblastoma[39] and breast cancer; or the inheritance of lesions which have a high propensity for becoming malignant, such as familial colonic polyps. How many gene loci are there in man which, in some way, control resistance or susceptibility to a particular malignancy? We have no way of knowing at present. Surely these genes exert an important influence on the types of chromosomal changes that may be present in the malignant cell.

The second factor affecting the karyotypic pattern relates to the different cells that are at risk of becoming malignant, and the varying states of maturation of these cells. The catalog of the nonrandom changes in various tumors maintained by Drs. Mitelman and Levan[20] provides clear evidence that the same chromosomes may be affected in a variety of tumors; No. 8 is a good example. On the other hand, some chromosomes seem to be involved in neoplasia involving a particular tissue; the involvement of No. 14 in lymphoid malignancies, and of No. 20 in red-cell abnormalities, might be suitable examples. Given the great genetic diversity, the number of different cell types that might become malignant, and the variety of carcinogens to which these cells are exposed, it is surprising that nonrandom karyotypic changes can be detected at all.

When one considers the number of nonrandom changes that are seen in a single malignancy such as ANLL, it is clear that not just one gene, but rather a class of genes is involved. Our knowledge of the human gene map has been developing concurrently with our understanding of chromosomal changes in leukemia.[40] It is now possible to try to correlate the chromosomes that are affected with the genes that they carry. Clearly, these efforts must be very preliminary since relatively few genes have been mapped, and since some of the chromosomes that are most frequently abnormal have few genetic markers. In regard to No. 8, for example, the only gene mapped is glutathione reductase,[41] which is located on the short arm, and from several patients whom I have studied who had a structural rearrangement affecting No. 8, I believe that it is the long arm of No. 8 that provides cells with a proliferative advantage. Nonetheless, we must look at the chromosome-gene correlations if we are to gain further insight into the question "why nonrandom changes."

I undertook such an analysis because of the observations of McDougall,[42] who noted that the sites of adenovirus-12 (AD-12) that induced breaks on chromosome No. 1 were correlated with genes related to nucleic acid metabolism. The current gene map contains the location of about 75 enzymes whose functions can be classified; 42 of them are related to carbohydrate or intermediary metabolism, and 16 to nucleic acid metabolism.[40] The other enzymes are either related to general functions or to lipid or protein metabolism, for which only a few are mapped. The 42 related to carbohydrate or intermediary metabolism are located on 18 chromosomes, abnormalities in 9 of which are involved in hematologic malignancies.[43] The 16 enzymes related to nucleic acid metabolism are located on 10 chromosomes, 7 of which are involved in hematologic malignancies; these 7 chromosomes carry 12 of the 16 enzymes.[43] One of the 3 chromosomes which I considered to be uninvolved is the X, which carries hypoxanthine-guanine phosphoribosyl transferase; an abnormality of this chromosome has been noted, however, in selected patients with AML. A summary of the chromosomal abnormalities and the genes related to nucleic acid metabolism is given in ref. 43.

Not only chromosomes that carry genes related to nucleic acid biosynthesis, but also the specific chromosomal regions associated with these genes may frequently be involved in rearrangements associated with hematologic malignancies. Thus, the most frequent abnormalities of No. 17 result either in an isochromosome for the long arm, or in a translocation with No. 15 in which the break in No. 17 is in band 17q21. This region of No. 17 contains genes for thymidine kinase, galactokinase, and a site particularly vulnerable to AD-12-induced breakage.[44] Furthermore, induction of host cell thymidine kinase and a high frequency of breaks in 17q21 are early functions of the virus, as is the synthesis of a tumor antigen which may have a role in control of DNA synthesis.

Thus it is possible that nonrandom chromosomal aberrations, when they occur, change the level of some enzymes related to nucleic acid metabolism, either through a change in location or through duplication of gene loci. Nonrandom chromosomal changes, particularly consistent, specific translocations, now seem clearly to be an important component in the proliferative advantage of the mutant cell in neoplasia. The challenge is to decipher the meaning of these changes.

SUMMARY

The consistent occurrence of nonrandom chromosomal changes in human malignancies suggests that they are not trivial epiphenomena. Whereas we do not understand their significance at present, one possible role they may fulfill is to provide the chromosomally aberrant cells with a proliferative advantage as the result of alteration of the number and/or location of genes related to nucleic acid biosynthesis. It would be expected that, the proliferative advantage provided by varying chromosomal aberrations differs in patients with different genetic constitutions.

ACKNOWLEDGEMENTS

Supported by the National Foundation-March of Dimes, the National Institutes of Health (CA 16910), the Leukemia Research Foundation, and an Otho S. A. Sprague institutional grant. The Franklin McLean Memorial Research Institute is operated by The University of Chicago for the United States Energy Research and Development Administration under contract EV-76-C-02-0069.

REFERENCES

1. Boveri, T. (1912) Beitr. Pathol., 14, 249-260.
2. Rowley, J.D. (in press) in Clinics in Haematology.
3. Rowley, J.D. (in press) in Seminars in Hematology.
4. Whang-Peng, J., Canellos, G.P. and Carbone, P.P. (1968) Blood, 32, 755-766.
5. Nowell, P.C. and Hungerford, D.A. (1960) Science, 132, 1197.
6. Caspersson, T., Gahrton, G., Lindsten, J., et al. (1970) Exp. Cell Res., 63, 238-244.
7. O'Riordan, M.L., Robinson, J.A., Buckton, K.E., et al. (1971) Nature (Lond.), 230, 167-168.
8. Rowley, J.D. (1973) Nature (Lond.), 243, 290-293.
9. Mitelman, F. (1974) Hereditas, 76, 315-316.
10. Sonta, S., Oshimura, M. and Sandberg, A.A. (1976) Blood, 48, 697-705.
11. Ishihara, T., Kohno, S.-I. and Kumatori, T. (1974) Brit. J. Cancer, 29, 340-342.
12. Sonta, S. and Sandberg, A.A. (in press) Blood.
13. Rowley, J.D. and Potter, D. (1976) Blood, 47, 705-721.
14. Trujillo, J.M., Cork, A., Hart, J.S., et al. (1974) Cancer, 33, 824-834.
15. Rowley, J.D. (1973) Ann. Genet., 16, 109-112.
16. Lindgren, V. and Rowley, J.D. (1977) Nature, 266, 744-745.
17. Rowley, J.D., Golomb, H.M. and Dougherty C. (1977) Lancet, I, 549.
18. Okada, M., Miyazaki, T. and Kumota, K. (1977) Lancet, I, 961.
19. Kaneko, Y. and Sakurai, M. (1977) Lancet, I, 961.
20. Mitelman, F. and Levan, G. (1976) Hereditas, 82, 167-174.
21. Zankl, H. and Zang, K.D. (1972) Humangenetik, 14, 167-169.
22. Mark, J. (1977) in Advances in Cancer Research, Vol. 24, Academic Press, New York, pp. 165-222.
23. Zech, L., Haglund, V., Nilsson, K., et al. (1976) Int. J. Cancer 17, 47-56.
24. McCaw, B.K., Epstein, A.K., Kaplan, H.S., et al. (1977) Int. J. Cancer 19, 482-486.
25. Rowley, J.D. (in press) in Molecular Human Cytogenetic, ICN-UCLA Symposia on Molecular and Cellular Biology, Vol. III, Sparkes, R.S., Comings, D. and Fox, C.F., eds., Academic Press, New York.
26. McClintock, B. (1961) Am. Naturalist, 95, 265-277.

27. Fincham, J.R.S. and Sastry, G.R.P. (1974) Ann. Rev. Genet., 8, 15-50.
28. Green, M.M. (1973) Genetics, 73, 187-194.
29. Cohen, S.N. (1976) Nature, 263, 731-738.
30. Hozumi, N. and Tonegawa, S. (1976) Proc. Nat. Acad. Sci. USA 70, 3628-3632.
31. Nevers, P. and Saedler, H. (1977) Nature, 268, 109-115.
32. Gahrton, G., Lindsten, J. and Zech, L. (1974) Blood, 48, 837-840.
33. Fialkow, P.J. (1974) New Engl. J. Med., 291, 26-35.
34. Huggins, C., Grand, L.C. and Brillantes, F.P. (1961) Nature, 189, 204.
35. Huggins, C., Grand, L. and Oka, H. (1970) J. Exper. Med., 131, 321-330.
36. Rowe, W.P. (1973) Cancer Res., 33, 3061-3068.
37. German, J. (1972) Prog. Med. Genet., 8, 61-101.
38. Mulvihill, J.J., Miller, R.W. and Fraumeni, Jr., J.F., eds. (1977) Progress in Cancer Research and Therapy, Vol. 3, Genetics of Human Cancer, Raven Press, New York.
39. Knudsen, A.G. (1973) Advan. Cancer Res., 17, 317-352.
40. McKusick, V.A. and Ruddle, F.H. (1977) Science, 196, 390-405.
41. De La Chapelle, A., Vuopio, P. and Icen, A. (1976) Blood, 47, 815-826.
42. McDougall, J.K. (1971) J. Gen. Virol., 12, 43-51.
43. Rowley, J.D. (in press) Proc. Nat. Acad. Sci. USA.
44. McDougall, J.K., Kucherlapati, R.S. and Ruddle, F.H. (1973) Nature (New Biol.), 245, 172-175.

Chromosomes Today Volume 6, A. de la Chapelle and M. Sorsa eds.

CHROMOSOMES AND GENES IN HUMAN CANCER CELLS: MULTIDISCIPLINARY APPROACHES TO A UNITARY GENODEMOGRAPHIC HYPOTHESIS

FREDERICK HECHT AND BARBARA KAISER-McCAW
Department of Pediatrics, University of Oregon Health Sciences Center, Portland, Oregon 97201 (USA)

ABSTRACT

The usual sequence of chromosome changes in human cancer cells appears to be (1) chromosome breakage, (2) "balanced" translocation (primary clone), (3) addition or subtraction of chromosomes, one at a time, to produce imbalance (secondary clone) and (4) further changes producing more imbalance (tertiary clone).

Chromosome regions in cancer cells can be classified as (a) restricted (no change allowed), (b) semi-restricted ("balanced" change or addition permitted) and (c) unrestricted (anything goes). These classes may correspond to the density of key genes and inversely to the density of repetitious DNA. This genodemographic hypothesis can be tested as more is learned about the locations of key genes and repetitive DNA in Man.

INTRODUCTION

Characteristic chromosome changes occur in a number of types of human cancer. This brief paper is designed to classify the usual sequence of chromosome changes in cancer and to propose an explanation: the genodemographic hypothesis. This hypothesis is testable, with time and multidisciplinary data.

SEQUENCE OF CHROMOSOME CHANGES IN CANCER CELLS

Chronic myeloid leukemia (CML) is the best studied malignancy as regards chromosomes. Most patients with CML have the Philadelphia translocation (9q+; 22q-); this is a "balanced" translocation in that apparently no net change in total material has occurred. In the blastic phase of CML the most common changes occurring next are the addition of another 22q- chromosome, producing imbalance, and then the addition of a chromosome 8, making for more imbalance. A similar sequence of changes occur in acute myeloid leukemia.

Burkitt's lymphoma is the best studied human lymphoid cancer. In Burkitt's lymphoma there is a "balanced" translocation between chromosomes 8 and 14 (8q-; 14q+). Then often there is loss of the 14q- to produce imbalance. And then additional changes occur.

From these and similar studies of other cancers, the following sequence of chromosome changes is proposed (Table 1).

TABLE 1

SEQUENCES OF CHROMOSOME CHANGES IN HUMAN CANCER

Step	Change	Balance	Class of clone
1	Original cell	+	Original
2	Balanced translocation	+	Primary
3	Addition or subtraction	-	Secondary
4	Further changes	-	Tertiary

ARE "BALANCED" TRANSLOCATIONS BALANCED?

The very specific patterns of chromosome change in tumor cells may reflect hot spots for breakage (for which there is no evidence) or nonrandom reunion (for which there is no evidence either).

A more tenable explanation is cell selection. According to this concept, certain chromosome changes confer an advantage on the cell.

POSITION EFFECT

The mechanism by which an advantage is conferred is unknown. Material may be lost (no evidence) or gained (no evidence). Intragenic breaks may occur (no evidence).

The most likely explanation appears to be position effect: a change in gene activity due to repositioning of chromosome material.

We have preliminary evidence for this using acridine orange to stain the 8q-; 14q+ translocation in Burkitt's lymphoma. However, such evidence is purely histochemical and not on the level of gene action.

CLASSES OF CHROMOSOME REGIONS IN CANCER CELLS

There appear to be three classes of chromosome regions in cancer cells (Table 2). One class consists of regions where no change is permitted: restricted regions. A second class comprises regions where "balanced" translocations or, less often, addition of material is permitted: semi-restricted regions. The third and last class is where anything goes: unrestricted regions.

TABLE 2

CLASSES OF CHROMOSOME REGIONS IN CANCER CELLS

Type of region	Change permitted	Example
Restricted	None	6p
Semi-restricted	"Balanced" translocation or addition	14q
Unrestricted	Anything	8p and q

THE GENODEMOGRAPHIC HYPOTHESIS

Up to this point one is on relatively firm ground. Past this point one can but speculate.

We wish to speculate that the sequence of chromosome changes and the classes of chromosome regions in cancer cells reflect genodemography, that is the demography of genes in Man. More precisely, it may be that genes of importance to somatic cells are not evenly distributed in the human genome. The density of key genes may vary. For example, the histocompatability loci and a large number of related immune loci are known to be tightly linked on 6p. Other areas such as on chromosome 8 appear semi-arid with very few genes.

The predictions from the genodemographic hypothesis are presented in Table 3.

TABLE 3

PREDICTIONS FROM GENODEMOGRAPHIC HYPOTHESIS

Type of region	Key gene density	Reiterative DNA
Restricted	High	Low
Semi-restricted	Medium	Medium
Unrestricted	Low	High

This hypothesis is now clearly amenable to preliminary testing and later to more definitive testing.

PRELIMINARY TEST OF GENODEMOGRAPHIC HYPOTHESIS

The relative length of each autosome in Man is known. Data in this regard were presented for banded chromosomes by three laboratories at the Paris Conference (1971): Standardization in Human Cytogenetics. We have taken the mean of the means presented there by the three laboratories. For example, chromosome 1 measures 8.9% of the total haploid length compared to 5.0% for chromosome 8 and 1.7% for chromosome 21.

The human gene map is still in an embryonic stage. Nonetheless, one can for purposes of example use the data summarized by V.A. McKusick for the International Workshop on Human Gene Mapping held in Winnipeg in August, 1977 (to be published by the National Foundation in its Original Article Series and concurrently by the Journal _Cytogenetics and Cell Genetics_).

The derivation of key gene density is as follows: number of genes mapped = A; relative length of chromosome (or region) = B; key gene density = A/B. Examples are given in Table 4.

TABLE 4

GENE DENSITIES

Chromosome number (no.)	1	8
No. genes mapped (A)	26	2
Relative length (B)	8.9	5.0
Gene density (A/B)	2.9	0.4

Note that chromosome 1, the first autosome to be mapped, has now about seven times the density of genes as chromosome 8. Further, chromosome 1 has, among its genes, the locus for Rh; chromosome 8 has no such distinguished gene locus yet.

Chromosome 1 is rarely involved in human neoplastic cell rearrangements, whereas chromosome 8 can be rearranged, added or lost with seeming impunity.

The gene densities for seven chromosomes are given, by way of example, in Table 5.

TABLE 5

GENE DENSITIES (SOME EXAMPLES)

Chromosome	Gene density (GD)	Mean GD
8, 14, 22	0.4, 0.6, 0.6	0.5
1, 12, 19	2.9, 2.2, 2.0	2.4
6	3.5	3.5

The correlation appears strikingly inverse between gene density and involvement in cancer cell rearrangements (and other chromosome changes). Chromosomes 8, 14 and 22 have low gene densities and are often involved in cancer cell changes. Chromosomes 1, 12 and 19 have moderate gene densities and are less often involved. Chromosome 6 has a high gene density and is sacrosanct (restricted) at least as regards 6p where the immune loci reside.

DISCUSSION: A SELF-CRITIQUE

Gene densities, as we have calculated them, are obviously subject to tremendous error, since the number of genes now mapped to a given chromosome may not reflect accurately the real gene density. Chromosome 1 may have many areas mapped on it now simply because historically it was the first autosome to be mapped. Chromosome 8 might prove to have as many genes someday.

Further, who is to say which genes are "key" genes?

And finally, it may not be gene density per se which matters, but rather tight linkage of key gene loci (such as the immune complex on 6p).

The better test of the genodemographic hypothesis may come from the location and extent of reiterative ("non-key") DNA. Biochemistry, cytology, genetics and other disciplines will allow this concept to be tested.

SUMMARY

Certain malignancies in Man are associated with certain chromosome changes. The usual sequence of changes appears to be: (1) chromosome breakage; (2) chromosome rearrangement to form a "balanced" translocation (primary clone); (3) addition or subtraction of chromosomes, often those in the rearrangement, to produce chromosome imbalance (secondary clone); and (4) further morphologic and numerical changes resulting in more imbalance (tertiary clone).

The patterns of chromosome change in malignant cells suggest a genodemographic hypothesis. Chromosome regions can be classified as: (a) Restricted regions where no change is permitted, (b) Semi-restricted regions where "balanced" changes or addition of material is permitted, and (c) Unrestricted regions where addition or subtraction of material is permitted. This classification presumably corresponds to the relative density of key genes: (a) restricted: dense clusters of key genes; relatively little redundancy, (b) semi-restricted: fewer key genes; and more redundancy, (c) unrestricted: few or no key genes; and much redundancy.

The genodemographic hypothesis will be subject to critical testing as more genes are precisely mapped to specific chromosome regions and as areas of redundancy are delineated.

The same hypothesis applies to constitutive chromosome changes.

ACKNOWLEDGEMENTS

The authors' research, including the present theoretical paper, is supported by grant CA16747 from the National Cancer Institute for work on the "Cytogenetics of Clonal Neoplasias" and by grant HD08236 from the National Institute of Child Health and Human Development for work on "Chromosome Structure, Function and Behavior".

The authors thank their colleagues Alan L. Epstein and Henry S. Kaplan (Stanford University) and Shiva Patil, Herman E. Wyandt, Kathleen Overton and Rosemary Milbeck (Oregon) for stimulating interactions.

Chromosomes Today Volume 6, A. de la Chapelle and M. Sorsa eds.

CHROMOSOMES AND THE ETIOLOGY OF CANCER

GÖRAN LEVAN and FELIX MITELMAN
Institute of Genetics and Department of Clinical Genetics, University of Lund,
S-223 62 LUND, Sweden

ABSTRACT

The correlation in experimental tumors between chromosome pattern and etiology has shown that only a few specific chromosomes are involved in changes during oncogenesis. Conditions may be similar in human tumors, but our ignorance of their etiology prevents direct conclusions. Indirect evidence may be expected from analyses of geographic and epidemiologic groups of neoplasms.

The discrepancy between the immediate chromosome lesions induced by a carcinogen and the breaks leading to marker formation suggests that the malignant transformation is often associated with a submicroscopic event, whereas the visible changes are secondary and cooperate to adjust the genic balance in favor of the transformed allele.

EXPERIMENTAL NEOPLASMS IN THE RAT

In a series of papers we have shown that experimental rat sarcomas exhibit distinctly nonrandom patterns of chromosome aberrations. Sarcomas induced by Rous sarcoma virus (RSV) often display trisomy for chromosome No. 7 and, subsequently, trisomies for No. 13 and No. 12[1-3], whereas histologically identical sarcomas induced by 7,12-dimethylbenz(α)anthracene (DMBA) most frequently display trisomy for No. 2[4,5]. Other chromosome types are involved only rarely and in a more random fashion. Thus, it appears in these cases that the etiologic factor plays an important role in determining the chromosome evolution of the tumor[6].

The finding that trisomy No. 2 is a common aberration in DMBA rat sarcomas became the more interesting in view of the results obtained by Kurita and coworkers[7,8] in rat leukemias induced by intravenous injections of the same chemical, DMBA. These workers have found that trisomy No. 2 is a characteristic feature of DMBA leukemias and that the aberration is seen in frequencies comparable to those we observed in the sarcomas (Table 1). Furthermore, Ahlström in our group found trisomy No. 2 in epithelial tumors obtained by painting ears of rats with DMBA[9]. Thus, in malignancies derived from quite different tissues, it again appears that the inducing agent is a deciding factor determining the chromosome pattern of the malignant cell population.

TABLE 1

CHROMOSOME ABERRATIONS IN RAT MALIGNANCIES

Average numerical deviation per 100 cells in 4 different induced rat malignancies

Neoplasm	Chromosome No. 2	7	12	13	16-20	Reference
RSV sarcoma		+29	+4	+15		1
DMBA sarcoma	+31				+18	4,5
DMBA leukemia	+28				+ 9	7,8
DMBA carcinoma	+24				+46	9

It is of special interest to note that DMBA has been shown to interact preferentially with chromosome No. 2 both in vivo and in vitro, as evidenced by selective increases in chromosome breakage and in sister chromatid exchange. Sugiyama and co-workers[10,11] have studied the effect in the bone marrow at various times after a single, intravenous injection of DMBA. At 6 hours post injection, the No. 2 chromosome contained 5 times more breaks per unit length than the remaining chromosomes and 2.5 times more than the pair among the latter that was broken most often. These workers also showed that the breaks induced by DMBA in chromosome No. 2 were distributed over the length of the chromosome in a distinctly nonrandom fashion, indicating that there are "hot spots" for the interaction between DMBA and the DNA of the No. 2 chromosome. Utilizing the sister chromatid exchange technique on rat bone marrow cells in vitro, the same workers have shown that DMBA induces an increase in nonrandom sister chromatid exchanges along the length of the No. 2 chromosome. The locations of the most frequent sister chromatid exchanges roughly coincide with the points most susceptible to chromosome breakage. Popescu and DiPaolo[12] have made similar experiments with embryonic rat fibroblasts in vitro and have obtained similar results.

It seems likely that there is a connection between the selective involvement of No. 2 in numerical and structural change in DMBA-induced malignancies, on the one hand, and the preferential chromosome damage in the No. 2 chromosome by acute DMBA treatment, on the other. There appears to be no immediate relationship, however, between the points vulnerable to breakage and the translocation points in various markers (Fig. 1), the latter points more often falling in areas not specifically vulnerable to breakage. This indicates that the primary effect of interaction between DMBA and the No. 2 chromosome is not to form the No. 2 derived markers sometimes seen in DMBA tumors. The selective involvement in breakage can rather be viewed as an indication of a preferential interaction between sites in the No. 2 chromosome and the carcinogen. This interaction can sometimes cause chromosome breakage or sister chromatid exchange, but more often results in submicroscopic

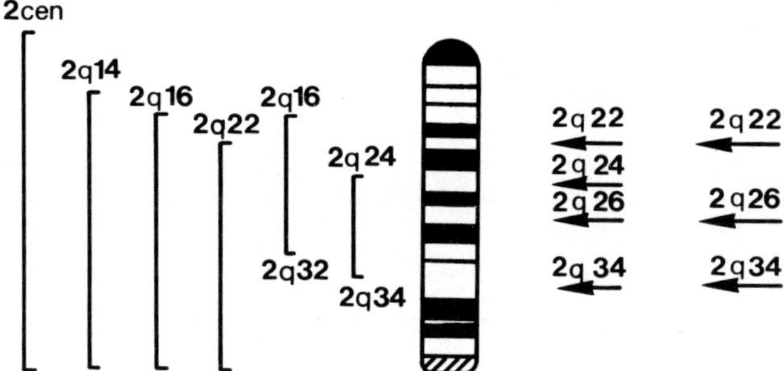

Translocation points in tumors induced with polycyclic hydrocarbons

Break points and points of SCE in cells treated with DMBA

Fig. 1. Comparison of translocation points in chromosome No. 2 derived markers of rat sarcomas and the points selectively vulnerable to breakage and sister chromatid exchange efter DMBA treatment. Left set of arrows: vulnerable points according to Popescu and DiPaolo[12]; right set of arrows: vulnerable points calculated from data by Ueda et al.[11]. The translocation points often fall in areas between the vulnerable points, indicating that the translocations are not formed at the time of tumor induction.

genetic change that may subsequently lead to malignant transformation. The presence in DMBA malignancies of translocations and trisomies involving the No. 2 chromosome may reflect a secondary step, viz. the amplification of the primary lesion to produce cells more malignant than those originally transformed.

Sugiyama and collaborators[13] have studied a large number of rat leukemias induced by two closely related agents, DMBA and 7,8,12-trimethylbenz(α)anthracene (TMBA). They find that both agents frequently induce leukemias involving trisomy No. 2. The frequency with which trisomy No. 2 leukemias occur is, however, distinctly different in the two inductions (Table 2). The fact that two related agents induce reproducible and significantly different rates of the same chromosomal aberration emphasizes the preciseness of the mechanism underlying these events and manifests strikingly the significance of the inducing agent.

TABLE 2

RAT LEUKEMIAS WITH TRISOMY No. 2

Frequency of animals displaying trisomy No. 2 in the stemline of leukemias induced by DMBA and TMBA[a]

Inducing agent	Per cent leukemias with No. 2 trisomy			No of leukemias analyzed
	Males	Females	Males+Females	
DMBA	29.6	28.9	29.4	197
TMBA	17.9	17.5	17.7	164

[a]Data from Sugiyama et al.[13]

HUMAN NEOPLASMS

Since several years we have been collecting, from our own experience and from the literature, karyotypic data from, as far as possible, all cases of human neoplasms analyzed by banding technique. These data have been utilized for an overall compilation of chromosome aberrations found[14,15]. For our latest survey of this

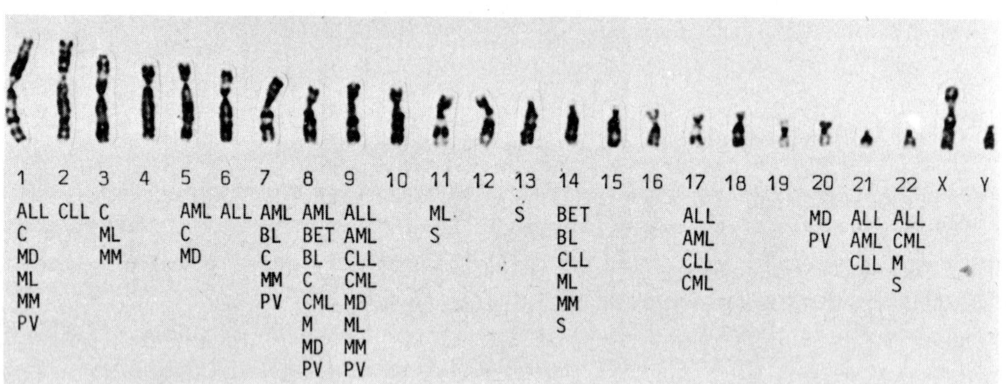

Fig. 2. Clustering of aberrations to specific chromosomes in human neoplasms. At the top, the 24 human chromosome types are represented; below 15 of them, one or more of 13 classes of neoplasms are given, viz. those with which each chromosome type has been most frequently involved in aberrations.

ALL: acute lymphocytic leukemia
AML: acute myeloid leukemia
BET: benign epithelial tumors
BL: Burkitt's lymphoma
C: carcinoma
CLL: chronic lymphocytic leukemia
CML: chronic myeloid leukemia
M: meningioma
MD: various myeloproliferative disorders
ML: malignant lymphoma
MM: multiple myeloma
PV: polycythemia vera
S: sarcoma

material[16] we are obliged to numerous colleagues who have cooperated by informing us of their unpublished cases. This survey contains data from 706 cases of human neoplasms exhibiting chromosome aberrations, not counting several hundred cases in which a t(9;22) is reported as the only aberration in Ph[1]-positive CML. The 706 cases are divided into 13 classes of neoplastic disease. In selecting about 5 chromosomes that are most commonly involved in aberrations in each of the 13 classes, it becomes obvious that they are distributed nonrandomly over the different human chromosome types (Fig. 2). The aberrations cluster to several specific chromosomes, mainly to Nos 1, 7, 8, 9, 14, 17, 21 and 22.

THE CHROMOSOMES IN MALIGNANT TRANSFORMATION - A HYPOTHESIS

The observation that but a few chromosomes are regularly affected in the development of human malignancies, agrees well with the findings in the experimental materials, and it appears reasonable to conclude that similar laws govern chromosome changes in the two systems. We now arrive at the key question: What is the role of chromosome aberrations in malignant transformation and tumor formation? Unfortunately, this question can still only be approached with more or less wild guesses.

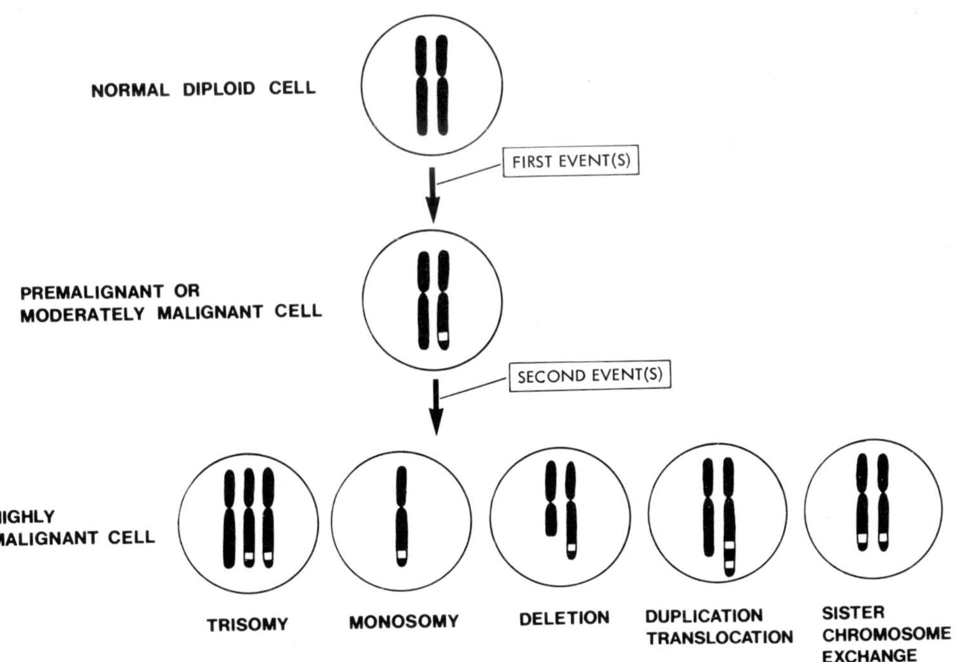

Fig. 3. A hypothetic model to explain the role of chromosome aberrations in neoplasms. For details, see text.

We have adopted a working hypothesis, the essence of which was first suggested by Östergren[17]. It is clearly vulnerable to criticism but may be useful as a basis for further experiments and discussion (Fig. 3). Most cancers originate in several steps. Clearly not all of these steps need to be on the level of genes or chromosomes, but the nature of the chromosomal steps might be that outlined in the model. The first event is assumed to be a direct interaction between a carcinogenic agent and the genetic material of a chromosome involved in transformation, the site in the chromosome of this material reacting specifically with the inducing agent. The first lesion might in itself not be enough to transform the cell completely, a premalignant or moderately malignant cell arising in the process. In order to increase in importance, the affected segment may need amplification relative to the normal segment still present in the unaffected homologue. One way to obtain this is by nondisjunction leading to trisomy with two copies of the affected chromosome. In such a trisomic cell there would be an increase in the relative amount of the aberrant gene product, possibly resulting in a more malignant cell population. Other processes to the same end are indicated in Fig. 3, such as loss of the normal allele by nondisjunction or deletion or by duplication of the affected allele by translocation or sister chromosome exchange. Each of these events would tend to increase the relative amount of the aberrant genetic material. Which specific process would be preferred in each individual case might depend on the genetic material of the affected chromosome as a whole. Clearly a cell cannot always tolerate trisomy or monosomy for any particular chromosome.

When the types of chromosome changes in human malignancies are examined, it is clear that different chromosomes are preferentially involved in different types of changes. In Table 3, the nature of the change in the chromosomes affected most com-

TABLE 3

CHROMOSOME ABERRATION IN AML

Nature of change in the chromosomes affected in at least 15% of the 186 cases recorded

Chromosome affected	Type of aberration (per cent)						Total No. of aberrations
	del	-1	t	+1	i	other	
No. 5	42	32	23	-	-	3	31
No. 7	33	56	-	6	-	6	36
No. 8	2	6	32	56	1	2	82
No. 9	9	3	34	49	-	6	35
No. 17	3	39	33	3	12	9	33
No. 21	-	26	43	26	-	5	58

monly in AML have been listed. Evidently, there is usually loss of material in chromosomes No. 5 and 7, whereas Nos. 8 and 9 are almost always trisomic or translocated. Chromosome No. 21 exhibits monosomy and trisomy at equal frequency. This may be in reflection of the small size of the No. 21; monosomy or trisomy for such a small chromosome might be equally tolerable to a cell, while with a larger chromosome one of the aberrations might have a distinct advantage before the other.

THE ETIOLOGIC FACTOR IN HUMAN NEOPLASMS

From the experimental data there was the indication that the etiologic factor affects the chromosome picture in the fully developed malignancy. Naturally, the possibility exists that the same is true in the human materials. However, the etiologic factor is seldom known in human malignancy, and therefore it is impossible at present to put this notion to a direct test. An indirect method to approach the question would be to study the geographic distribution of the chromosome aberrations of different neoplasms. It may be that different etiologic agents predominate in different regions of the world, leading to geographic variation in the chromosome aberrations of tumors.

There are some indications that this could be the case. Thus, there is the 5q- anomaly seen at high frequency in a restricted area of Belgium[18], and trisomy No. 9 in AML seen mainly in Australia[19]. The question is whether such differences are accidental and will even out when larger materials become available. We have calculated the frequencies of some common chromosome aberrations in AML and CML in different geographic areas. Table 4 represents the relative frequencies of the three

TABLE 4

CHROMOSOME ABERRATION IN CML

Comparison of frequencies of the 3 most common secondary aberrations in Ph^1-positive CML in two different geographic regions

Geographic region	Frequency (per cent)			Total No. of cases
	+No.8	i(17q)	$2Ph^1$	
USA	43	28	39	46
Europe	39	30	46	76

most common secondary aberrations in Ph^1-positive CML in the areas of USA and Europe. Clearly the frequencies are very similar in the two. On the other hand, there are in AML some clear-cut differences between geographic regions. The frequency of the specific translocation of t(8;21), for instance, varies considerably and appears to be very high in Japan and low in Europe, especially in Sweden, where only one case out of 39 has displayed this particular abnormality as compared to 8 out of 29 in

Japan. One must remember, however, that these differences may easily be fortuitous. Also, the data may be biased, since many published cases represent selections from larger series analyzed.

ACKNOWLEDGEMENTS

Work in our laboratory, discussed in the present article, has been supported by grants from the Swedish Cancer Society and from the John and Augusta Persson Foundation for Medical Research.

REFERENCES

1. Mitelman, F. (1971) The chromosomes of fifty primary Rous rat sarcomas. Hereditas, 69, 155-186.
2. Mitelman, F. (1972) Predetermined sequential chromosome changes in serial transplantation of Rous rat sarcomas. Acta Pathol. Microbiol. Scand. 80A, 313-328.
3. Levan, G. and Mitelman, F. (1976) G-banding in Rous rat sarcomas during serial transfer: Significant chromosome aberrations and incidence of stromal mitoses. Hereditas, 84, 1-14.
4. Mitelman, F. and Levan, G. (1972) The chromosomes of primary 7,12-dimethylbenz(α)anthracene-induced rat sarcomas. Hereditas, 71, 325-334.
5. Levan, G., Ahlström, U. and Mitelman, F. (1974) The specificity of chromosome A2 involvement in DMBA-induced rat sarcomas. Hereditas, 77, 263-280.
6. Mitelman, F., Mark, J., Levan, G. and Levan, A. (1972) Tumor etiology and chromosome pattern. Science, 176, 1340-1341.
7. Kurita, Y., Sugiyama, T. and Nishizuka, Y. (1968) Cytogenetic studies on rat leukemia induced by pulse doses of 7,12-dimethylbenz(α)anthracene. Cancer Res., 28, 1738-1752.
8. Kurita, Y., Sugiyama, T. and Nishizuka, Y. (1969) Cytogenetic analyses of cell populations in rat leukemia induced by pulse doses of 7,12-dimethylbenz(α)anthracene. Gann, 60, 529-535.
9. Ahlström, U. (1974) Chromosomes of primary carcinomas induced by 7,12-dimethylbenz(α)anthracene in the rat. Hereditas, 78, 235-244.
10. Sugiyama, T. (1971) Specific vulnerability of the largest telocentric chromosome of rat bone marrow cells to 7,12-dimethylbenz(α)anthracene. J. Nat. Cancer Inst., 47, 1267-1275.
11. Ueda, N., Uenaka, H., Akematsu, T. and Sugiyama, T. (1976) Parallel distribution of sister chromatid exchanges and chromosome aberrations. Nature, 262, 581-583.
12. Popescu, N. C. and DiPaolo, J. A. (1977) Vulnerability of specific rat chromosomes to in vitro chemically induced damage. Int. J. Cancer, 19, 419-433.
13. Sugiyama, T., Uenaka, H., Ueda, N., Fukuhara, S. and Maeda, S. (1977) Reproducible chromosomal changes of polycyclic hydrocarbon-induced rat leukemia; Incidence and chromosome banding pattern. J. Nat. Cancer Inst., in press.
14. Levan, G. and Mitelman, F. (1975) Clustering of aberrations to specific chromosomes in human neoplasms. Hereditas, 79, 156-160.
15. Mitelman, F. and Levan, G. (1976) Clustering of aberrations to specific chromosomes in human neoplasms. II. A survey of 287 neoplasms. Hereditas, 82, 162-174.

16. Levan, G. and Mitelman, F. (1977) Clustering of aberrations to specific chromosomes in human neoplasms. III. Incidence and geographic distribution of aberrations in 706 cases. Hereditas, in press.

17. Östergren, G. Personal communication.

18. Sokal, G., Michaux, J. L., Berghe, H. van den, Cordier, A., Rodhain, J., Ferrant, A., Moriau, M., Bruyère, M. de and Sonnet, J. (1975) A new hematologic syndrome with a distinct karyotype: the 5q- chromosome. Blood, 46, 519-533.

19. Ford, J. H., Pittman, S. M., Singh, S., Wass, E. J., Vincent, P. C. and Gunz, F. W. (1975) Cytogenetic basis of acute myeloid leukemia. J. Nat. Cancer Inst., 55, 761-765.

Chromosomes Today Volume 6, A. de la Chapelle and M. Sorsa eds.
© 1977 Elsevier/North-Holland Biomedical Press, Amsterdam, The Netherlands

CHROMOSOME FINDINGS IN EFFUSIONS FROM PATIENTS WITH HODGKIN'S DISEASE

DIETER K. HOSSFELD
University of Essen, Medical Clinic (Tumor Research), 43 Essen (Germany)

ABSTRACT

Effusion cells from 5 patients with advanced Hodgkin's disease were studied. Although no Hodgkin cells could be identified cytologically in 4 of 5 cases, striking chromosome anomalies of clonal nature, among them a number of recurrent marker chromosomes, were demonstrated in all cases. It is concluded that the effusion cells were intimately related to the pathogenesis of Hodgkin's disease.

INTRODUCTION

Our knowledge of the chromosome constitution in Hodgkin's disease is primarily based on the study of lymph nodes. The cytologic and histologic examinations of effusions into pleural or peritoneal cavities rarely yielded cells commonly considered diagnostic for Hodgkin's disease[1]. Hence, such material appeared unsuitable for chromosome analysis. Irrespective of cytologic findings I studied the chromosomes of all Hodgkin-patients who came into our clinic with effusions during the last 3 years. The results of 5 such findings shall be reported here.

MATERIAL AND METHODS

At the time the effusion cells were obtained for chromosome analysis all 5 patients had stage 4 Hodgkin's disease. Patient No. 2 had mixed cellularity type, the others had lymphocytic depletion type. All had received radiotherapy and combined chemotherapy. None of them responded to further chemotherapeutic trials, and they died within 6 months after the demonstration of the effusion.

At the time of chromosome analysis the leading manifestation of Hodgkin's disease was diffuse involvement of both lungs in patient No. 1, at autopsy, too, the pleura was not affected. In patients No.2 and 4 the stage of the disease was characterised by enlargement of the mediastinum, invasion of the right lung (site of the pleural effusions), and massive subdiaphragmal lymphadenopathy. Patient No. 3 had intensive bone marrow and liver infiltration, discrete subpleural lung infiltration, and, at autopsy, subserosal involvement of the small intestine. In patient No. 5 both lungs and the left pleura (site of the effusion) were massively invaded.

Cytologically and histologically the effusions of patients 2 to 5 were described to contain lymphocytes, macrophages and mesothelial cells, but no cells resembling Hodgkin or Sternberg cells. In the pleural effusion of patient No. 1, however, the latter cells were identified.

In patients No. 1, 2, 4 and 5 cells for chromosome analysis were derived from pleural effusions, while in patient No. 3 the cells came from a peritoneal effusion. Chromosome preparations were made 2 hours after the cells had been obtained from the patients. Techniques to achieve G- and C-banding have been described previously[2].

RESULTS

Numerical chromosome findings are summerized in Tables 1 and 2.

TABLE 1

Summary of numerical chromosome findings in cases 1-4. Numbers in parenthesis indicate the number of metaphases.

Case No.	Chromosome Number						Mode	No.Metaphases counted
	45	46	47	48	49			
1	2	13	17	56		88-96(11) 160(1)	48	100
2	2	12	10	84	7	94-98(24) 130(1) 170(1)	48	148
3	4	91	2			92(3)	46	100
4	4	66	3			56(2) 70-85(11) 89-93(10) 121-130(4)	46	100

TABLE 2

Summary of numerical chromosome findings in case 5

Case No. 5	Chromosome Number												Mode	No.Metaphases counted
	58	58	59	60	61	62	63	64	65-66	117-118	171	214		
	7	13	14	30	7	6	4	6	3	8	1	1	60	100

Case No. 1. The modal chromosome number was 48. The modal karyotype was 48, XY, +9, +12, 17p+ (Fig. 1). The majority of metaphases with 47 chromosomes lacked the extra No. 9. All the karyotyped metaphases with 46 chromosomes were normal. The marker chromosome 17p+ was found in 70% of the hyperdiploid metaphases. The extra material was probably derived from the short arm of one X chromosome. 75% of the polyploid metaphases had 96 chromosomes.

Cases No. 2. The modal chromosome number was also 48. No satisfactory banding could be obtained. However, the following karyotypic features could be identified in all hyperdiploid metaphases (Figs.2,3): 1q+, iso 1q, 1q-, 3q-, -10,t (3q;11), +14, +15, -17, +19, +20, -22, -Y. 60% of the hyperdiploid metaphases contained a 14q+ in addition. A large metacentric chromosome one arm of which resembled No. 13 was seen in 59% of the metaphases. Occasional cells disclosed abnormal chromosomes resembling a No. 3q- and 5q-. Metaphases with 47 and 49 chromosomes revealed an inconstant pattern. This applied also to the polyploid metaphases which, however, always demonstrated one or more sets of the marker chromosomes. Among the diploid cells 8 were normal and the others exhibited various characteristics of the modal karyotype.

Case No. 3. The modal number was 46. The modal karyotype was 46, XY, 14q+, 15p+, -16, +marker (Figs. 4,5). The marker was a medium-sized submetacentric chromosome with a dark band in the middle of the large arms. The extra material of 14q+ probably was from the distal part of the long arms of No. 10. 80% of the metaphases corresponded to the modal karyotype. 20% of the metaphases were normal. The number of polyploid cells was unusually small in this case.

Case No. 4. The modal number was 46. The modal karyotype was 46, XY, +2q+, t(3q+;11q), t (4q-;11q), -13 (Fig. 6). The origin of the additional band of 2q+ could not be identified. On the basis of C-banding pattern it was concluded that a non-centromeric band was intercalated in t(3q+;11q-). About 60% of the diploid metaphases showed the modal karyotypic theme, while about 30% had a missing No. 16 and an extra No. 17 in addition. Not a single normal metaphase was found. About one fourth of the metaphases had polyploid chromosome numbers. Except for the marker chromosomes no constant karyotypic picture emerged from the study of a large number of such metaphases, even when the chromosome number was identical. In only 40% of the tetra- and pentaploid metaphases all three markers could be detected in duplicate or triplicate.

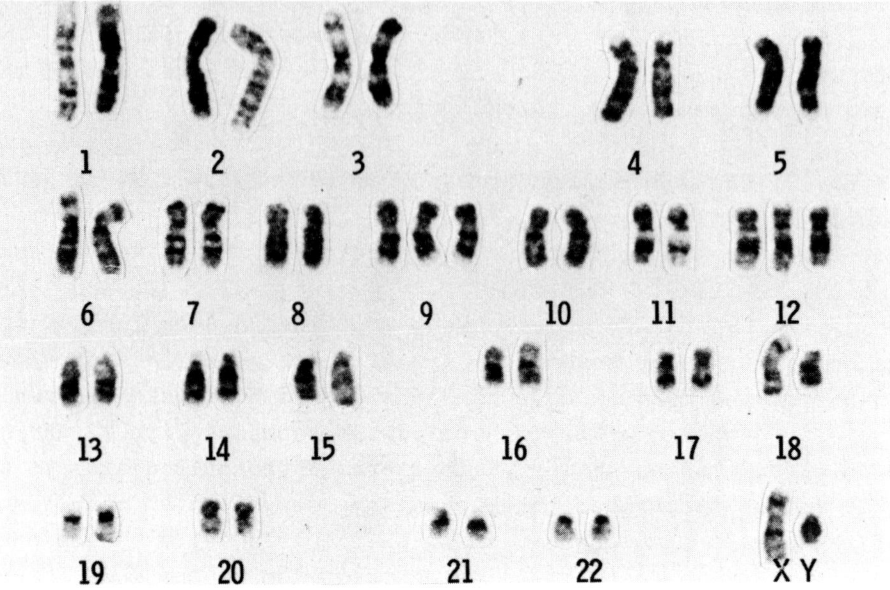

Fig. 1. Representative karyotype of case No. 1

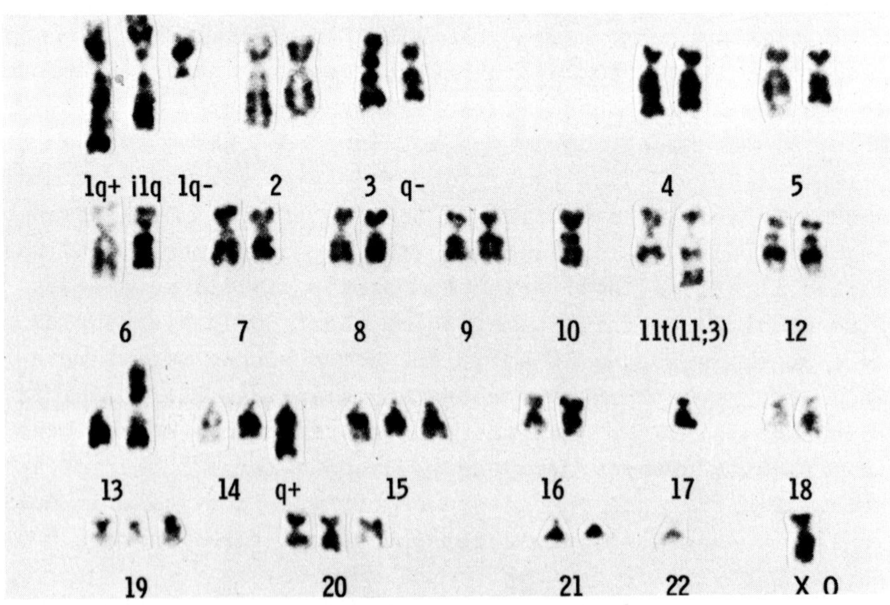

Fig. 2. Representative karyotype of case No. 2

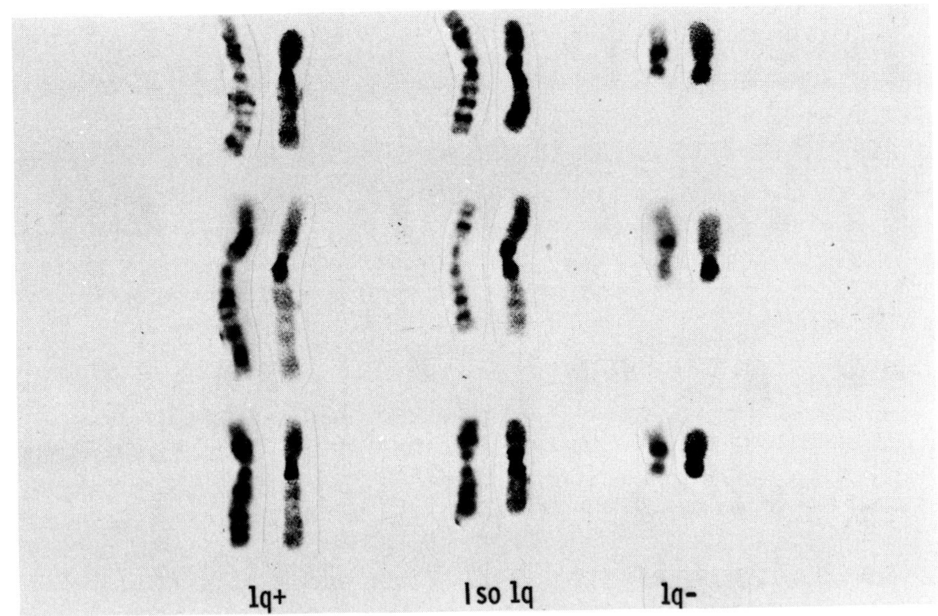

Fig. 3. G-and C-banding patterns of markers i1q,1q+, and 1q- in case No. 2

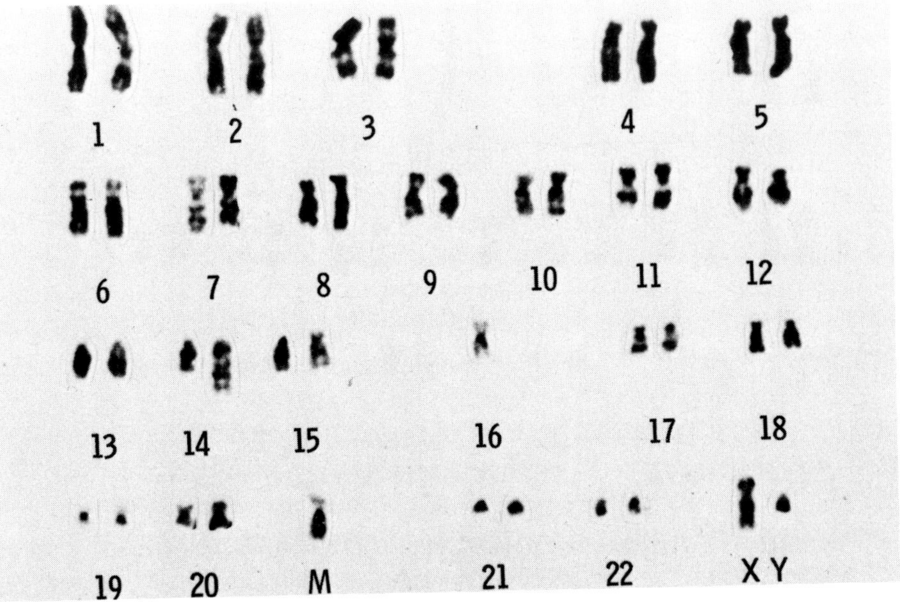

Fig. 4. Representative karyotype of case No. 3

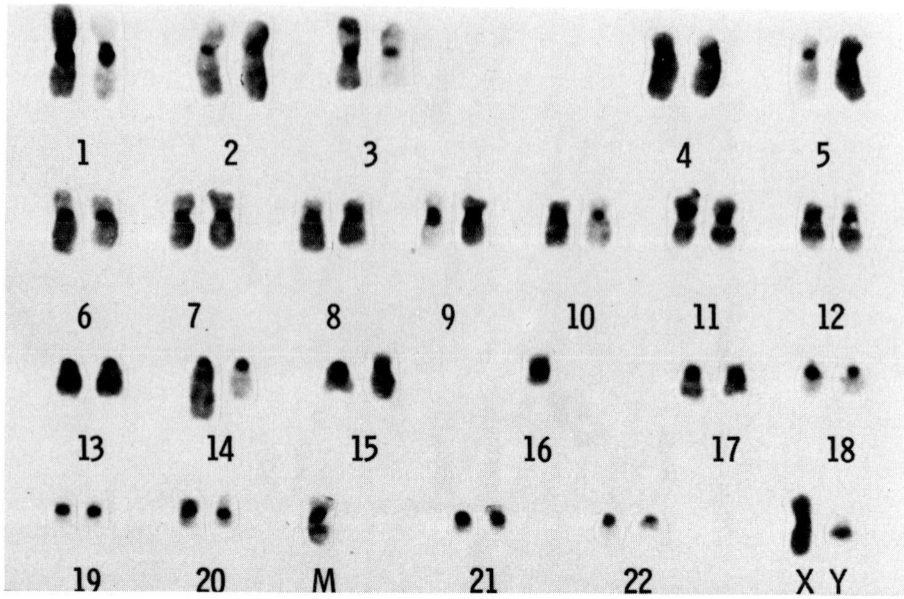

Fig. 5. C-bands in case No. 3

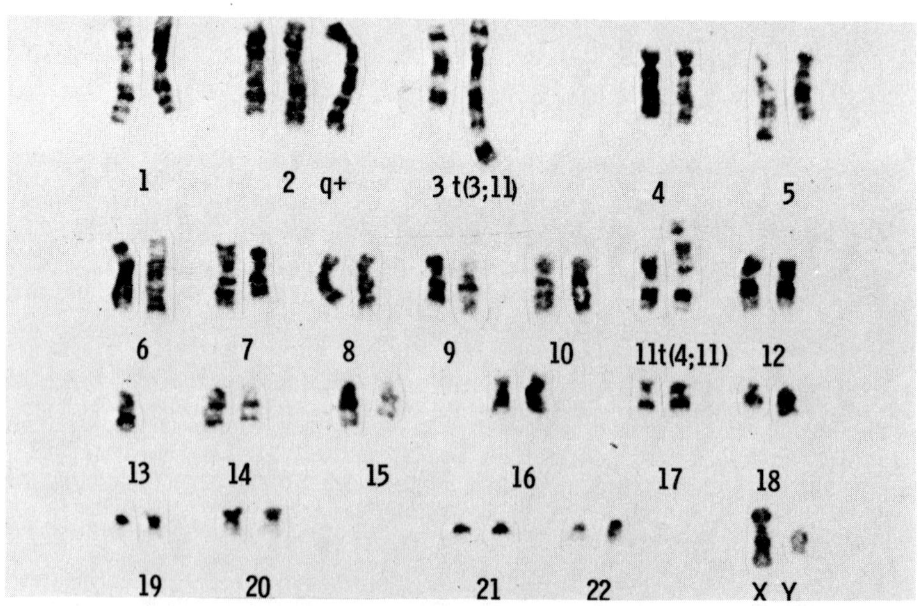

Fig. 6. Representative karyotype of case No. 4

Case No. 5. The modal chromosome number was 60. Compared to the other cases this mode was rather flat comprising only 30% of the metaphases. About 35% of the cells had chromosome numbers between 55 and 59, and in about 25% of the cells the numbers ranged between 61 and 68. Chromosome numbers in the pentaploid range were seen in 8%.Karyotypic analysis revealed at least 17 marker chromosomes (Figs. 7, 8). It appeared that except for chromosomes 4, 10 and 14 all the other pairs contributed to structural and numerical changes. Even among the cells with the modal chromosome number a great karyotypic diversity was noted. In addition, the description of the modal karyotype was hampered by a banding quality less than optimal. Nevertheless, it could be demonstrated that none of the metaphases was normal, and that all the cells had at least 10 karyotypic changes in common. Such changes were 1p-; t(1p;5q);t(1;5);6q-;t(?;9q);t(5q;12);iso 17q;+19; +19;-21;-22. The nature of other markers remained unclarified or questionable (see Figures). Karyotypes of near-pentaploid metaphases disclosed the characteristic anomalies usually but not invariably to be present in duplicate. Preliminary chromosomal data of cases 4 and 5 have been published previously[3].

TABLE 3

Summary of karyotypic findings

Case 1	48,XY,+9,+12,17p+
Case 2	48,XO, 1q+,i1q,+1q-,3q-,-10,t(3q;11),+14,+15,-17,+19,+20,-22
Case 3	46,XY,14q+,15p+,-16,M
Case 4	46,XY,+2q+,t(3q;11q),t(4q;11q),-13
Case 5	60,XY,1p-,+t(1p;5q),+t(1;5),6q-,t(?;9q),+t(5q;12),+i17q,+19, +19,-21,-22

DISCUSSION

This study provided 3 interesting findings. Firstly, except for case No. 1, no malignant cells could be identified in the effusions, and yet, striking numerical and structural chromosome changes were found. Secondly, compared to chromosomal data in Hodgkin diseased lymph nodes, the percentage of abnormal metaphases was considerably higher. Thirdly, a number of marker chromosomes appeared to correspond to markers commonly seen in various other malignancies.

Obviously, the crucial question concerns to nature of the cells investigated. That they are malignant is borne out by the chromosomal findings, particularly the recurrent marker chromosomes, and the fact

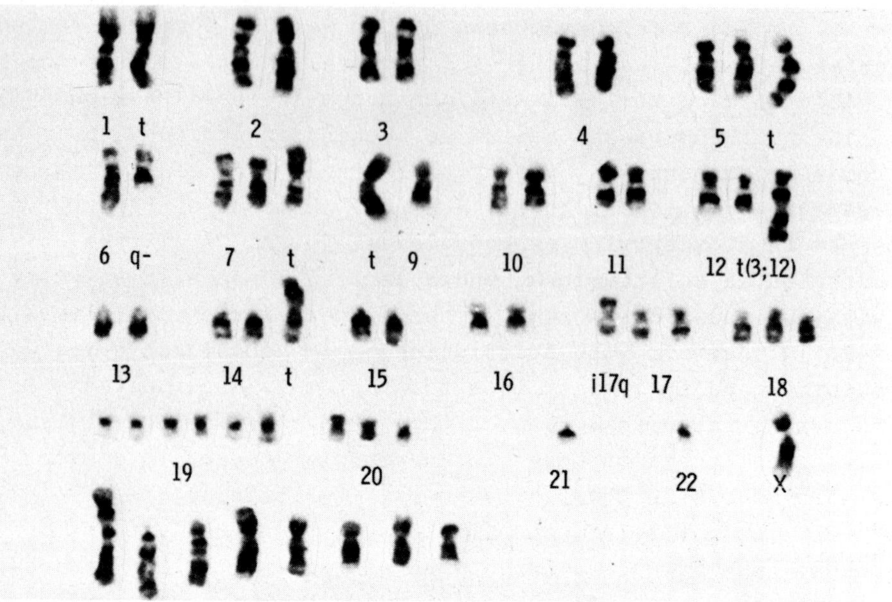

Fig. 7. Representative karyotype of case No. 5

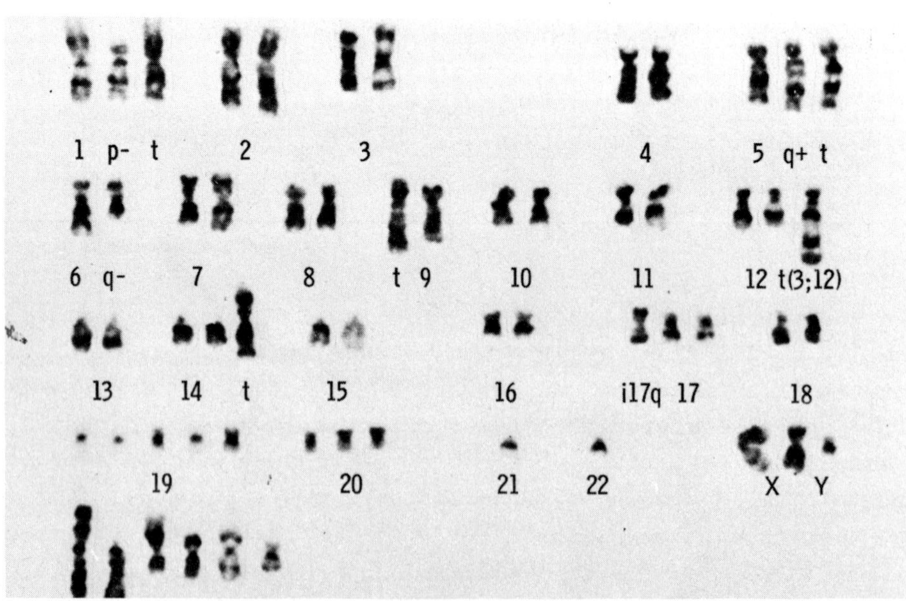

Fig. 8. Representative karyotype of case No. 5

that these cells are capable of continuous growth in diffusion chambers[3]. In a paper to be published[4] the growth characteristics in diffusion chambers, and the cytochemical and immunological reactions will be described. Presently it appears that the cells are neither typical T or B lymphocytes nor macrophages. We believe that the cells are intimately related to the pathogenesis of Hodgkin's disease, and that the inability to identify them cytologically may be due to the unusual environment of the cells and other yet unknown factors. It is known that the pathologist needs an appropriate stromal reaction to diagnose Hodgkin cells.

Our view would be strengthened by the demonstration of identical chromosomal changes in effusion cells and lymph nodes. So far I did not have the opportunity for such a comparative study. A lymph node, removed immediately after death from patient No. 4, yielded only one analysable metaphase, which, however, clearly contained the markers found in the effusion cells.

It is not surprising to find a considerably smaller percentage of normal metaphases or none in effusions than in lymph nodes. It is believed that in lymph nodes the normal metaphases are derived from lymphocytes, immunoblasts and plasma cells which comprise the largest fraction of the cell mass, while the number of Hodgkin cells is small. No such stromal reactions occur in effusions. In addition, own data suggest that also in lymph nodes the number of normal metaphases decreases if the histology is lymphocytic depletion.

Unexpected was the finding that 4 out 5 effusions had modes in the diploid range. In lymph nodes it is common to find modes in the triploid range[5]. Sinks and Clein[6] reported a case with "Hodgkin-leukemia" in whom the circulating Hodgkin cells were also shown to have a mode at 48. Whether this represents tissue specific differences or a chance phenomenom remains to be seen.

A chromosome anomaly characteristic for Hodgkin-effusions could not be identified. Of interest, however, was the demonstration of a number of marker chromosomes which have been described previously not only in Hodgkin's disease[7] but also in non-Hodgkin lymphomas[8,9,10] and carcinomas[11]. Data from the literature suggest that the markers 6q-, 14q+ and structural rearrangements involving chromosomes 3, 5, 11 and 12 might be of some specificity in Hodgkin but also non-Hodgkin lymphomas. The markers i1q, 1p-, i17q, and trisomy in groups 9, 14 and 19, on the other hand, appear to characterize malignancy in general. The fact that such markers could be detected in our material underscores that the cells were truly malignant and, possibly, of lymphocytic origin.

Finally, it was an important observation that in each case all abnormal metaphases shared a number of chromosomal characteristics, irrespective of the ploidy value. This strongly indicates monoclonality, and that the diploid pre-Hodgkin cells are precursors of the polyploid Sternberg cells.

ACKNOWLEDGEMENT

Supported by the Deutsche Forschungsgemeinschaft. The technical help of Mrs. E. Wendehorst is greatfully acknowledged.

REFERENCES

1. Kaplan, H.S. (1972) Hodgkin's Disease, Harvard University Press, Cambridge, MA, pp. 1-452.
2. Hossfeld, D.K. (1974) Humangenetik, 23, 111.
3. Boecker, W.R. et al. (1975) Nature 258, 235.
4. Boecker, W.R., Hossfeld, D.K. to be published
5. Sandberg, A.A., Hossfeld, D.K. (1974), In Handbuch der allgemeinen Pathologie (E. Grundmann, Ed.), Springer, Berlin-Heidelberg-New York, Vol. 6, V, pp. 141-287
6. Sinks, L.F., Clein, G.P. (1966), Brit. J. Haemat., 12, 447.
7. Reeves, B.R. (1973), Humangenetik, 20, 231.
8. Fleischmann, E.W., Prigogina, E.L. (1977), Humangenetik, 35, 269.
9. Mark, J. (1975), Hereditas, 81, 289.
10. Reeves, B.R., Stathopoulos, G. (1976), Hum Genet., 31, 203.
11. Kakati, S. et al. (1976), Cancer 38, 770.

Chromosomes Today Volume 6, A. de la Chapelle and M. Sorsa eds.

THE CYTOGENETICS OF HUMAN LYMPHOMAS: CHROMOSOME 14 IN BURKITT'S, DIFFUSE HISTIOCYTIC AND RELATED NEOPLASMS.

*KAISER-MCCAW, B., +EPSTEIN, A. L. *OVERTON, K. M., +KAPLAN, H. S. and *HECHT, F.
*Department of Pediatrics, University of Oregon Health Sciences Center, Portland, Oregon: +Cancer Research Laboratory, Stanford University School of Medicine, Stanford, California (USA).

ABSTRACT

Rearrangements of chromosome 14 have been observed in benign and malignant lymphocyte clones in ataxia telangiectasia (AT) and in lymphoid neoplasms such as African and North American Burkitt's lymphomas (BL). To characterize other lymphoid neoplasms, established cell lines from 10 patients with diffuse histiocytic lymphoma (DHL) were analyzed. An AT patient who subsequently developed North American Burkitt's lymphoma was also studied.

Various rearrangements of chromosome 14 were noted, plus other non-random changes, especially involving chromosome 6 in the DHL cells. The chromosome 14 rearrangements in neoplastic cells consistently involve two specific regions: a "donor" site at band 14q12 and a "receptor" site at band 14q31. These rearrangements arise as a result of either intra- or interchromosomal exchange between chromatids.

INTRODUCTION:

Investigations of chromosome 14 in lymphoproliferative disorders became feasible with the development of banding techniques. The addition of an extra band onto the long arm (q) of one chromosome in the no. 14 pair was observed in both tumor biopsies and cell lines established from African Burkitt's lymphoma.[1] Subsequently, it was demonstrated that the extra band was due to a translocation from the long arm of chromosome 8 in African Burkitt's lymphoma[2] and in the North American Burkitt's lymphoma as well.[3] Chromosome 14 rearrangements have also been observed in other lymphoid neoplasms, such as lymphosarcomas and reticulosarcomas.[4] In patients with ataxia telangiectasia, a rare autosomal recessive disorder predisposing to leukemias and lymphomas, benign clones of lymphocytes are marked by rearrangements of chromosome 14.[5] The proportion of these clones in vivo usually increases with time. We monitored an A-T patient who subsequently developed a T-cell leukemia and noted that the clone of cells containing the 14q+ translocation correlated with the progression of a premalignant to a malignant state. We hypothesize that structural rearrangements of 14q may be an important step in the development of lymphoid malignancies.

MATERIALS AND METHODS

Pleural or peritonial effusions were obtained from 10 patients with diffuse histiocytic lymphomas. With the aid of a tissue culture method which permits the rapid screening of nutrient requirements, the tumor cells were established in continuous culture.[6] The lines are designated SU-DHL-1 through SU-DHL-10. SU-DHL-1 and -2 are pure histiocytic, SU-DHL-3 through -7 and -10 are B cell, and SU-DHL-8 and -9 are null cells. A type of C RNA virus has been detected in culture fluids of SU-DHL-1[7] and possibly SU-DHL-10 as well.

Cytogenetic analysis was done with quinacrine banding.

RESULTS

The results are summarized in Table I.

Table I: Diffuse Histiocytic Lymphomas

Line	Cell Type*	Ig	Karyotype	Chromosome Changes 14	Other
SU-DHL-1	H	--	79,XY	No	Hyperdiploid, 6q-, Y Trans., Mult. Markers
SU-DHL-2	H	--	52,XX	No	6q-, Trisomy 9, 11, 13
SU-DHL-3	B	G	47,XY	14q+	+8, 3p+
SU-DHL-4	B	G	47,XY	14q+	+8, 3p+
SU-DHL-5	B	G	47,XX	No	+Markers
SU-DHL-6	B	G	N.C.		
SU-DHL-7	B	A	43,XX	14q+	Hypodiploid, Mult. Markers
SU-DHL-8	N	--	N.C.		
SU-DHL-9	N	G	50,XX	No	6q-, Trisomy 9, 11, 13
SU-DHL-10	B	G	46-92,XY	No	Tetraploid

*H = Histiocytic, B = B Lymphocyte, N = Null Cell
N.C. = Study Not Complete

The DHL cell lines do not have stable chromosome numbers, especially SU-DHL-1 and -10. (Fig. 1) Perhaps by coincidence, both these lines are infected by a C-type virus. All the lines tended toward hyperdiploidy, except SU-DHL-7, which is undergoing a reduction in chromosome number.

There were numerous chromosome rearrangements, often unique to single cells. Not all the rearranged chromosomes or marker chromosomes could be identified. The 6q- observed in SU-DHL-1, -2 and -9 was similar in morphology to that reported in the acute lymphoblastic lymphomas.[8] To date, three of the 10 lines have rearrangements in 14q. (Fig. 2) The cell lines also continue to acquire new chromosome rearrangements with time in culture. A repeat study of SU-DHL-3 after a six month

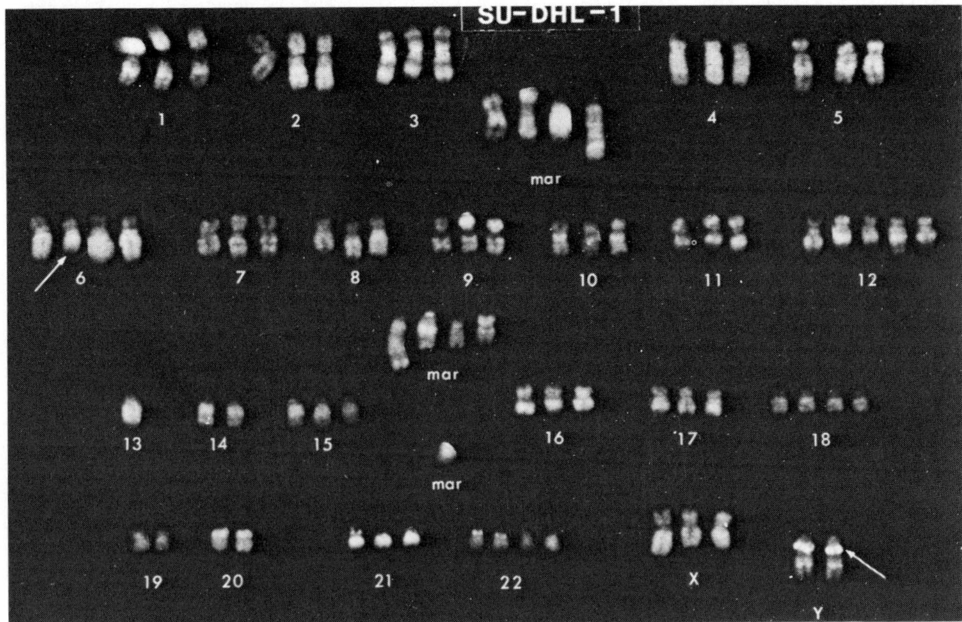

Figure 1

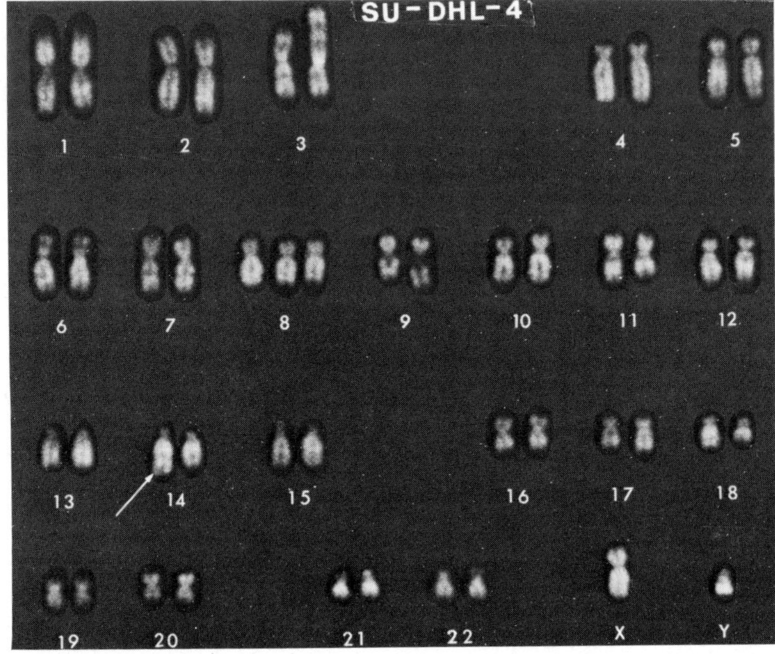

Figure 2: Karyotype of line SU-DHL-4 showing extra material translocated onto the long arm of chromosome 14 (arrow), trisomy 8 and $3p^{+}$.

interval demonstrated that this line is now monosomy 8 and 14 and has several new large marker chromosomes.

We previously reported our observations on 2 North American and three African Burkitt's lymphoma cell lines.[9] Recently, a patient with ataxia telangiectasia, a genetic disease predisposing to leukemia and lymphoma, was found to also have North American Burkitt's lymphoma. The cell line derived from this individual has been designated as SU-AmB-3. Our overall findings are summarized in Table II.

Table II: Burkitt's Lymphoma

Type	Line	Cell Type	EBV	Karyotype	Chromosome Changes 14 Other
Afr.	EB3	B	+	46,XY	T(8;14)
Afr.	HR-1	B	+	47,XY	T(8;14) +Marker
Afr.	RAJI	B	+	48,XY	T(8;14) +Markers
N.A.	SU-AmB-1	B	-	46,XY	T(8;14)T(4;5;7)
N.A.	SU-AmB-2	B	+	46,XY	T(8;14)
N.A.*	SU-AmB-3	B	+	46,XY	14q+

* Ataxia Telangiectasia

Afr. = African, N.A. = North American

DISCUSSION

Although some chromosome rearrangements are observed in the lymphoproliferative disorders which do not involve 14q, it is increasingly clear that 14q+ markers commonly occur. After reviewing our data and the literature, most neoplastic rearrangements of chromosome 14 can be placed in one of the following categories (Table III).

The first 3 categories result in a balanced karyotype. Category (I) is a normal pair of 14 chromosomes. Category (II) is a balanced rearrangement between the two homologs. In other words, a 14;14 balanced translocation. Category (III) is a balanced translocation of chromosome 14 with another chromosome. The 8;14 translocation in Burkitt's lymphoma and the 11;14 translocations observed in lymphosarcomas[10] would be examples.

The last 3 categories result in an unbalanced karyotype. A 14q+ only (category IV) probably arises through the same mechanism as (II) above, with the subsequent loss of the smaller, deleted 14. We have, in fact, observed this during clonal evolution in ataxia telangiectasia.[5] The short arm and part of the long arm of one No. 14 are now lost. In category V, there is one normal 14 and a metacentric chromosome composed of two 14's which may represent an iso-14. Category VI includes a normal 14 and a 14q+, which is a tandem repeat of the long arm of 14. A ring composed of two 14's is included in category VI.

Table III: Chromosome 14 Rearrangements

Class	Number 14 Chromosomes	Morphology Of One No. 14	Morphology Of Other No. 14	Mechanism	Balanced Rearrangement	Result
I	2	N	N	----	Yes	N
II	2	14q+	14q-	Translocation Between 14's	Yes	N
III	2	14q+	N	Translocation Between 14 and Other Chromosome	Yes	N
IV	1	14q+	Lost	Translocation Between 14's	No	Del (14p)& Prox (14q)
V	2	Iso14q/ T(14q;14q)	N	Isochromosome/ Translocation	No	Del (14p)& Dup (14q)
VI	2	14q+	N	Tandem Dup 14q	No	Dup (14q)

N = Normal, p = Short Arm, q = Long arm
Del = Deletion, Dup = Duplication

There appear to be only two sites on chromosome 14 which are involved in all of these rearrangements; 14q12 is a "donor" site contributing material either to its 14 homolog or as in some A-T clones, another autosome, usually the 7.[5] 14q31 acts as a "receptor" site, receiving material either from its 14 homolog or from another chromosome, usually the 8 in Burkitt's lymphoma.

All these rearrangements of chromosome 14 could arise from either (1) an inter-chromosomal event if there is somatic pairing, misalignment, chromatid breakage at 14q12 and 14q31 followed by a refusion or (2) an intrachromosomal event if there is breakage between sister chromatids, crossing-over, followed by refusion, perhaps promoted by a loop. An intrachromosomal scheme must be invoked to explain a normal chromosome 14 remaining in the cell.

Heteromorphisms, such as satellites, can sometimes be used to designate each no. 14 chromosome. For example, one chromosome 14 may be marked by large or bright satellites, and the other, by very dull or small satellites. In a benign lymphocyte clone marked by a 14;14 translocation (category II), the donor 14 and the receptor 14 could be correlated with the 14 chromosomes found in the patient's fibroblasts. Normal fibroblasts could only be obtained from SU-DHL-4; heteromorphisms are being used to try to identify the 14 involved in the translocation.

Cells with a rearrangement of chromosome 14 may have a proliferative advantage. However, there may also be selection <u>against</u> cells not carrying a rearrangement of chromosome 14. For example, spontaneous chromosome breakage in lymphocytes from normal people occur randomly throughout the karyotype. Only

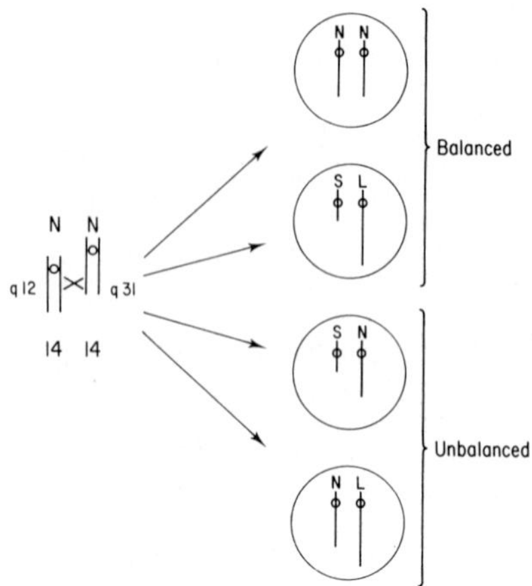

Rearrangements of Chromosome 14
Interchromosomal

Figure 3: Possible mechanism for the origin of chromosome 14 rearrangements. N = normal, S = short, deleted 14, L = long, duplicated 14.

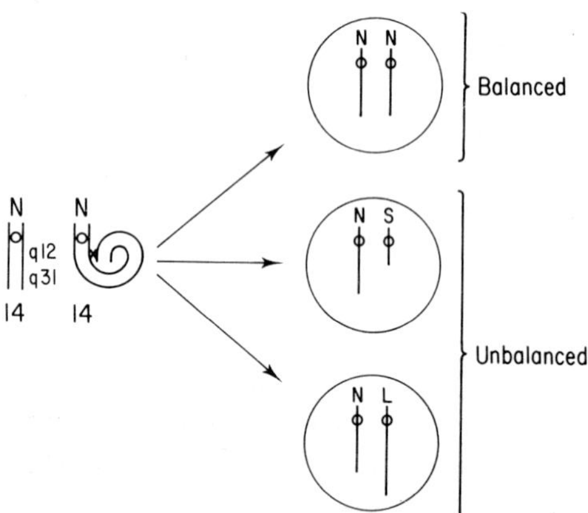

Rearrangements of Chromosome 14
Intrachromosomal

Figure 4: Possible mechanism for the origin of chromosome 14 rearrangements. (See Figure 3).

about 1% of all these random breaks are destined to be involved in balanced chromosome rearrangements, but instead remain as single chromatid or iso-chromatid breaks. By contrast, over 60% of breaks in 14q result in a balanced chromosome rearrangment, usually with the other 14[11] or the 7. The frequency of 7;14 balanced translocations in cultured lymphocytes is 4 x 10^{-4}.[12] This means that cells with random chromosome breakage may not survive because this breakage results in monosomy lethal to the cell. Breaks in chromosome 14 usually result in a balanced translocation. Since genetic material is not lost, the cell survives.

The proliferative advantage of neoplastic cells with a balanced rearrangement of chromosome 14 may also be due to "position effect"; in other words, the effect of a gene or genes may be dependent upon its location and its position in respect to neighboring genes. The chromosome material translocated onto the end of 14 in Burkitt's lymphoma frequently shows altered fluorochromatic staining with acridine orange. If this can be verified, it will constitute cytochemical evidence for position effect in malignant lymphoid cells with a balanced translocation. (See discussion by Hecht and Kaiser-McCaw, this volume).

Congenital rearrangements of chromosome 14 have been reported for every band[13] whereas neoplastic rearrangements due to a somatic mutation occur only at bands 14q12 and q31. This discrepancy cannot yet be explained. The genes assigned to chromosome 14 are ribosomal RNA (14p12), nucleoside phosphorylase (14q11-q21) and tryptophanyl tRNA synthetase (14q21-qter).[14] Neoplastic rearrangements of chromosome 14 may be useful to confirm these gene assignments as well as map additional genes to this chromosome.

The specificity of rearrangements of chromosome 14 in lymphoproliferative disorders has been compared to the specificity of the 9;22 translocation most frequently seen in patients with CML. Rearrangements of chromosome 14 may yet prove to be of diagnostic value. However the similarity between 14 rearrangements and the Philadelphia rearrangement appears to end there. Chromosome 14 rearrangements consist of a variety of alternative rearrangements which may be either balanced or unbalanced. The instability of chromsome 14 and its lability to various rearrangements must ultimately be explored and explained at the molecular level. It is possible that these rearrangements can be correlated with an arrest of the normal differentiation pathway for lymphocytes, resulting in abnormal proliferation and disease.

SUMMARY

(1) There are two "active" regions on chromosome 14 (q12 and q31) in neoplastic cells.

(2) There may be homology between these regions.

(3) This homology promotes inter- and intra-chromosomal exchanges at these two points.

(4) The regions may be polymorphic in the general population, with more than one alternative structure at the two sites on each chromosome 14.

(5) Chromosome 14 is the only chromosome in the karyotype which behaves in this fashion.

(6) The resultant rearrangements often appear to be balanced.

(7) There is selection for cells carrying a rearrangement of 14q.

(8) This selection or proliferative advantage may be due to position effect.

ACKNOWLEDGEMENTS

We would like to thank Mr. Michael Pauly for technical assistance and Mrs. Dorothy Conze and Miss Rosemary Milbeck for preparing the manuscript.

This work was supported by grant 1 R01 HD 08236 from the National Institute of Child Health and Development, by grant 5 R01 CA 16747 and contract N01-CP-43228 from the National Cancer Institute, National Institutes of Health. Alan Epstein is a medical and graduate student supported by Medical Scientists Program grant GM-01922 from the National Institutes of Health.

REFERENCES

1. Manolov, G., Manolova, Y. (1971) A Marker Band in One Chromosome No. 14 in Burkitt Lymphomas, Hereditas, 69, 300.

2. Zech, L., Haglund, U., Nilsson, K., Klein, G. (1976) Characteristic Chromosomal Abnormalities in Biopsies and Lymphoid Cell Lines from Patients with Burkitt and Non-Burkitt Lymphomas, Int. J. Cancer 17, 47-56.

3. Kaiser-McCaw, B., Epstein, A.L., Kaplan, H.S. and Hecht, F. (1977) Chromosome 14 Translocation in African and North American Burkitt's Lymphoma, Int. J. Cancer 19, 482-486.

4. Mark, J. (1977) Chromosome Abnormalities and Their Specificity in Human Neoplasms. An Assessment of Recent Observations by Banding Techniques. In: Advances in Cancer Research, Vol. 74, Klein, G and Weinhouse, S. eds., New York: Academic Press, pp. 165-222.

5. McCaw, B.K., Hecht, F., Harnden, D.G., and Teplitz, R.L. (1975) Somatic Rearrangement of Chromosome 14 in Human Lymphocytes, Proc. Natl. Acad. Sci. USA 72, 2071-2075.

6. Epstein, A.L., Kaplan, H.S. (1974) Biology of Human Malignant Lymphomas. Establishment in Continuous Cell Culture and Heterotransplantation of Diffuse Histiocytic Lymphomas, Cancer 34, 6.

7. Kaplan, H.S., Goodenow, R.S., Epstein, A.L., Gartner, S., Decleve, A., and Rosenthal, P.N. (1977) Isolation of a Type C RNA Virus from an Established Human Histiocytic Lymphoma Cell Line, Proc. Natl. Acad. Sci., USA 74, 2564-2568.

8. Oshimura, M., Sandberg, A.A. (1976) Chromosomal 6q - Anomaly in Acute Lymphoblastic Leukaemia, Lancet, 12-25.

9. McCaw, B.K., Epstein, A.L., Kaplan, H.S., and Hecht, F. (1977) Chromosome 14 Translocation in African and North American Burkitt's Lymphoma, Int. J. Cancer, 482-486.

10. Fleischman, E.W. and Prigogina, E.L. (1977) Karyotype Peculiarities of Malignant Lymphomas, Hum. Gen., 269-279.

11. Ayme, S., Mattei, J.F., Mattei, M.G., Aurran, Y. and Giraud, F. (1976) Non-random Distribution of Chromosome Breaks in Cultured Lymphocytes of Normal Subjects, Hum. Gen., 161-175.

12. Hecht, F., McCaw, B.K. Peakman, D., Robinson, A. (1975) Non-random Occurrence of 7-14 Translocations in Human Lymphocyte Cultures, Nature, 243-244.

13. Borgaonkar, D.S. (1975) Chromosomal Variation in Man, The John Hopkins University Press.

14. Francke, U., Denney, R.M., Ruddle, F.H. (1977) Intrachromosomal Gene Mapping in Man: The Gene for Tryptophanyl-tRNA Synthetase Maps in Region q21--qter of Chromosome 14, Somatic Cell Gen., 381-389.

Chromosomes Today Volume 6, A. de la Chapelle and M. Sorsa eds.

CHROMOSOME ABERRATIONS, DNA POST-REPLICATION REPAIR AND LETHALITY OF TUMOUR CELLS WITH A DIFFERENTIAL SENSITIVITY TO ALKYLATING AGENTS

D. SCOTT,
Paterson Laboratories, Christie Hospital and Holt Radium Institute, Withington, Manchester M20 9BX, U.K.

ABSTRACT

Rat tumour cells (Yoshida lymphosarcoma) sensitive to bifunctional alkylating agents sustain much more chromosome damage than resistant Yoshida tumour cells when exposed to sulphur mustard (SM). The amount of chromosome damage is almost sufficient to account for the degree of cell killing. The capacity of the sensitive (YS) cells for DNA post-replication repair, which can be inhibited by caffeine, is less than that of the resistant (YR) cells as measured by the magnitude of caffeine enhancement of SM-induced lethality and chromosome damage. Failure to perform post-replication repair (PRR) results in chromosome aberrations which lead to cell death. The difference is PRR capacity between YR and YS cells is not paralleled by a difference in their spontaneous sister chromatid exchange frequencies.

INTRODUCTION

Many of the agents used in cancer therapy, both radiation and chemical, have been shown to induce chromosome structural aberrations[26,28]. After X-irradiation these are probably the major cause of cell death[4,7,14,27] but with other agents their contribution to cell lethality has not been quantitated. We have previously shown[27] that cells from the Yoshida lymphosarcoma of rats which are very sensitive to bifunctional alkylating agents[10] sustain much more chromosome damage than cells from resistant Yoshida sarcomas when exposed to sulphur mustard (SM) in vitro (Fig.1) in spite of the fact that the SM is incorporated to the same extent into DNA, RNA and protein of sensitive (YS) and resistant (YR) cells[27].

These studies have now been extended in an attempt to answer the following questions:

1) To what extent is chromosome structural damage responsible for the lethal effect of SM on these cells? This has been done by comparing, in the same experiment, the frequency of cells with chromosome aberrations with the proportion of cells killed.

2) Does the differential sensitivity of YS and YR cells to cell killing and chromosome damage result from differences in DNA repair capacity? We previously found[27] that YS and YR cells do not differ in their capacity for excision repair after SM, a process whereby, after treatment of cells with, for example, ultraviolet-light (UV) or alkylating agents, single-strand lesions in DNA are excised and replaced by new

DNA which is synthesised outside the normal S phase using the intact DNA strand as template[5]. If, however, DNA is synthesised during the S phase on parental (template) DNA containing unexcised lesions the molecular weight of the newly-synthesised DNA is lower than that of untreated cells but later increases to control levels[3,19]. The usual interpretation of this observation is that gaps are left in newly-synthesised DNA opposite lesions in parental DNA and that these gaps are later filled by post-replication repair[3,19] (PRR). The capacity of YS and YR cells for PRR has now been investigated and is reported in this paper.

PRR can be inhibited by caffeine[6,11,29] which binds to single-stranded DNA[8] presumably preventing the action of the enzymes involved in the gap-filling process[20]. If cells treated with UV or alkylating agents are post-treated with a non-toxic dose of caffeine the amount of cell killing[11,30] and chromosome damage[17,23] is usually increased compared with cells not receiving caffeine. The implication is therefore that caffeine inhibits PRR and this leads to an increase in chromosome structural aberrations and hence an increase in cell killing[17]. In cells which lack PRR capacity, caffeine post-treatment would not be expected to increase the amount of cell death or chromosome damage. In the present study the degree of enhancement by caffeine of chromosome damage and cell killing after SM treatment of YS and YR cells has, therefore, been used as a measure of their respective PRR capacities as in the experiments of Roberts and Ward[22] comparing Chinese hamster and HeLa cells.

The frequency of sister chromatid exchanges (SCEs) in untreated YS and YR cells has also been measured using the harlequin-banding method of Perry and Wolff[21] since it has been suggested[2,15] that SCE formation may involve DNA post-replication repair. On this hypothesis any difference in PRR capacity between YS and YR cells would be reflected in a difference in their spontaneous SCE frequencies.

MATERIALS AND METHODS

Cell culture methods[10], cytogenetic techniques[27] and SM treatments[27] have been described previously. Briefly, cells were grown as suspension cultures in Fischer's medium + 20% horse serum in an atmosphere of 5% CO_2 and 95% air at 37°C and SM treatments were always for 1hr at 37°C.

Cell survival was determined by back-extrapolation of the exponential region of growth curves[1,27] from cell counts made on control and treated cells at 2 day intervals. Regression lines with a common slope were fitted to plots of log cell number against time after treatment using a computerised least squares analysis. Best-fitting values of D_o and extrapolation number (n) for the survival data were obtained using the computer programme of Gilbert[12].

Caffeine was given immediately after 1h SM treatments and always for a period of 2 days, after which cells were transferred to caffeine-free medium. Cell cycle times were obtained from labelled mitosis curves after pulse labelling with a non-toxic dose of tritiated thymidine (^{3}HTdR, 0.01µCi/ml, S.A. 9Ci/mM) for 1h, washing cells twice in medium containing cold thymidine (2μg/ml) and resuspending cells in fresh medium. Autoradiographs were prepared with Ilford L4 emulsion, exposed for 1-5 months and developed (Kodak D19) for only 30 secs at 20°C to obtain small silver grains which did not obscure the underlying chromosomes. SCE frequencies were determined after a 32h exposure of cells to BUdR (5μg/ml), the last hour with vinblastine sulphate (0.1μg/ml), and then fixed and stained by the method of Perry and Wolff[21]. Chromosome structural aberrations were classified according to the system described by Savage[24].

RESULTS

A) Aberration Frequency in Relation to Cell Killing after SM Treatment

In this experiment ^{3}HTdR was given in conjunction with the SM for 1hr in order to determine the cell cycle time. Controls received ^{3}HTdR only. After treatment the cells were kept at room temperature for 4½hrs to prevent progression through the cycle while cells were washed, counted and placed in culture vessels for later sampling which, for cytogenetic studies, began 3½hrs after the cells had been returned to 37°C. At each fixation time vinblastine was added to those cultures to be fixed at the next sampling time (usually 3-4 hrs later) in order to sample the entire mitotic population up to 68 hrs post-treatment (Figs 2-5).

SM doses were given to YS and YR cells in an attempt to achieve similar levels of cell killing in the two lines. Thus 7 or 9ng/ml were given to YS cells and 50 or 80ng/ml to YR cells. The lethality achieved was: for YS7 cells 76 ± 3%; YS9, 95 ± 3%; YR 50, 47 ± 7% and YR 80, 88±2%.

In both YS and YR cells almost all of the aberrations observed were of the chromatid type[9,24], approx. 75% being chromatid exchanges, 15% terminal deletions and a few isochromatid and interstitial deletions. Aberrations occurred at mitosis for several cell cycles after treatment (Figs 2-5).

In comparing aberration frequencies with cell killing the following assumptions have been made:

1. Cell death results from loss of chromosome material usually as acentric fragments[14]. When symmetrical chromatid exchanges segregate at anaphase a proportion (assumed to be 50%) will result in chromatin loss (and associated chromatin duplication) from the daughter cells[18]; this has also been assumed to be lethal. A lethal effect (LE) was attributed to each aberration; those resulting in chromatin loss from both daughter cells (e.g. isochromatid deletions) having LE = 1.0 and those affecting only one daughter cell (e.g. chromatid deletions) having LE = 0.5.
2. The aberrations frequencies (before adjusting for lethal effect) can be represented by the curves drawn between the points in Figs 2-5.
3. The probability of chromosome aberrations arising in a cell during a particular interphase is independent of whether the cell had chromosome damage at its previous mitosis. Only aberrations which have arisen during the previous interphase are

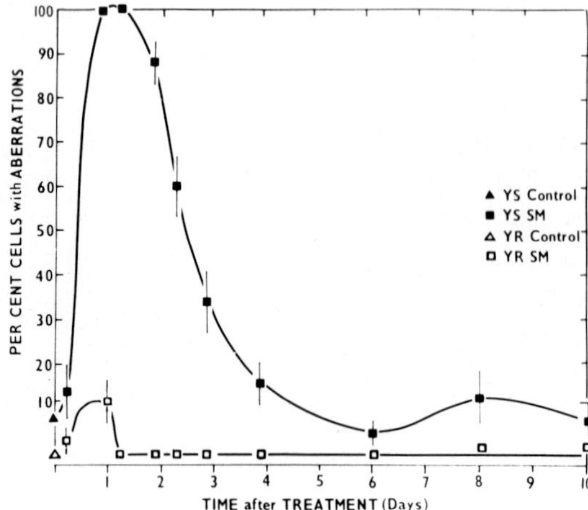

Fig.1. Aberrations in YS and YR cells after 20ng/ml SM.[27]

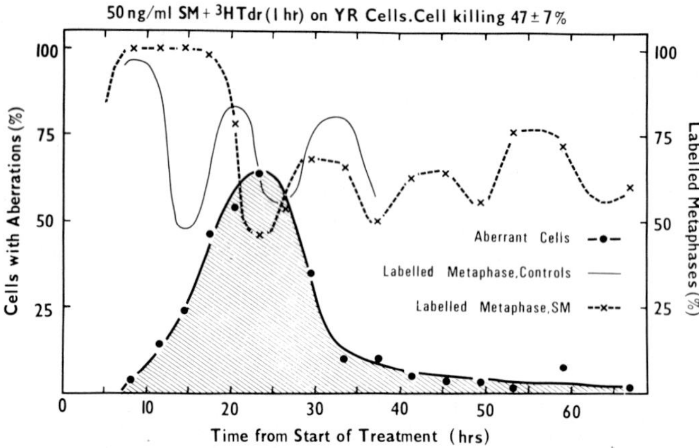

Fig.2. 50 cells scored for each aberration frequency, 100 for labelling index.

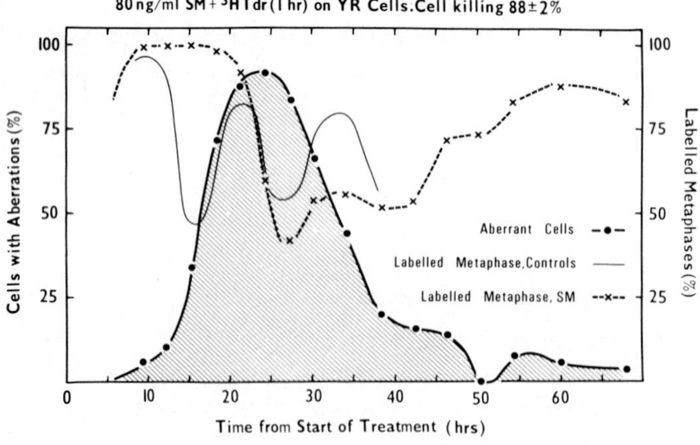

Fig.3. Legend as for Fig.2.

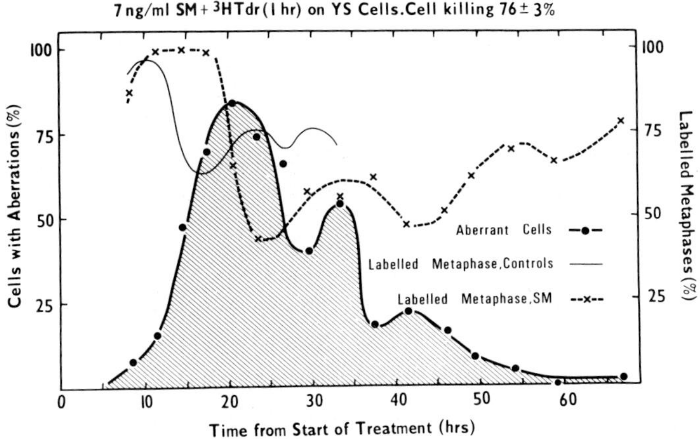

Fig.4. Legend as for Fig.2.

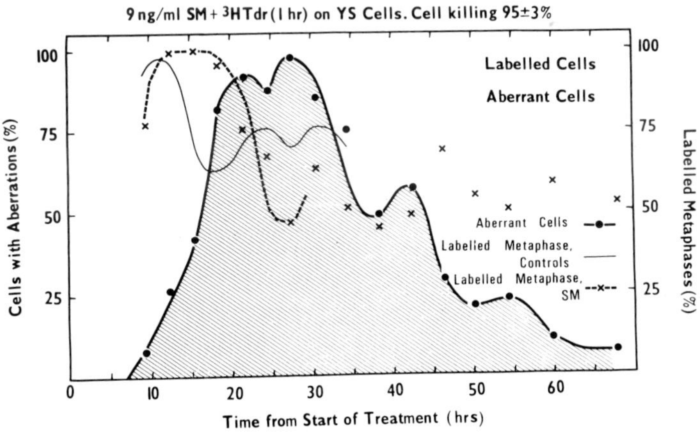

Fig.5. Legend as for Fig.2.

counted as lethal events, the few "derived"[25] chromosome-type aberrations are not.
4. The first trough in the labelled mitosis curve after SM + ^{3}HTdR is taken to be the time of transition from the first to the second post-treatment cycle. Cell cycles after the first are assumed to be of 12 hrs duration, the normal cycle time for these cells[10]. This is clearly the case for YR50 cells as seen from the labelled mitosis curve (Fig.2) but is not clear after more toxic doses of SM (Figs 3-5) when synchrony is lost at later cell cycles. Roberts and Ward[22] found that in SM-treated Chinese hamster cells the second post-treatment cycle was of normal duration.

The total amount of chromosome damage occurring after treatment is calculated as in the following example. Suppose the average aberration frequency over the first cell cycle is such as to give a lethal effect to 40% of the cells, then 60% of cells passing into the next cycle will not have sustained lethal damage. If the mean lethal aberration frequency during the second cell cycle is 50%, then 50% of the above 60% (i.e. 30%) of cells will sustain second cycle damage so a total of 40 + 30% (70%) of the original population will have been lethally damaged, leaving 30% (100-70%) of cells to pass undamaged into the third cell cycle, and so on. Using this method and the above assumptions there is remarkable agreement between the aberration frequencies and cell killing in YR50 cells (Table 1) whereas with more toxic SM levels the aberration frequencies are not quite sufficient to account for the proportion of cells killed and an additional lethal mechanism may be involved.

TABLE 1

COMPARISON OF PREDICTED CELL KILL FROM CHROMOSOME DAMAGE WITH OBSERVED KILL

Cell Cycle	Lethal effect of aberrations (% of cells)			
	YR50	YR80	YS7	YS9
1	24	43	41	50
2	26	39	32	63
3	4	5	9	25
4	1	2	1	6
Calculated kill*	47	68	67	87
Observed kill	47	88	76	95

* calculated as described in text

B. Effect of SM ± Caffeine on YS and YR Cells

1) Cell killing The toxic effect of caffeine alone on YS and YR cells was variable from experiment to experiment and had to be assessed each time; in general YS cells were more sensitive than YR cells (Figs 6-10).

In the first experiment in which SM-induced cell killing with or without

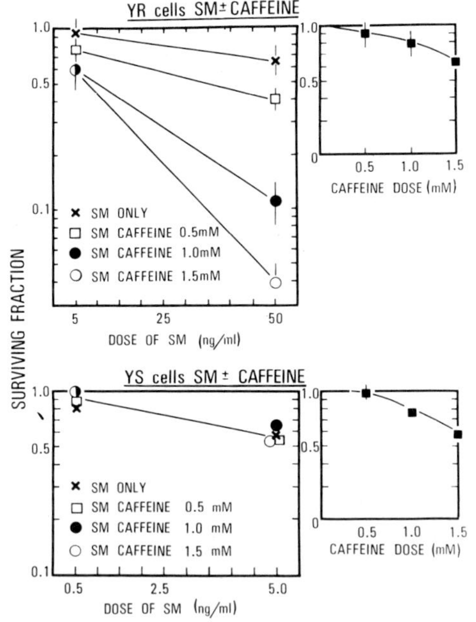

Fig.6. Enhancement of SM-induced lethality by caffeine in YR but not in YS cells. SM doses chosen to give approx. similar killing of YS and YR cells.

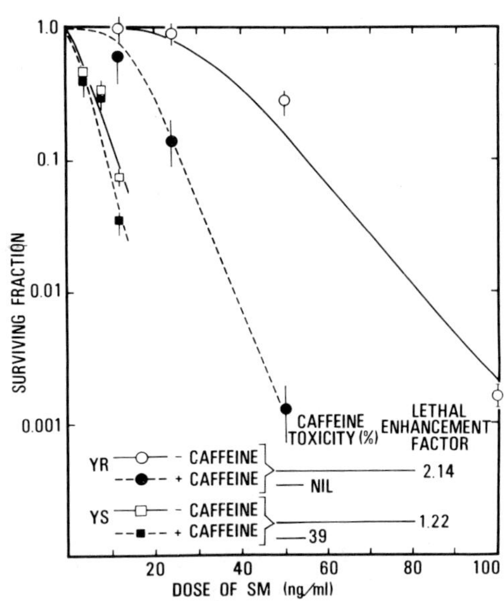

Fig.7. Survival after SM± caffeine (0.5mM) of YS and YR cells. For YR cells common n = 13.8 ± 4.6 (D_o SM = 11.4, SM + caff. 5.3) and for YS cells common n = 1.8 ± 0.9 (D_oSM = 4.0, SM + caff. 3.3).

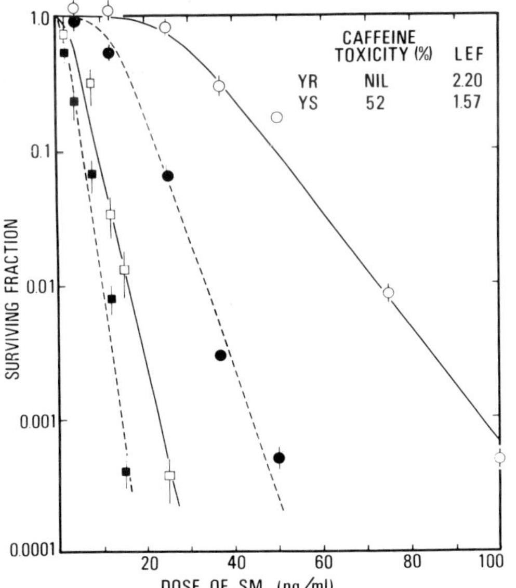

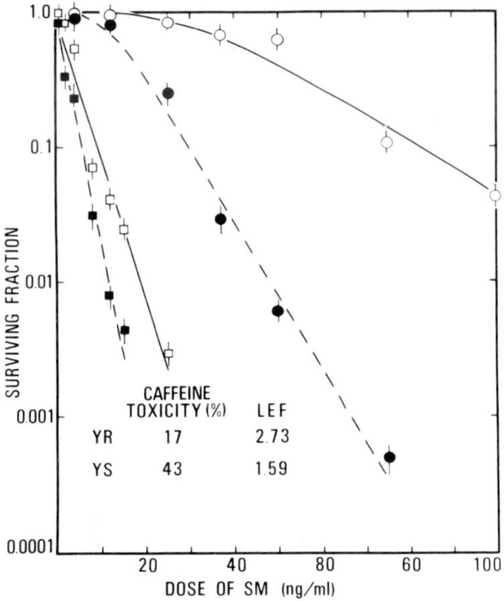

Fig.8. Survival after SM± caffeine (0.5mM) of YS and YR cells. For YR cells common n = 15.8 ± 8.8 (D_o SM = 9.9, SM + Caff. 4.5) and for YS cells common n = 2.7 ± 0.9 (D_oSM = 2.8, SM+ caff. = 1.8).

Fig.9. Survival after SM± caffeine (1.0mM) of YS and YR cells. For YR cells common n = 5.0 ± 1.3 (D_oSM = 21.3, SM+ caff. = 7.8) and for YS cells common n = 0.9 ± 0.2 (D_oSM = 4.1, SM+ caff. = 2.6).

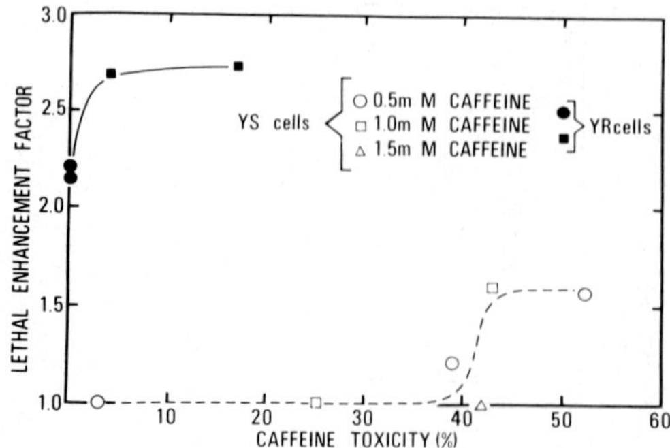

Fig.10. LEF values for YS cells in experiments shown in Figs 6-9 and YR cells in Figs 7-9 plus an additional experiment on YR cells where LEF = 2.7, caffeine toxicity 4%.

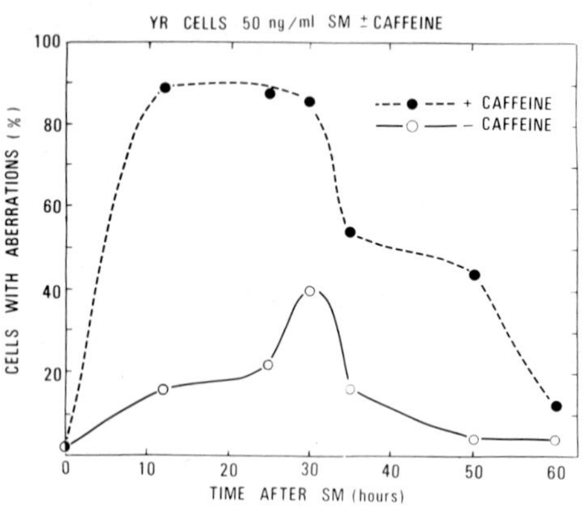

Fig.11. Percent YR cells with aberrations after 50ng/ml SM± caffeine (1.0mM). 50 cells scored per point. Cell kill was 54±3% after SM and 99.7±0.5% after SM+ caffeine.

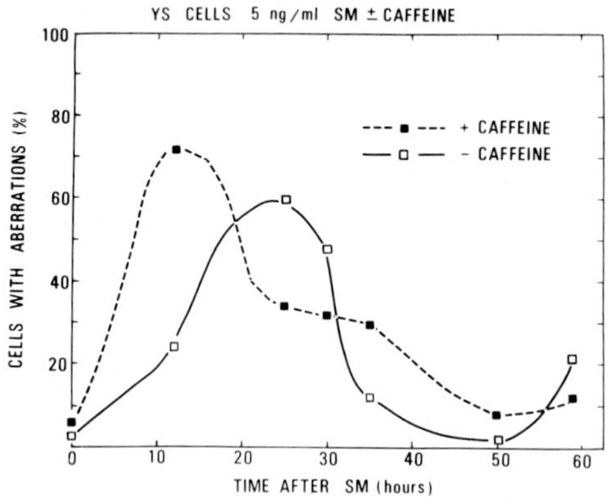

Fig.12. Percent YS cells with aberrations after 5ng/ml SM± caffeine (1.0mM). 50 cells scored per point. Cell kill was 29±7% after SM and 43±3% after SM+ caffeine.

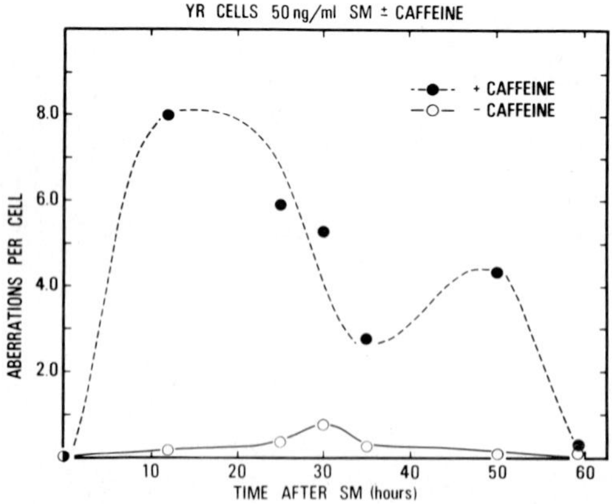

Fig.13. Same experiment as Fig.11 (YR cells) but aberration frequencies expressed per cell.

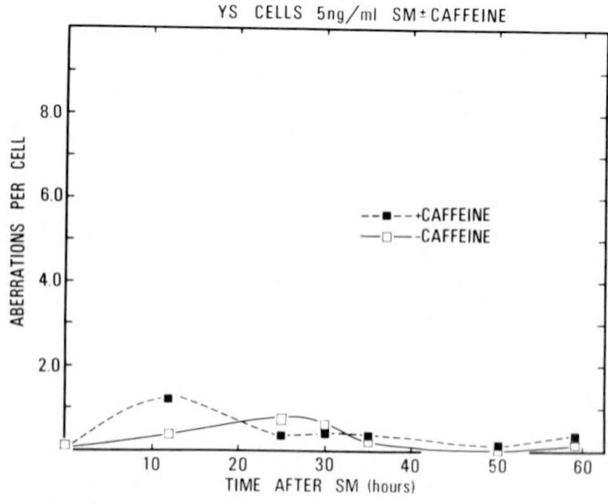

Fig.14. Same experiment as Fig.12 (YS cells) but aberration frequencies expressed per cell.

caffeine post-treatment was measured (Fig.6) the degree of enhancement of lethality by caffeine increased with increasing caffeine concentration in YR cells but no enhancement was seen in YS cells even at a caffeine dose which alone killed 40% of YS cells. In later experiments (Figs 7-9) some degree of enhancement of lethality (lethal enhancement factor, LEF) by caffeine was achieved in YS cells but only at caffeine concentrations which alone killed 40% or more of cells, whereas considerably higher LEF values could be achieved in YR cells with completely non-toxic doses of caffeine (data summarised in Fig 10).

Computer analysis[12] of the survival data showed that both the slope (D_o) of the exponential regions of the survival curves and the shoulder size (given by the extrapolation number, n) were reduced by caffeine post-treatment of SM-treated YS and YR cells (see also reference 11). However, in no case was the reduction in n statistically significant so, for the convenience of quoting a single value for the degree of enhancement, the data with and without caffeine were fitted to common value of n and the LEF taken as the ratio of the resulting fitted D_o values. These best-fitting values of D_o and common n are given in the legend to Figs 7-9.

2) Chromosome aberrations YS and YR cells were exposed to 5ng/ml and 50ng/ml SM, respectively, to achieve approximately similar levels of cell lethality. Aberration frequencies were determined at various intervals after treatment and compared with frequencies after SM + caffeine (1.0mM) post-treatment. Caffeine markedly enhanced the amount of SM-induced cell killing and chromosome damage in YR cells (Fig.11) but had a much lesser effect on YS cells (Fig.12). This contrast was most obvious when average aberration frequencies per cell were compared (Figs 13, 14). In both YS and YR cells after SM± caffeine over 70% of aberrations were chromatid exchanges.

Again, caffeine alone was more toxic to YS than YR cells (18% and 37% respectively) and appeared to induce some aberrations (see reference 16) in YS cells, i.e. 4% of untreated YS cells had aberrations whereas the frequency was 8% at 24hrs after caffeine treatment; YR cells had 2% aberrations with or without caffeine.

C) Sister-chromatid Exchange Frequencies

Sister-chromatid exchange frequencies were equal in untreated YS and YR cells at 0.23 ± 0.01 and 0.22 ± 0.02 per chromosome, respectively, from an analysis of 25 cells per cell line.

DISCUSSION

A. Aberrations in Relation to Lethality

Although many assumptions were necessary in the quantitation of SM-induced lethal aberrations, these are considered to be reasonable assumptions and lead to the conclusion that most of the lethality observed could have resulted from loss of chromosome material via structural aberrations. The best evidence for the lethal effect of chromatin loss comes from the experiments of Grote and Revell[14] who observed directly that X-irradiated diploid hamster cells which lost chromosome fragments at mitosis failed to form normal surviving colonies. Aneuploid cells could tolerate some loss of chromatin and still survive[13].

B. DNA Repair Capacity of YS and YR Cells

Caffeine post-treatment of SM-treated Yoshida cells enhances lethality in YR cells to a greater extent than in YS cells, the latter requiring fairly toxic doses of caffeine to obtain any enhancement at all (Fig.10). Although there are factors other than PRR capacity which can determine the magnitude of the 'caffeine effect' (e.g. i. PRR is caffeine-resistant[11,20] in some cells and no lethal enhancement is obtained, ii. the effectiveness of caffeine inhibition may depend on the rate of PRR)[20] the simplest and most reasonable interpretation of the present result is that YS cells have a lesser capacity for PRR than YR cells and that this is the reason for the ultrasensitivity of YS cells to SM and perhaps to other alkylating agents.

The 'caffeine effect' on YS cells after a toxic dose of caffeine may result from a mechanism other than PRR inhibition, perhaps involving the synergistic interaction of chromosome aberrations induced by SM with those induced by the toxic caffeine dose (see RESULTS and reference 16).

Since caffeine enhances both chromosome damage and lethality to a greater extent in YR than in YS cells after SM, the importance of chromosome aberrations in cell killing is again indicated. Failure to perform the gap-filling process of PRR appears to lead to chromosome structural aberrations which in turn lead to cell death; this repair process is less efficient in YS than in YR cells.

C. Sister Chromatid Exchange Frequencies

Sister chromatid exchange frequencies in YS and YR cells were equal, an observation which, like those of Wolff _et al_[31], does not support the hypothesis that PRR is involved in SCE formation.

ACKNOWLEDGEMENTS

The expert and devoted technical assistance of Mrs Christine Blease is gratefully acknowledged. This work was supported by grants from the Cancer Research Campaign and the Medical Research Council.

1. Alexander, P. and Mikulski, Z.B. (1961) Biochem.Pharmacol. 5, 275.
2. Bender, M.A., Griggs, G.H. and Bedford, J.S. (1974) Mutation Res., 24, 117.
3. Buhl, S.N., Setlow, R.B. and Regan, J.D. (1972) Int.J.Radiat.Biol. 22, 417.
4. Carrano, A.V. (1973) Mutation Res., 17, 355.
5. Cleaver, J.E. (1974) Adv.Radiat.Biol., 4, J.T. Lett, H. Adler and M. Zelle, eds., Academic Press, New York and London, pp 1-75.
6. Cleaver, J.E. and Thomas, G.H. (1969) Biochem.Biophys.Res.Comm., 36, 203.
7. Dewey, W.C., Stone, L.E., Miller, H.H. and Giblak, R.E. (1971) Radiat.Res. 47, 672.
8. Domon, M., Barton, B., Porte, A. and Rauth, A.M. (1970) Int.J.Radiat.Biol. 17, 395.
9. Evans, H.J. and Scott, D. (1969) Proc.Roy.Soc.B. (London) 173, 491.
10. Fox, M. and Fox, B.W. (1971/72) Chem. Biol.Interactions. 4, 363.
11. Fujiwara, Y. and Tatsumi, M. (1976) Mutation Res., 37, 91.
12. Gilbert, C.W. (1969) Int.J.Radiat.Biol., 16, 323.
13. Grote, S.J. (1972) Ph.D. Thesis, University of London.
14. Grote, S.J. and Revell, S.H. (1972) Current Topics in Radiation Res., 83, 55.
15. Kato, H. (1973) Exptl.Cell Res., 82, 383.
16. Kihlman, B.A. (1977) Caffeine and Chromosomes, Elsevier, North-Holland.
17. Kihlman, B.A., Sturelid, B., Hartley-Asp, B. and Nilsson, K. (1974) Mutation Res., 26, 105.
18. Lea, D.E. (1955) Actions of Radiations on Living Cells, Cambridge University Press, England, pp 195.
19. Lehmann, A.R. (1972) J.Molec.Biol. 66, 319.
20. Lehmann, A.R., Kirk-Bell, S., Arlett, C.F., Paterson, M.C., Lohman, P.H.M., de Weerd-Kastelein, E.A. and Bootsma, D. (1975) Proc.Nat.Acad.Sci.U.S.A. 72, 219.
21. Perry, P. and Wolff, S. (1974) Nature, 251, 156.
22. Roberts, J.J. and Ward, K.N. (1973) Chem.-Biol.Interactions, 7, 241.
23. Roberts, J.J., Sturrock, J.E. and Ward, K.N. (1974) Mutation Res., 26, 129.
24. Savage, J.R.K. (1975) J.Medical.Genet., 12, 103.
25. Scott, D. (1969) Cell and Tissue Kinet., 2, 295.
26. Scott, D., (1976) in Chemotherapy, Proc IXth Int.Chemotherapy Cong., 7, K. Hellmann and T.A. Connors, eds., Plenum, London, pp 95-103.
27. Scott, D., Fox, M. and Fox, B.W. (1974) Mutation Res., 22, 207.
28. Sieber, S.M. and Adamson, R.H. (1975) Adv.Cancer Res., 22, 57.
29. Trosko, J.E., Frank, P., Chu, E.H.Y. and Becker, J.E. (1973) Cancer Res., 33, 2444.
30. Walker, I.G. and Reid, B.D. (1971) Mutation Res., 12, 101.
31. Wolff, S., Bodycote, J., Thomas, G.H. and Cleaver, J.E. (1975) Genetics, 81, 349.

Chromosomes Today Volume 6, A. de la Chapelle and M. Sorsa eds
© 1977 Elsevier/North-Holland Biomedical Press, Amsterdam, The Netherlands

POSTSCRIPT

FRANCIS CRICK
The Salk Institute for Biological Studies, P.O. Box 1809, San Diego, California 92112, U.S.A.

The organisers have asked me, as an outsider from molecular and cell biology, to make a few retrospective comments on the very interesting and successful 6th International Chromosome Conference.

My most general impression is that these meetings, which started as rather informal family gatherings, have (now that the attendance has grown) become somewhat isolated from the main stream of recent biological developments. Evolution, guided by natural selection, must always remain the grandest theme in biology, but if we are to arrive at a deep and quantitative knowledge of it we must understand developmental biology in far greater depth than we do now. To do this it is essential (among other things) for us to know exactly what constitutes a eukaryotic "gene" and how it acts. In prokaryotes we already know, in broad outline, the answers to these questions. It is one of the scandals of molecular biology that we still cannot answer them for eukaryotes.

From this broader point of view the metaphase chromosome, which is the object of study of many of the Conference members, is the dullest form of chromosome: an inert package needed to make orderly mitosis possible. The meiotic chromosome, in its less active phases is not much better. For pure developmental biology (as opposed to evolutionary biology) meiotic recombination is almost irrelevant, except as part of the mechanism needed to use genetics as an experimental tool. Without any doubt the most important form of the chromosome is the active form found in interphase. Unfortunately this is the least rewarding form to study microscopically. But even here, in the extremely favourable case of the giant polytene chromosomes (especially of *Drosophila*, with its excellent genetics) cytological studies have, from the point of view of the molecular biologist, been somewhat disappointing. Puffs are certainly revealing and their detailed study increasingly rewarding, but the meaning of bands and interbands, the exact location, nature and control of the genes for most ordinary enzymes (such as those, say, for the Krebs cycle) or the genes for fundamental developmental processes (such as the homoeotic mutants) are not only unknown but are unlikely to be discovered by cytological methods unassisted by molecular biology.

There is, of course, no doubt that the new banding techniques for metaphase chromosomes are a great step forward, since they have not only allowed the decisive identification of individual chromosomes but also those parts of chromosomes involved in the larger deletions and translocations. Their medical importance has allowed the subject to be adequately supported financially. They are also responsible for a considerable advance in genetic mapping, especially for the human genome, which in the long run is indispensable. Nevertheless, one must realize that the results are on a very coarse scale. It will be a good time before a resolution as high as that already obtained in *Drosophila* will be reached and, as I have already explained, even this is not enough. Nevertheless these studies on metaphase chromosomes are yielding interesting results, both for medicine and for evolutionary studies and should certainly be pursued. My point is that they are only the merest beginnings to a full-scale attack on the problems confronting us. Important as the study of cancer is in a medical context, from the wider view of developmental biology cancer is a freak. Sometimes freaks are extremely useful and give the game away. Sometimes they only complicate rather simple underlying mechanisms. It remains to be seen which aspect of cancer research falls into which category.

One feature of the meeting, to the molecular biologist like myself, was how little was said about the many promising recent developments in molecular biology. Some of these were briefly touched on in the evening panel on the "Past and future of chromosome research" by one or two speakers, in particular by Dr. J. Wahrman's talk on "Near future trends in chromosome studies". The two main advances are fairly well known. The first is "genetic engineering" - the ability to take a fairly long stretch of DNA (say 10^4 base-pairs long), put it into a micro-organism and multiply it up so that much more of it is available for various experimental studies - in DNA terms, to be able to obtain biochemically useful amounts of a "pure gene" rather than the usual semi-random mixtures. The second is the ability to characterize the broad features of such lengths of DNA by restriction enzyme mapping (combined with hybridization) and, even more powerfully, by precise DNA sequencing. The two methods which are available for this are both so rapid that DNA sequencing is now considerably faster than either RNA or protein sequencing. Moreover, the methods may become even faster. This detailed information about genes in higher organisms, though laborious to obtain and, at present, difficult to interpret with precision, will, I am sure, prove essential for establishing the structure and function of eukaryotic genes.

In addition we have had an important advance in two-dimensional protein mapping - as many as two or three thousand proteins can now be distinguished on a single two-dimensional pattern - which will prove invaluable for the study of non-histone proteins as well as for many other problems in developmental biology.

The study of the three-dimensional structure of chromosomes, after passing through a very confused period, has also made a spectacular jump though it has still very far to go. More than 80% (perhaps almost 100%) of the DNA in the chromosomes is coiled onto nucleosomes. Some 10 to 20% of these have a looser structure and at present are little understood. The "transcriptionally active" regions of DNA probably have such nucleosomes. (The genes for rRNA are probably so active that few, if any, nucleosomes remain on them). The more inert nucleosomes, which make up the majority, have been better characterised.

As touched on by both Dr. Comings and Dr. Varshavsky, the "core nucleosomes" - the particles consisting of 140 base pairs of DNA plus two molecules each of the four major histones - are very much the same in almost all eukaryotes so far studied, from fungi to man. They have not only been crystallized, but the crystals are now sufficiently big to be studied by single-crystal X-ray diffraction. A Fourier projection (to a resolution of about 20 Å) along the shortest axis shows the core nucleosome to be a flatish cylinder, about 110 Å in diameter and a little under 60 Å in height. Klug has called this a "platysome". The DNA is almost certainly wound round the outside, around a more central protein core, the axis of this DNA probably forming about 1 3/4 turns of a rather shallow helix.

Between each platysome is a linker region, often about 60 base pairs long but very variable in length, both between species, between different tissues of the same species and also within a single type of cell. With this linker uncoiled one sees "beads-on-a-string". When the linker is coiled up the so-called "100 Å" fibre is formed. This is coiled again to produce the "250 Å" fibre visible in the electron microscope, but the exact details of this fold are not yet clear. It may be fairly regular (in which case it is called a "solenoid") or it may be more irregular (forming a string of "super-beads"). The higher levels of coiling are at the moment even less understood.

A neglected branch of the subject which is just beginning to make progress is the packaging and processing of the primary gene product, the nascent RNA, but there are now signs that this, too, may well advance rapidly. This is partly because of the great effort, by very able experimentalists which have been put into the study of various small eukaryotic viruses (usually oncogenic, such as SV 40). Such viruses may be useful models for eukaryotic genes, since their DNA is associated with histones and forms nucleosomes. They are at present far easier to study, by genetics, sequencing, by hybridization and by other methods than almost any set of eukaryotic genes so far available.

Some of the unexpected results already obtained on small viruses, such as "gene compaction" - the use of one stretch of DNA to code in part (each in its own phase) for two distinct proteins, may simply reflect the limited amount of DNA a viral capsid can contain. Other results, such as the avoidance of the duplication of nucleic acid sequences, may be due to the much higher rate of recombination found for such viruses (per length of DNA) compared to the host chromosome, in which, in eukaryotes, repetition is not at all uncommon. But the other recently discovered phenomenon, the production of a single messenger RNA from non-adjacent stretches of the DNA genome may perhaps reflect an unexpected and fundamental process in the treatment of hnRNA (heterogeneous nuclear RNA). Rather than merely cutting up hnRNA to make short lengths of mRNA it may be that certain stretches are looped out, excised and the remaining pieces spliced together. This has yet to be established but it makes an attractive hypothesis and the mechanism is in any case likely to be known before the 7th International Chromosome Conference in three years time.

I feel that chromosome workers will ignore these coming advances in molecular biology at their peril. It is not enough, in order to understand the Book of Nature, to turn over the pages looking at the pictures. Painful though it may be, it will also be necessary to learn to read the text. Only with the assistance of molecular biology will this be possible.

REFERENCES

In addition to several of the papers in the present volume, in particular the three referred to in the text above, I would recommend to the reader the Cold Spring Harbor volume for the 1977 meeting on "Chromatin".

AUTHOR INDEX

SUBJECT INDEX